Software Reliability

Springer
Singapore
Berlin
Heidelberg
New York
Barcelona
Hong Kong
London
Milan
Paris
Tokyo

Software
Reliability

Hoang Pham

Rutgers, The State University of New Jersey, USA

 Springer

Dr. Hoang Pham
Department of Industrial Enginering
Rutgers, The State University of New Jersey
96 Frelinghuysen Road
Piscataway, NJ 08854-8018
United States of America

Library of Congress Cataloging-in-Publication Data
Pham, Hoang
    Sofware reliability / Hoang Pham.
      p.   cm.
    Includes bibliographical references and index.
    ISBN 9813083840
    1. Computer software -- Reliability   I. Title.
QA76.76.R4454   2000
005. 1--dc21   99-40446
                                      CIP

ISBN 981-3083-84-0

© Springer-Verlag Singapore Pte. Ltd. 2000
Printed in Singapore

The publisher makes no representation, express or implied, with regard to the accuracy of the information contained in this book and cannot accept any legal responsibility or liability for any errors or omissions that may be made.

The use of general descriptive names, registered names, trademarks, etc. in this publication does not imply, even in the absence of a specific statement, that such names are exempt from the relevant protective laws and regulations and therefore free for general use.

Typesetting: Expo Holdings Sdn. Bhd, Malaysia
SPIN 10691552   5 4 3 2 1 0

for
Michelle
Hoang, Jr.
and
David

Preface

Today, there are countless software systems controlling the operation of chemical plants, refineries, nuclear power plants, hospital patient control rooms, commercial aircraft, and air traffic control. They are also used to design bridges, buildings, and other complex structures. But writing the code for such high-performance systems is not in any way software engineering stand alone. Modern software systems are becoming more complex and sophisticated in their demand for performance, and thus, modern software or programs are designed using the entire engineering software processes and understanding to solve sophisticated problems. Systems must also be well-documented, tested, and verified to ensure maximum performance as designed.

Software reliability is the probability that the software will not fail for a specified period of time under specified conditions. The greatest problem facing the software industry today is how to quantitatively assess software reliability characteristics.

This book aims to deal with the modeling and analysis of software systems that provides (1) assurance that the software has achieved safety goals, and (2) a means of rationalizing whether further testing is warranted or whether the software has been sufficiently tested to allow its release or unrestricted use. Such predictions provide a quantitative basis for achieving reliability, risk, and cost goals.

The text is suitable for a one-semester or two-quarter graduate course on software reliability in engineering, operations research, computer science and engineering, mathematics, statistics, and management. A prerequisite is a basic course in probability, statistics, or an introduction to reliability engineering. It is intended that the book will be a valuable reference for practising software and reliability engineers, statisticians, safety engineers, and researchers in the field. It is also intended that the individual, after having utilized this book, will be thoroughly prepared to pursue advanced studies in software reliability engineering and research in this field.

A 3.5″ diskette containing the recently developed computer software, NHPP Software™, is also included. It provides useful software reliability tools for reliability estimation, mean value function, cost estimation, and a wide range of the non-homogeneous Poisson process (NHPP) software reliability models as described in Chapters 5 and 6.

The topics covered are organized as follows:

- **Chapter 1** gives a brief introduction to software reliability and basic terminologies used throughout the book. This chapter also provides the literature available in the area of software reliability engineering.

- **Chapter 2** discusses the fundamental definitions and concepts of reliability. These concepts provide the basis for quantifying the reliability of a system.

- **Chapter 3** provides the basic concepts of software engineering assessment including software lifecycle, software development process and its applications, software verification and validation, and data analysis.

- **Chapter 4** presents various software reliability models and methods for estimating software reliability and other performance reliability measures, such as software complexity and the number of remaining errors.

- **Chapter 5** comprehensively covers software reliability models for the failure phenomenon based on the NHPP. The generalized NHPP model and the software mean time between failures are also discussed.

- **Chapter 6** discusses some recent software cost models based on the NHPP software reliability functions. In addition to the cost of traditional models, the models mentioned here include the warranty and risk costs due to software failures. Various optimal release policies of the software system, that is, when to conclude testing and release the software, are also presented.

- **Chapter 7** is devoted to the basic concepts of fault-tolerant software techniques and several other advanced techniques including self-checking schemes. The reliability analysis of fault-tolerant software schemes such as recovery block, *N*-version programming, and hybrid fault-tolerant systems are presented.

- **Chapter 8** discusses recent advances and research directions in software reliability engineering. Several new software reliability models that incorporate environmental factors are also discussed.
- **Appendices A** and **B**, which are optional for the more statistically inclined readers, provide basic mathematical statistics and an introduction to probability theory used for deriving relationships in the body of the text. **Appendix C** provides a survey form which engineers may adopt when interviewing clients, in order to obtain a better understanding of the latter's priorities. **Appendix D** contains various distribution tables.

A list of references for further reading and problems are included at the end of each chapter.

Solutions to selected problems are also provided towards the end of the book.

Acknowledgments

I would like to thank Sung Hoon Hong of the Chonbuk National University, Korea, for his tireless effort in reading several drafts of this manuscript. I would also like to thank the following reviewers: Min Xie of the National University of Singapore, Shigeru Yamada of the Tottori University, Japan, and Ming Zhao of the University of Gavle, Sweden, who provided numerous comments and suggestions that have extensively improved the book.

Feedback from Taghi Khoshgoftaar of the Florida Atlantic University, Ming-Wei Lu of Chrysler Corporation, Auburn Hills, Eric Wong of the BellCore, Jai-Hyun Byun of Gyeongsang National University in Korea, and Hongzhou Wang of the Lucent Technologies are much appreciated. I am also indebted to Xuemei Zhang of Rutgers University who developed the software, Software Reliability & Cost Estimation™ that accompanies this book.

I am also grateful to the students of the Department of Industrial Engineering at Rutgers University who have used preliminary versions of this book during the past three years and provided numerous comments and suggestions.

Gillian Chee, Senior Editor, and her associates at Springer-Verlag deserve significant praise for their patience and understanding when deadlines were missed.

Finally and most importantly, I want to thank my wife, Michelle, and my sons, Hoang Jr. and David, for their love, patience, understanding, and support. It is to them that this book is dedicated.

Hoang Pham
June 1999

Contents

Preface. vii
Acknowledgments. xi

1 Introduction. 1
  1.1 The Need for Software Reliability. 1
  1.2 Software Reliability Engineering 4
  1.3 Why Does Software Cost So Much?. 5
  1.4 Basic Definitions and Terminologies 7
    Further Reading 9
    Problems 11
    References. 11

2 Reliability Engineering Measures 13
  2.1 Reliability Definitions 13
  2.2 System Mean Time to Failure 16
  2.3 Failure Rate Function 18
  2.4 Reliability Function for Common Distributions 19
    2.4.1 Binomial Distribution. 19
    2.4.2 Poisson Distribution. 20
    2.4.3 Exponential Distribution 21
    2.4.4 Normal Distribution. 22
    2.4.5 Log Normal Distribution 24
    2.4.6 Weibull Distribution. 26
    2.4.7 Gamma Distribution. 28
    2.4.8 Beta Distribution. 29
    2.4.9 Hazard Rate Model I. 29
    2.4.10 Hazard Rate Model II 29
  2.5 Maintainability and Availability 30
    2.5.1 Maintainability 30

2.5.2 Availability 32
Problems 33
References 34

3 Software Engineering Assessment 37
3.1 Introduction 37
3.2 Software Versus Hardware Reliability 38
3.3 Software Reliability and Testing Concepts 44
3.4 Software Lifecycle 48
3.4.1 Analysis Phase 50
3.4.2 Design Phase 53
3.4.3 Coding Phase 56
3.4.4 Testing Phase 56
3.4.5 Operating Phase 60
3.5 Software Development Process and Its Applications . . . 62
3.5.1 Analytic Hierarchy Process 62
3.5.2 Evaluation of Software Development Process 63
3.6 Software Verification and Validation 67
3.7 Data Collection and Analysis 69
Further Reading 71
Problems 71
References 71

4 Software Reliability Modeling 73
4.1 Introduction 73
4.2 Halstead's Software Metric 74
4.3 McCabe's Cyclomatic Complexity Metric 78
4.4 Error Seeding Models 81
4.4.1 Mills' Error Seeding Model 81
4.4.2 Cai's Model 83
4.4.3 Hypergeometric Distribution Model 85
4.5 Failure Rate Models 87
4.5.1 Jelinski–Moranda Model 87
4.5.2 Schick–Wolverton Model 90
4.5.3 Jelinski–Moranda Geometric Model 92
4.5.4 Moranda Geometric Poisson Model 92
4.5.5 Negative-Binomial Poisson Model 93
4.5.6 Modified Schick–Wolverton Model 93
4.5.7 Goel–Okumoto Debugging Model 93

4.6 Curve Fitting Models. 94
    4.6.1 Estimation of Errors Model. 94
    4.6.2 Estimation of Complexity Model 95
    4.6.3 Estimation of Failure Rate Model 95
4.7 Reliability Growth Models 95
    4.7.1 Coutinho Model 96
    4.7.2 Wall and Ferguson Model 96
4.8 Non-Homogeneous Poisson Process Models. 96
4.9 Markov Structure Models 97
    4.9.1 Markov Model with Imperfect Debugging 98
    4.9.2 Littlewood Markov Model. 98
    4.9.3 Software Safety Model 99
    Problems 100
    References. 102

5 NHPP Software Reliability Models. 105
5.1 Introduction 105
5.2 Parameter Estimation 106
5.3 NHPP Models 108
    5.3.1 NHPP Exponential Model. 108
    5.3.2 NHPP S-Shaped Model. 115
    5.3.3 NHPP Imperfect Debugging Model 120
    5.3.4 NHPP Imperfect Debugging S-Shaped Model . . . 128
5.4 Applications 134
5.5 Imperfect Debugging Versus Perfect Debugging 145
5.6 A Generalized NHPP Software Reliability Model 148
5.7 Mean Time Between Failures for NHPP 152
    Problems 155
    References. 157

6 Software Cost Models 159
6.1 Introduction 159
6.2 A Software Cost Model with Risk Factor. 162
6.3 A Generalized Software Cost Model 165
6.4 A Cost Model with Multiple Failure Errors 169
    6.4.1 Cost Subject to Reliability Constraint 171
    6.4.2 Cost Subject to the Number of Remaining Errors
         Constraint 172
    6.4.3 Software Reliability Subject to Cost Constraint. . 173

6.5 Applications.................................. 173
    Problems 176
    References 177

7 Fault-Tolerant Software 179
    7.1 Introduction............................... 179
    7.2 Basic Fault-Tolerant Software Techniques 181
        7.2.1 Recovery Block Scheme.................... 182
        7.2.2 N-Version Programming 182
        7.2.3 Other Advanced Techniques................. 183
    7.3 Self-Checking Duplex Scheme 184
    7.4 Reliability Modeling 186
        7.4.1 Recovery Block 187
        7.4.2 N-Version Programming 187
        7.4.3 Hybrid Fault-Tolerant Scheme 189
    7.5 Reduction of Common-Cause Failures 192
    7.6 Summary 193
        Problems 194
        References 195

8 Software Reliability Models with
   Environmental Factors 199
    8.1 Introduction................................ 199
    8.2 Definition of Environmental Factors 199
    8.3 Environmental Factors Analysis 205
    8.4 A Generalized Model with Environmental Factors 210
        8.4.1 Environmental Factors Estimation Using
              Maximum Likelihood Estimation 211
        8.4.2 Environmental Factors Estimation Using
              Maximum Partial Likelihood Approach 212
    8.5 Enhanced Proportional Hazard Jelinski–Moranda
        (EPJM) Model............................. 213
    8.6 An Application with Environmental Factors 216
        Problems 219
        References 220

Appendix A   Theory of Estimation.................. 221
    A.1 Point Estimation 221
        A.1.1 Maximum Likelihood Estimation Method 223
        A.1.2 Method of Moments 230

A.2 Goodness of Fit Techniques 230
    A.2.1 Chi-Squared Test 231
    A.2.2 Kolmogorov–Smirnov d Test 232
    A.2.3 Least Squared Estimation 233
A.3 Interval Estimation. 235
    A.3.1 Confidence Intervals for the Normal
        Parameters 235
    A.3.2 Confidence Intervals for the Exponential
        Parameters 238
    A.3.3 Confidence Intervals for the Binomial
        Parameters 240
    A.3.4 Confidence Intervals for the Poisson
        Parameters 241
A.4 Tolerance Limits 242
    A.4.1 Tolerance Limits for Normal Populations. 242
    A.4.2 Tolerance Limits for Exponential
        Populations 243
A.5 Non-Parametric Tolerance Limits 244
A.6 Sequential Sampling. 245
A.7 Bayesian Methods. 253
    Problems. 259
    References 260

Appendix B   Stochastic Processes. 261
    B.1 Introduction 261
    B.2 Markov Processes 262
        B.2.1 System Mean Time Between Failures 271
    B.3 Poisson Processes 282
    B.4 Renewal Processes 284
    B.5 Quasi-Renewal Processes 287
    B.6 Non-Homogeneous Poisson Processes. 290
        Further Reading 292
        Problems. 293
        Reference 293
        LaPlace Transform 293

**Appendix C   Survey of Factors that Affect
                Software Reliability. 297**

Appendix D   Distribution Tables   301

Solutions to Selected Problems.   325

Index   337

1

Introduction

*"If you don't test your software,
how do you know it works."*

1.1 The Need For Software Reliability

In our modern society, computers are used in diverse areas for various applications, for example, air traffic control, nuclear reactors, aircraft, real-time military, industrial process control, automotive mechanical and safety control, and hospital patient monitoring systems. As the functionality of computer operations becomes more essential and complicated and critical software applications increase in size and complexity, there is a greater need for computer software reliability. Faults in software design thus become more subtle.

Let us define the terms such as "software error", "fault" and "failure" (IEEE Std. 610.12, 1990). An error is a mental mistake made by the programmer or designer. A fault is the manifestation of that error in the code. A software failure is defined as the occurrence of an incorrect output as a result of an input value that is received with respect to the specification.

A computer system consists of two major components: hardware and software. Although extensive research has been carried out on hardware reliability, the growing importance of software dictates that the focus shift to software reliability. Software reliability is different from hardware reliability in the sense that software does not wear out or burn out. The software itself does not fail, unless flaws within the software result in a failure in its dependent system.

In recent years, the cost of developing software and the penalty cost of software failure have become the major expenses in a system (Pham, 1992). Failure of the software may result in an intended system state or course of action. A loss event could ensue in which

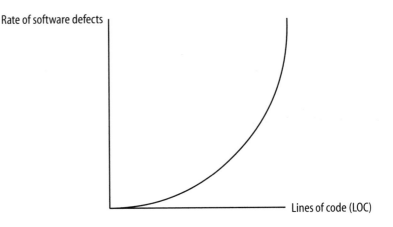

FIG. 1.1. The rate of software defect changes.

property is damaged or destroyed, people are injured or killed, and/ or monetary costs are incurred. A quantitative measure of loss is called the *risk cost of failure* (Pham, 1996, 1999). In other words, risk cost is a quantitative measure of the severity of loss resulting from a software failure. A research study has shown that professional programmers average six software defects for every 1,000 lines of code (LOC) written. At that rate, a typical commercial software application of 350,000 LOCs may contain over 2,000 programming errors including memory-related errors, memory leaks, language-specific errors, errors calling third-party libraries, extra compilation errors, standard library errors, etc. As software projects become larger, the rate of software defects increases geometrically (see Fig. 1.1). Table 1.1 shows the defect rates of several software applications per 100 LOC.

Locating software faults is extremely difficult and costly. A study conducted by Microsoft showed that it takes about 12 programming hours to locate and correct a software defect. At this rate, it can take more than 24,000 hours (or 11.4 man-years) to debug a program of 350,000 LOC with a cost of over US$1 million.

Software errors have caused spectacular failures, some with dire consequences, such as the following examples. On 31 March 1986, a Mexicana Airlines Boeing 727 airliner crashed into a mountain because the software system did not correctly negotiate the mountain position. Between March– June 1986, the massive Therac-25 radiation therapy machines in Marietta, Georgia; Boston, Massachusetts; and Tyler, Texas, overdosed cancer patients due to flaws in the computer program controlling the highly automated devices. During the period of 2–4 November 1988, a computer virus infected

TABLE 1.1. Defect rates.

Application	Number of systems	Fault density (per 100 LOC)
Airborne	8	1.28
Strategic	18	0.66
Tactical	6	1.00
Process control	2	0.18
Production	9	1.30
Developmental	2	0.40

software at universities and defense research centers in the United States causing system failures. On 10 December 1990, Space Shuttle Columbia was forced to land early due to computer software problems. On 17 September 1991, a power outage at the AT&T switching facility in New York City interrupted service to 10 million telephone users for nine hours. The problem was due to the deletion of three bits of code in a software upgrade and failure to test the software before its installation in the public network. During the 1991 Gulf War, a software problem resulted in the failure of the Patriot missile system to track an Iraqi Scud missile causing the deaths of 28 American soldiers. On 26 October 1992, the computer-aided dispatch system of the Ambulance Service in London, which handles more than 5,000 requests daily in the transportation of patients in critical condition, failed after installation. This led to serious consequences for many critical patients.

A recent inquiry revealed that a software design error and insufficient software testing caused an explosion that ended the maiden flight of the European Space Agency's (ESA) Ariane 5 rocket, less than 40 seconds after lift-off on 4 June 1996. The problems occurred in the flight control system and were caused by a few lines of Ada code containing three unprotected variables. One of these variables pertained to the rocket launcher's horizontal velocity. A problem occurred when the ESA used the software for the inertial-reference flight-control system in the Ariane 5, similar to the one used in the Ariane 4. The Ariane 5 has a high initial acceleration and a trajectory that leads to a horizontal velocity acceleration rate five times greater than that found in Ariane 4. Upon lift-off, the Ariane 5's horizontal velocity exceeded a limit that was set by the old software in the backup inertial-reference system's computer.

This stopped the primary and backup inertial-reference system's computers, causing the rocket to veer off course and explode. The ESA report revealed that officials did not conduct a pre-flight test of the Ariane 5's inertial-reference system, which would have located the fault. The companies involved in this project had assumed that the same inertial-reference-system software would work in both Ariane 4 and Ariane 5. The ESA estimates that corrective measures will amount to US$362 million.

Generally, software faults are more insidious and much more difficult to handle than physical defects. In theory, software can be error-free, and unlike hardware, does not degrade or wear out but it does deteriorate. The deterioration here, however, is not a function of time. Rather, it is a function of the results of changes made to the software during maintenance, through correcting latent defects, modifying the code to changing requirements and specifications, environments and applications, or improving software performance. All design faults are present from the time the software is installed in the computer. In principle, these faults could be removed completely, but in reality, the goal of perfect software remains evasive (Friedman, 1995). Computer programs, which vary for fairly critical applications between hundreds and millions of lines of code, can make the wrong decision because the particular inputs that triggered the problem were not tested and corrected during the testing phase. Such inputs may even have been misunderstood or unanticipated by the designer who either correctly programmed the wrong interpretation or failed to identify the problem. These situations and other such events have made it apparent that we must determine the reliability of the software systems before putting them into operation.

1.2   Software Reliability Engineering

Research on software reliability engineering have been conducted during the past 25 years, and more than 50 statistical models have been proposed for estimating software reliability. Most existing models for predicting software reliability are based purely on the observation of software product failures. These models also require a considerable amount of failure data to obtain an accurate reliability prediction. However, information concerning the development of the software product, the method of failure detection, environmental factors, etc., are ignored.

On the other hand, not many software practitioners, developers, or users utilize these models to evaluate computer software reliability as they do not know how to select and apply them. A survey conducted in the late 1990s by the American Society for Quality (ASQ) reported that only 4% of the participants responded positively when asked if they could use a software reliability model.

Many researchers are currently pursuing the development of statistical models that can be used to evaluate the reliability of real-world software systems. To develop a useful software reliability model and be able to make sound judgments when using the model, one needs to understand how software is produced and tested, the types of errors, and how errors are introduced. Environmental factors can help us in justifying the usefulness of the model and its applicability in a user environment. In other words, these models would be valuable if practitioners, software developers and users could use the information about the software development process, incorporating the environmental factors, thus giving greater confidence in estimates based on small numbers of failure data.

1.3   Why Does Software Cost So Much?

In the early 1970s when computers were first used in the business world, storage space was at a premium and the use of a two-digit convention to represent the year seemed appropriate. For example, a date such as 20 April 1997 is typically represented in software as YY/MM/DD, or 97/04/20, and 1 January 2000 will look like 00/01/01 on our computers, but at that time, the year 2000 was a long way off. However, the time has arrived causing the year 2000 to be a major software concern of the 20th century. It has been called the Year 2000 Problem, Y2K Problem or the Millennium Bug. The Year 2000 problem involves either or all of the following: (i) the year 2000 represented as a two-digit number causes failures in arithmetic, comparisons, and input/output to databases or files when manipulating date data, (ii) using an incorrect algorithm to recognize leap years for years divisible by 400, and (iii) system date data types that may roll over and fail due to the storage register becoming full.

Incorrect software programs will assume that the maximum value of a year field is "99" and will roll systems over to the year 1900 instead of 2000, resulting in negative date calculations and the creation of many overnight centenarians. Incorrect leap year calculations will therefore incorrectly assume that the year 2000 has only 365 days instead of 366. The Year 2000 Problem is widespread. It affects hardware, embedded firmware, languages and compilers, operating systems, nuclear power plants, air-traffic control, security services, database-management systems, communications systems, transaction processing systems, banking systems and medical services. Several examples are as follows. In nuclear power plants, the Nuclear Regulatory Commission states that the "bugs" may affect security control, radiation monitoring and accumulated burn-up programs which involve calculations to estimate the hazard posed by radioactive fuel. In air-traffic control, one possible danger is computer lockup. While planes continue to fly at 12:01 am on 1 January 2000, the screens monitoring them, if not upgraded, might lock. Even if the computers may keep track of the planes, they may mix them up with flights recorded at the same time on a previous day. The worst scenario is that, at 12:00 am, the nation's air-traffic control systems will all go dead and some planes will lose the ability to navigate effectively.

Today, testing the Year 2000 Problem software comprises approximately 45% of the entire Problem project effort. One way of minimizing the test effort is to use testing tools to develop automated and repetitive test methods. Test scenarios provide a reasonable way to conduct effective stress tests of various problem dates. The Year 2000 Problem requires testing at numerous points in time, each requiring supporting test data and test scenarios. Following are some possible dates that may need to be considered during testing: 01/01/1998; 01/01/1999; 09/09/1999; 01/01/2000; 01/1/2000; 10/10/2000; 02/29/2000 (leap year) and 12/31/2000 (366th day of the year).

In 1997, United States federal officials estimated the cost of analysis and modification of the Year 2000 Problem to be US$2 per line. As an estimated 15 billion lines of code have to be changed to cope with the problem, the work may cost up to US$30 billion and the worldwide cost at US$600 billion. This estimate, however, reflects only conversion costs and may not include the cost of replacing hardware, testing and upgrading the systems.

The year 2000 is, however, not the only date dangerous to software applications. Over the next 50 years, at least 100 million applications around the world will need modification because of formatting problems with dates or related data. The total cost of the remedies in the United States could exceed US$5 trillion. These problems include: the "nine" end-of-file, the global positioning system's date roll over; the social insecurity identity numbers, the date on which Unix and C libraries rollover and some date-like patterns used for data purposes. In the early 21st century, the nine digits assigned to the United States social security numbers and the ten digits allotted to long-distance telephones will no longer suffice for American citizens and required phone lines, respectively. Changes to these numbering systems will also affect software. There is a solution to such problems but it will need a global agreement. A good starting point would be to examine all known date and data-like problems, and how computers and application software basically handle dates.

1.4   Basic Definitions and Terminologies

Let us define the following terms:

Bug. A design flaw that will result in symptoms exhibited when an object is subjected to an appropriate test.

Clean test. A test with the primary purpose of validation, i.e., test designed to demonstrate the software's correct working.

Dirty test. A test with the primary purpose of falsification i.e., test designed to break the software.

Integration test. Test that explore the interaction and consistency of successfully tested units.

Object. A generic term that includes data and software.

Operational profile. The set of operations that the software can execute, given the probability of their occurrence.

Risk. The combination of the frequency or probability, and the consequence of a specified hazardous event.

Safety-related system. Those systems that enable, independently of other systems, the tolerable risk level to be met.

Safety integrity. The likelihood of a safety-related system achieving its required functions under all stated conditions within a time period.

Safety validation. The process of determining the level of conformance of the final operating system to safety requirements specification.

Software availability. The probability that a system has not failed due to a software fault.

Software defect. A generic term referring to a fault or a failure.

Software error. An error made by a programmer or designer, e.g., a typographical error, an incorrect numerical value, an omission, etc.

Software fault. An error that leads to a software fault. Software faults can remain undetected until software failure results.

Software failure. A failure that occurs when the user perceives that the software has ceased to deliver the expected result with respect to the specification input values. The user may need to identify the severity of the levels of failures, e.g., catastrophic, critical, major or minor, depending on their impact on the systems. Severity levels may vary from one system to another, and from application to application. Typically, the severity of a software system effect is classified into four categories:

> **Category 1**: *Catastrophic* This category is for disastrous effects, e.g., loss of human life or permanent loss of property, the effect of an erroneous medication prescription or an air-traffic controller error.

> **Category 2**: *Critical* This category is for disastrous but restorable damage. It includes damage to equipment without lost of human life or where there is major but curable illness or injury.

> **Category 3**: *Major* This category is for serious failures of the software system where there is no physical injury to people or other systems. This may include erroneous purchase orders or the breakdown of a vehicle.

> **Category 4**: *Minor* This category is for faults that cause marginal inconveniences to a software system or its users. Examples might be a vending machine that momentarily cannot provide change or a bank's computer system that is not working when a consumer requests an account balance.

Software debugging. Activity to isolate faults and eliminate underlying error.

Software MTTF. The expected time when the next failure is observed due to software faults.

Software Maintainability. The probability that a program will be restored to working condition in a given period of time when it is being changed, modified or enhanced.

Software MTTR. The expected time to restore a system to operation upon a failure due to software faults.

Software reliability. The probability that software will not fail for a specified period of time under specified conditions.

Software testing. A verification process for software quality evaluation and improvement.

Software validation. The process of ensuring that the software is executing the correct task.

Software verification. The process of ensuring that the software is executing the task correctly.

System availability. The probability that a system is available when needed.

System testing. Test that explores system behavior that cannot be done by unit, component, or integration testing. System testing presumes that all components have been previously and successfully integrated and is often performed by independent testers.

Test. A sequence of one or more subtests executed as a sequence because the result of one subtest is the input of the next.

Test design. The process of specifying the input and predicting the result for that input.

Test strategy. A systematic method used to select and/or generate tests to be included in a test suite.

Test suite. A set of one or more tests, usually aimed at a single input, with a common purpose and database, usually run as a set.

Acronyms

cdf	cumulative distribution function
GO	Goel and Okumoto
JM	Jelinski and Moranda
ln	natural logarithm
MLE	maximum likelihood estimate
NHPP	non-homogeneous Poisson process
pdf	probability density function
SW	Schick and Wolverton
WF	Wall and Ferguson

Further Reading

There are many survey papers on software reliability. Interested readers are referred to the review papers by Ramamoorthy and Bastani (1982), Goel (1985), Yamada and Osaki (1985), Yamada (1991) and Pham (1999).

The *Handbook of Software Reliability Engineering* edited by M. R. Lyu (McGraw-Hill, New York and IEEE Computer Society Press, Los Angeles, 1996), the book *Software-Reliability-Engineered Testing Practice* by J. D. Musa (McGraw-Hill, New York, 1997), *Software Assessment: Reliability, Safety, Testability* by M.A. Friedman and J.M.

Voas (John Wiley & Sons, New York, 1995), *Introduction to Software Management Model* by S. Yamada and M. Takahashi (Kyoritsu Shuppan, Tokyo, 1993), and *Software Reliability Models* by S. Yamada (JUSE Press, Japan, 1994), are new and good textbooks for students and practitioners. *The State of the Art Report: Software Reliability* edited by A. Bendell (Pergamon Press, Oxford, 1986), *Software Reliability: Achievement and Assessment* edited by B. Littlewood (Alfred Waller Ltd., Oxford, 1987), *Software Reliability* by S. Yamada (Software Research Center, Tokyo, 1990), and *Software Reliability Handbook*, edited by P. Rook (Elsevier Applied Science, Amsterdam, 1990) are good reference books containing many interesting results. In addition, *Software Reliability Models: Theoretical Developments, Evaluation, and Application*, by Y.K. Malaiya and P.K. Srimani (IEEE Computer Society Press, Los Angeles, 1990), and *Software Reliability and Testing* by H. Pham (IEEE Computer Society Press, Los Angeles, 1995) are edited books which the reader may find useful.

Many research and tutorial papers on software reliability have been published in the *IEEE Transactions on Software Engineering*, *IEEE Transactions on Reliability*, and *IEEE Software Magazine*. Several special issues on software reliability are of practical interest:

Special issues of the *International Journal of Reliability, Quality and Safety Engineering* on "Software Reliability," Vol. 4(3), September 1997; on "Software Reliability Model and Applications," Vol. 6(1), March 1999.

Special issue of the *IEEE Computer* on "System Testing and Reliability," November, 1996.

Special Section of the *IEEE Transactions on Reliability* on "Software Reliability Engineering," December, 1994.

Special issue of the *IEEE Transactions on Software Engineering* on "Software Reliability," November, 1993.

Special issue of the *Journal of Computer and Software Engineering* on "Reliable Software," Vol. 1(4), 1993.

Special issue of the *IEEE Transactions on Reliability* on "Software Fault-Tolerant," June 1993.

Special issue of the *IEEE Transactions on Computers* on "Fault-Tolerant Computing," Vol. 41(5), 1992.

Special issue of the *IEEE Software* on "Reliability Measurement," July, 1992.

Special issue of the *Reliability Engineering and System Safety Journal* on "Software Reliability and Safety," Vol. 32(1 & 2), 1991.

Special issue of the *Information and Software Technology* on "Software Quality Assurance," Vol. 32(1), 1990.

Special issue of the *IEEE Transactions on Software Engineering* on "Software Reliability," Part I, Vol. SE-11(12), 1985; Part II, Vol. SE-12(1), 1986.

Special issue of the *Journal of Systems and Software* on "Software Reliability," Vol. 1(1), 1980.

Several other journals occasionally publish papers on the subject:

Journal of Systems and Software

International Journal of Reliability, Quality and Safety Engineering

Reliability Engineering and System Safety Journal

Microelectronics and Reliability — An International Journal.

IIE Transactions on Quality and Reliability Engineering.

IEICE Transactions on Fundamentals of Electronics, Communications and Computer Science.

There are also a great number of proceedings of international conferences where many interesting papers on software reliability and testing can be found, for example,

IEEE Annual Reliability and Maintainability Symposium

ISSAT International Conference on Reliability and Quality in Design

IEEE International Computer Software and Applications Conference

IEEE International Conference on Fault-Tolerant Computing

IEEE International Conference on Software Engineering

IEEE International Symposium on Software Reliability Engineering

This list is by no means exhaustive, but it will provide readers with basic knowledge of software reliability.

Problems

1.1. What is the difference between reliability and availability?

1.2 What are the differences between hardware failures and software failures?

1.3 Provide examples of software failures, software defects and software faults.

References

Friedman, M.A. and J.M. Voas, *Software Assessment — Reliability, Safety, Testability,* John Wiley & Sons, New York, 1995.

Goel, A.L., "Software reliability models: Assumptions, limitations, and applicability," *IEEE Trans. Software Engineering*, Vol. SE-2(12), 1985.

IEEE Standard Glossary of Software Engineering Terminology, IEEE Standard 610.12, 1990.

Pham, H., *Fault-Tolerant Software Systems: Techniques and Applications*, IEEE Computer Society Press, Los Angeles, 1992.

Pham, H., "A software cost model with imperfect debugging, random life cycle and penalty cost," *Int. J. Systems Science*, Vol. 27(5), 1996.

Pham, H., "Software reliability," in *Wiley Encyclopedia of Electrical and Electronics Engineering*, J. G. Webster (ed.), John Wiley & Sons, New York, 1999.

Pham, H. and X. Zhang, "A software cost model with warranty and risk costs," *IEEE Trans. Computers*, Vol. 48(1), 1999.

Ramamoorthy, C. V. and F. B. Bastani, "Software reliability status and perspective," *IEEE Trans. Software Engineering*, Vol. SE-8(4), 1982.

Yamada, S., "Software quality/reliability measurement and assessment: Software reliability growth models and data analysis," *J. Information Processing*, Vol. 14(3), 1991.

Yamada, S. and S. Osaki, "Software reliability growth modeling: Models and applications," *IEEE Trans. Software Engineering*, Vol. SE-11(12), 1985.

2

Reliability Engineering Measures

The analysis of the reliability of a system must be based on precisely defined concepts. Since it is readily accepted that a population of supposedly identical systems, operating under similar conditions, fail at different points in time, then a failure phenomenon can only be described in probabilistic terms. Thus, the fundamental definitions of reliability must depend on concepts from probability theory. This chapter describes the fundamental definitions and concepts of reliability. These concepts provide the basis for quantifying the reliability of a system. They allow precise comparisons between systems or provide a logical basis for improvement in a failure rate. Various examples reinforce the definitions as presented in Sections 2.1 to 2.3. Section 2.4 examines common distribution models useful in reliability. Several distribution models are discussed and the resulting hazard functions are derived.

In general, a system may be required to perform various functions, each of which may have a different reliability. In addition, at different times, the system may have a different probability of successfully performing the required function under stated conditions. In reliability engineering, the term *failure* means that the system is not capable of performing a function when required. The term *capable* is used here to define if the system is capable of performing the required function. However, the term *capable* is unclear and only various degrees of capability can be defined.

2.1 Reliability Definitions

Reliability is defined as the probability of success or the probability that the system will perform its intended function under specified design limits. Mathematically, reliability $R(t)$ is the probability that a system will be successful in the interval from time 0 to time t:

$$R(t) = P(T > t), \quad t \geq 0 \tag{2.1}$$

where T is a random variable denoting the time-to-failure or failure time.

Unreliability $F(t)$, a measure of failure, is defined as the probability that the system will fail by time t.

$$F(t) = P(T \leq t) \quad t \geq 0.$$

In other words, $F(t)$ is the failure distribution function. If the time-to-failure random variable T has a density function $f(t)$, then

$$R(t) = \int_t^\infty f(s)ds$$

or, equivalently,

$$f(t) = -\frac{d}{dt}[R(t)].$$

The density function can be mathematically described in terms of T:

$$\lim_{\Delta t \to 0} P(t < T \leq t + \Delta t).$$

This can be interpreted as the probability that the failure time T will occur between the operating time t and the next interval of operation, $t + \Delta t$.

Consider a new and successfully tested system that operates well when put into service at time $t = 0$. The system becomes less likely to remain successful as the time interval increases. The probability of success for an infinite time interval, of course, is zero. Thus, the system functions at a probability of one and eventually decreases to a probability of zero. Clearly, reliability is a function of mission time. For example, one can say that the reliability of the system is 0.995 for a mission time of 24 hours. However, a statement such as the reliability of the system is 0.995 is meaningless because the time interval is unknown.

Example 2.1 A computer system has an exponential failure time density function

$$f(t) = \frac{1}{9,000} e^{-\frac{t}{9,000}} \quad t \geq 0.$$

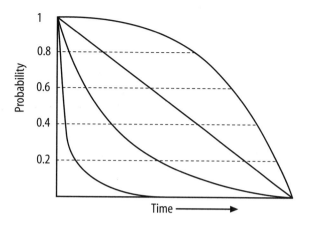

FIG. 2.1. Reliability function versus time.

What is the probability that the system will fail after the warranty (six months or 4,380 hours) and before the end of the first year (one year or 8,760 hours)?

Solution: From Eq. (2.1), we obtain

$$P(4,380 < T \leq 8,760) = \int_{4,380}^{8,760} \frac{1}{9,000} e^{-\frac{t}{9,000}} dt$$

$$= 0.237.$$

This indicates that the probability of failure during the interval from six months to one year is 23.7%.

If the time to failure is described by an exponential failure time density function, then

$$f(t) = \frac{1}{\theta} e^{-\frac{t}{\theta}} \quad t \geq 0, \theta > 0 \tag{2.2}$$

and this will lead to the reliability function

$$R(t) = \int_{t}^{\infty} \frac{1}{\theta} e^{-\frac{s}{\theta}} ds = e^{-\frac{t}{\theta}} \quad t \geq 0. \tag{2.3}$$

Consider the Weibull distribution where the failure time density function is given by

$$f(t) = \frac{\beta t^{\beta-1}}{\theta^{\beta}} e^{-(\frac{t}{\theta})^{\beta}} \quad t \geq 0, \theta > 0, \beta > 0.$$

Then, the reliability function is

$$R(t) = e^{-(\frac{t}{\theta})^{\beta}} \quad t \geq 0.$$

Thus, given a particular failure time density function or failure time distribution function, the reliability function can be obtained directly. Section 2.4 provides further insight for specific distributions.

2.2 System Mean Time to Failure

Suppose that the reliability function for a system is given by $R(t)$. The expected failure time during which a component is expected to perform successfully, or the system mean time to failure (MTTF), is given by

$$MTTF = \int_{0}^{\infty} tf(t)dt. \tag{2.4}$$

Substituting

$$f(t) = -\frac{d}{dt}[R(t)]$$

into Eq. (2.4) and performing integration by parts, we obtain

$$MTTF = -\int_{0}^{\infty} td[R(t)]$$

$$= [-tR(t)]\Big|_{0}^{\infty} + \int_{0}^{\infty} R(t)dt. \tag{2.5}$$

The first term on the right-hand side of Eq. (2.5) equals zero at both limits, since the system must fail after a finite amount of operating time. Therefore, we must have $tR(t) \to 0$ as $t \to \infty$. This leaves the second term, which equals

$$MTTF = \int_{0}^{\infty} R(t)dt. \tag{2.6}$$

Thus, in reliability engineering, MTTF is the definite integral evaluation of the reliability function. In general, if $\lambda(t)$ is defined as the failure rate function, then, by definition, MTTF is not equal to $1/(\lambda(t))$.

TABLE 2.1. Results of a twelve-component life duration test.

Component	Time to failure (hours)
1	4,510
2	3,690
3	3,550
4	5,280
5	2,595
6	3,690
7	920
8	3,890
9	4,320
10	4,770
11	3,955
12	2,750

The MTTF should be used when the failure time distribution function is specified because the reliability level implied by the MTTF depends on the underlying failure time distribution. Although the MTTF measure is one of the most widely used reliability calculations, it is also one of the most misused calculations. It has been misinterpreted as "guaranteed minimum lifetime." Consider the results as given in Table 2.1 for a twelve-component life duration test.

Using a basic averaging technique, the component MTTF of 3,660 hours was obtained. However, one of the components failed after 920 hours. Therefore, it is important to note that the system MTTF denotes the average time to failure. It is neither the failure time that could be expected 50% of the time, nor is it the guaranteed minimum time of system failure.

A careful examination of Eq. (2.6) will show that two failure distributions can have the same MTTF and yet produce different reliability levels. This is illustrated in a case where the MTTFs are equal, but with normal and exponential failure distributions. The normal failure distribution is symmetrical about its mean, thus

$$R(\mathrm{MTTF}) = P(Z \geq 0) = 0.5$$

where Z is a standard normal random variable. When we compute for the exponential failure distribution using Eq. (2.3), recognizing that $\theta = \mathrm{MTTF}$, the reliability at the MTTF is

$$R(\mathrm{MTTF}) = e^{-\frac{\mathrm{MTTF}}{\mathrm{MTTF}}} = 0.368.$$

Clearly, the reliability in the case of the exponential distribution is about 74% of that for the normal failure distribution with the same MTTF.

2.3   Failure Rate Function

The probability of a system failure in a given time interval $[t_1, t_2]$ can be expressed in terms of the reliability function as

$$\int_{t_1}^{t_2} f(t)dt = \int_{t_1}^{\infty} f(t)dt - \int_{t_2}^{\infty} f(t)dt$$

$$= R(t_1) - R(t_2)$$

or in terms of the failure distribution function (or the unreliability function) as

$$\int_{t_1}^{t_2} f(t)dt = \int_{-\infty}^{t_2} f(t)dt - \int_{-\infty}^{t_1} f(t)dt$$

$$= F(t_2) - F(t_1).$$

The rate at which failures occur in a certain time interval $[t_1, t_2]$ is called the *failure rate*. It is defined as the probability that a failure per unit time occurs in the interval, given that a failure has not occurred prior to t_1, the beginning of the interval. Thus, the failure rate is

$$\frac{R(t_1) - R(t_2)}{(t_2 - t_1)R(t_1)}. \tag{2.7}$$

Note that the failure rate is a function of time. If we redefine the interval as $[t, t + \Delta t]$, the expression in Eq. (2.7) becomes

$$\frac{R(t) - R(t + \Delta t)}{\Delta t R(t)}.$$

The rate in the above definitions is expressed as failures per unit time, when in reality, the time units might be in terms of miles, revolutions, etc.

The *hazard function* is defined as the limit of the failure rate as the interval approaches zero. Thus, the hazard function $h(t)$ is the instantaneous failure rate, and is defined by

$$h(t) = \lim_{\Delta t \to 0} \frac{R(t) - R(t + \Delta t)}{\Delta t R(t)}$$

$$= \frac{1}{R(t)}[-\frac{d}{dt}R(t)]$$

$$= \frac{f(t)}{R(t)}.$$

The quantity $h(t)dt$ represents the probability that a device of age t will fail in the small interval of time t to $t + dt$. The importance of the hazard function is that it indicates the change in the failure rate over the life of a population of components by plotting their hazard functions on a single axis. For example, two designs may provide the same reliability at a specific point in time, but the failure rates up to this point in time can differ. The death rate, in statistical theory, is analogous to the failure rate as the force of mortality is analogous to the hazard function.

Therefore, the hazard function or hazard rate or failure rate function is the ratio of the probability density function (pdf) to the reliability function. It is noted that all hazard functions must satisfy two conditions:

(1) $h(t) \geq 0$ for all $t \geq 0$

(2) $\int_{0}^{\infty} h(t)dt = \infty.$

2.4   Reliability Function for Common Distributions

This section presents some of the common distribution functions and their related hazard functions that have applications in reliability and system engineering. Several hazard models are presented in the following sections.

2.4.1   Binomial Distribution

The binomial distribution is one of the most widely used discrete random variable distributions in reliability and quality inspection. It has applications in reliability engineering, e.g., when one is dealing with a situation in which an event is either a success or a failure.

The pdf of the distribution is given by

$$P(X = x) = \binom{n}{x} p^x (1 - p)^{n-x} \quad x = 0, 1, 2, \ldots, n$$

$$\binom{n}{x} = \frac{n!}{x! \, (n - x)!}$$

where
n = number of trials
x = number of successes
p = single trial probability of success.

The reliability function, $R(k)$, (i.e., at least k out of n items are good) is given by

$$R(k) = \sum_{x=k}^{n} \binom{n}{x} p^x (1 - p)^{n-x}.$$

Example 2.2   Suppose in the production of lightbulbs, 90% are good. In a random sample of 20 lightbulbs, what is the probability of obtaining at least 18 good lightbulbs?

Solution   The probability of obtaining 18 or more good lightbulbs in the sample of 20 is

$$R(18) = \sum_{x=18}^{20} \binom{20}{18} (0.9)^x (1 - 0.9)^{20-x}$$

$$= 0.677.$$

2.4.2 Poisson Distribution

Although the Poisson distribution can be used in a manner similar to the binomial distribution, it is used to deal with events in which the sample size is unknown. This is also a discrete random variable distribution whose probability density function is given by

$$P(X = x) = \frac{(\lambda t)^x e^{-\lambda t}}{x!} \quad \text{for } x = 0, 1, 2, \ldots$$

where
λ = constant failure rate
x = is the number of events.

In other words, $P(X = x)$ is the probability of exactly x failures occurring in time t. Therefore, the reliability Poisson distribution, $R(k)$ (the probability of k or fewer failures) is given by

$$R(k) = \sum_{x=0}^{k} \frac{(\lambda t)^x e^{-\lambda t}}{x!}.$$

This distribution can be used to determine the number of spares required for the reliability of standby redundant systems during a given mission.

2.4.3 Exponential Distribution

Exponential distribution plays an essential role in reliability engineering because it has a constant failure rate. This distribution has been used to model the lifetime of electronic and electrical components and systems. This distribution is appropriate when a used component that has not failed is as good as a new component — a rather restrictive assumption. Therefore, it must be used diplomatically since numerous applications exist where the restriction of the memoryless property may not apply. For this distribution, we have reproduced Eqs. (2.2) and (2.3), respectively,

$$f(t) = \frac{1}{\theta} e^{-\frac{t}{\theta}} = \lambda e^{-\lambda t}, \quad t \geq 0$$

$$R(t) = e^{-\frac{t}{\theta}} = e^{-\lambda t} \quad t \geq 0$$

(2.8)

where $\theta = 1/\lambda > 0$ is an MTTF's parameter and $\lambda \geq 0$ is a constant failure rate.

The hazard function or failure rate for the exponential density function is constant, i.e.,

$$h(t) = \frac{f(t)}{R(t)}$$

$$= \frac{\frac{1}{\theta} e^{-\frac{t}{\theta}}}{e^{-\frac{t}{\theta}}} = \frac{1}{\theta} = \lambda.$$

The failure rate for this distribution is λ, a constant, which is the main reason for this widely used distribution.

We will now discuss some properties of the exponential distribution that are useful in understanding its characteristics, when and where it can be applied.

Property 1: Memoryless property   The exponential distribution is the only continuous distribution satisfying

$$P\{T \geq t\} = P\{T \geq t + s | T \geq s\} \quad \text{for } t > 0, s > 0.$$

This result indicates that the conditional reliability function for the lifetime of a component that has survived to time s is identical to that of a new component. This term is the so-called "used-as-good-as-new" assumption.

The lifetime of a fuse in an electrical distribution system may be assumed to have an exponential distribution. It will fail when there is a power surge causing the fuse to burn out. Assuming that the fuse does not undergo any degradation over time and that power surges that cause failure are likely to occur equally over time, then use of the exponential lifetime distribution is appropriate, and a used fuse that has not failed is as good as new.

Property 2 If $T_1, T_2, \ldots, T_n$ are independently and identically distributed exponential random variables (r.v.'s) with a constant failure rate λ, then

$$2\lambda \sum_{i=1}^{n} T_i \sim \chi(2n),$$

where $\lambda^2 (r)$ is a chi-squared distribution with degrees of freedom r. This result is useful for establishing a confidence interval for λ.

Example 2.3 Suppose that an electrical component has a failure rate of 10^{-4} failures per hour. What is the reliability of this component at 50 hours?

Solution From Eq. (2.8), the reliability is

$$R(50 \text{ hours}) = e^{-0.0001(50)} = 0.995.$$

2.4.4 Normal Distribution

Normal distribution plays an important role in classical statistics owing to the Central Limit Theorem. In reliability engineering, the normal distribution primarily applies to measurements of product susceptibility and external stress. This two-parameter distribution is used to describe systems in which a failure results due to some wearout effect for many mechanical systems.

The normal distribution takes the well-known bell shape. This distribution is symmetrical about the mean and the spread is measured by variance. The larger the value, the flatter the distribution. The pdf is given by

$$f(t) = \frac{1}{\sigma\sqrt{2\pi}} e^{-\frac{1}{2}\left(\frac{t-\mu}{\sigma}\right)^2} \quad -\infty < t < \infty$$

where μ is the mean value and σ is the standard deviation. In this case, the cumulative distribution function (cdf) is

$$F(t) = \int_{-\infty}^{t} \frac{1}{\sigma\sqrt{2\pi}} e^{-\frac{1}{2}\left(\frac{s-\mu}{\sigma}\right)^2} ds$$

and

$$R(t) = 1 - F(t).$$

There is no closed form solution for the above equation. However, tables for the standard normal density function are readily available (see Appendix D) and can be used to find probabilities for any normal distribution. If

$$Z = \frac{T - \mu}{\sigma}$$

is substituted into the normal pdf, we obtain

$$f(z) = \frac{1}{\sqrt{2\pi}} e^{-\frac{z^2}{2}} \quad -\infty < Z < \infty.$$

This is a so-called standard normal pdf, with a mean value of 0 and a standard deviation of 1. The standardized cdf is given by

$$\Phi(t) = \int_{-\infty}^{t} \frac{1}{\sqrt{2\pi}} e^{-\frac{1}{2}s^2} ds \tag{2.9}$$

where Φ is a standard normal distribution function. Thus, for a normal random variable T, with mean μ and standard deviation σ,

$$P(T \le t) = P\left(Z \le \frac{t - \mu}{\sigma}\right) = \Phi\left(\frac{t - \mu}{\sigma}\right)$$

where Φ yields the relationship necessary if standard normal tables are to be used.

The hazard function for a normal distribution is a monotonically increasing function of t. This can be easily shown by proving that $h'(t) \ge 0$ for all t. Since

$$h(t) = \frac{f(t)}{R(t)}$$

then

$$h'(t) = \frac{R(t)f'(t) + f^2(t)}{R^2(t)} \ge 0.$$

One can try this proof by employing the basic definition of a normal density function f.

Example 2.4   A component has a normal distribution of failure times with $\mu = 2{,}000$ hours and $\sigma = 100$ hours. Find the reliability of the component and the hazard function at 1,900 hours.

Solution   The reliability function is related to the standard normal deviate z by

$$R(t) = P\left(Z > \frac{t - \mu}{\sigma}\right)$$

where the distribution function for Z is given by Eq. (2.9). For this particular application,

$$R(1{,}900) = P\left[z > \frac{1{,}900 - 2{,}000}{100}\right]$$

$$= P[z > -1]$$

$$1 - \Phi(-1).$$

From the standard normal table in Appendix D, we obtain

$$R(1{,}900) = 1 - \Phi(-1) = 0.84135.$$

The value of the hazard function is found from the relationship

$$h(t) = \frac{f(t)}{R(t)} = \frac{\phi(z = \frac{t-\mu}{\sigma})}{\sigma R(t)}$$

where Φ is a pdf of standard normal density. Here

$$h(1{,}900) = \frac{\phi(-1.0)}{\sigma R(t)}$$

$$= \frac{0.84135}{100(0.15865)}$$

$$= 0.053 \text{ failures/cycle}.$$

2.4.5 Log Normal Distribution

This distribution, with its applications in maintainability engineering, is able to model failure probabilities of repairable systems and to model uncertainty in failure rate information. The log normal density function is given by

$$f(t) = \frac{1}{\sigma t \sqrt{2\pi}} e^{-\frac{1}{2}\left(\frac{\ln t - \mu}{\sigma}\right)^2} \quad t \geq 0$$

where μ and σ are parameters such that $-\infty < \mu < \infty$, and $\sigma > 0$. Note that μ and σ are not the mean and standard deviations of the distribution.

If a random variable X is defined as $X = \ln T$, then X is normally distributed with a mean of and a standard deviation of σ. That is,

$$E(X) = E(\ln T) = \mu$$

and

$$V(X) = V(\ln T) = \sigma^2.$$

Since $T = e^X$, the mean of the log normal distribution can be found by using the normal distribution. Consider that

$$E(T) = E(e^X) = \int_{-\infty}^{\infty} \frac{1}{\sigma\sqrt{2\pi}} e^{\left[x - \frac{1}{2}\left(\frac{x-\mu}{\sigma}\right)^2\right]} dx$$

and by rearrangement of the exponent, this integral becomes

$$E(T) = e^{\mu + \frac{\sigma^2}{2}} \int_{-\infty}^{\infty} \frac{1}{\sigma\sqrt{2\pi}} e^{-\frac{1}{2\sigma^2}[x - (\mu + \sigma^2)]^2} dx.$$

Thus, the mean of the log normal distribution is

$$E(T) = e^{\mu + \frac{\sigma^2}{2}}.$$

Proceeding in a similar manner,

$$E(T^2) = E(e^{2X}) = e^{2(\mu + \sigma^2)},$$

thus, the variance for the log normal is

$$V(T) = e^{2\mu + \sigma^2}[e^{\sigma^2} - 1].$$

The cumulative distribution function for the log normal is

$$F(t) \int_0^t \frac{1}{\sigma s \sqrt{2\pi}} e^{-\frac{1}{2}\left(\frac{\ln s - \mu}{\sigma}\right)^2} ds$$

and this can be related to the standard normal deviate Z by

$$F(t) = P[T \le t] = P(\ln T \le \ln t)$$
$$= P\left[Z \le \frac{\ln t - \mu}{\sigma}\right].$$

Therefore, the reliability function is given by

$$R(t) = P\left[Z > \frac{\ln t - \mu}{\sigma}\right], \tag{2.10}$$

and the hazard function would be

$$h(t) = \frac{f(t)}{R(t)} = \frac{\phi \frac{\ln t - \mu}{\sigma}}{\sigma t R(t)}. \tag{2.11}$$

Example 2.5   The failure time of a certain component is log normal distributed with $\mu = 5$ and $\sigma = 1$. Find the reliability of the component and the hazard rate for a life of 50 time units.

Solution   Substituting the numerical values of μ, σ, and t into Eq. (2.10), we compute

$$R(50) = P\left[Z > \frac{\ln 50 - 5}{1}\right]$$
$$= P[Z > -1.09]$$
$$= 0.8621$$

and using Eq. (2.11) for the hazard function,

$$h(50) = \frac{\phi\left(\frac{\ln 50 - 5}{1}\right)}{50(1)(0.8621)}$$
$$= \frac{\phi(-1.09)}{50(0.8621)} = 0.032 \text{ failures/unit.}$$

Thus, values for the log normal distribution are easily computed by using the standard normal tables.

2.4.6 Weibull Distribution

The exponential distribution, discussed in Subsection 2.4.3, is limited in applicability owing to the memoryless property. The Weibull distribution is a generalization of the exponential distribution and is often used to represent fatigue life, ball bearing life, and vacuum tube life. The Weibull distribution is extremely flexible and appropriate for modeling component lifetimes with fluctuating hazard rate functions and to represent various types of engineering applications.

The three-parameters probability density function is

$$f(t) = \frac{\beta(t - \gamma)^{\beta - 1}}{\theta^{\beta}} e^{-\left(\frac{t - \gamma}{\theta}\right)^{\beta}} \quad t \geq \gamma \geq 0 \tag{2.12}$$

where θ and β are known as the scale and shape parameters, respectively, and γ is known as the location parameter. These parameters are always positive. By using different parameters, this distribution can follow the exponential distribution, the normal distribution, etc.

It is clear that, for $t \geq \gamma$, the reliability function $R(t)$ is

$$R(t) = e^{-\left(\frac{t-\gamma}{\theta}\right)^{\beta}} \quad \text{for } t > \gamma > 0, \beta > 0, \theta > 0 \qquad (2.13)$$

hence,

$$h(t) = \frac{\beta(t-\gamma)^{\beta-1}}{\theta^{\beta}} \quad t > \gamma > 0, \beta > 0, \theta > 0. \qquad (2.14)$$

It can be shown that the hazard function is decreasing for $\beta < 1$, increasing for $\beta > 1$, and constant when $\beta = 1$.

Example 2.6   The failure time of a certain component has a Weibull distribution with $\beta = 4$, $\theta = 2,000$, and $\gamma = 1,000$. Find the reliability of the component and the hazard rate for an operating time of 1,500 hours.

Solution   A direct substitution into Eq. (2.13) yields

$$R(1,500) = e^{-\left(\frac{1,500-1,000}{2,000}\right)^4} = 0.996.$$

Using Eq. (2.14), the desired hazard function is given by

$$h(1,500) = \frac{4(1,500 - 1,000)^{4-1}}{(2,000)^4}$$
$$= 3.13 \ 10^{-5} \text{ failures/hour.}$$

Note that the Rayleigh and exponential distributions are special cases of the Weibull distribution at $\beta = 2$, $\gamma = 0$, and $\beta = 1$, $\gamma = 0$, respectively. For example, when $\beta = 1$ and $\gamma = 0$, the reliability of the Weibull distribution function in Eq. (2.13) reduces to

$$R(t) = e^{-\frac{t}{\theta}}$$

and the hazard function given in Eq. (2.14) reduces to $1/\theta$, a constant. Thus, the exponential is a special case of the Weibull distribution. Similarly, when $\gamma = 0$ and $\beta = 2$, the Weibull probability density function becomes the Rayleigh density function. That is

$$f(t) = \frac{2}{\theta} t e^{-\frac{t^2}{\theta}} \quad \text{for } \theta > 0, t \geq 0.$$

2.4.7 Gamma Distribution

The gamma distribution can be used as a failure probability function for components whose distribution is skewed. The failure density function for a gamma distribution is

$$f(t) = \frac{t^{\alpha-1}}{\beta^\alpha \Gamma(\alpha)} e^{-\frac{t}{\beta}} \quad t \geq 0, \quad \alpha, \beta > 0 \tag{2.15}$$

where α is the shape parameter and β is the scale parameter. Hence,

$$R(t) = \int_t^\infty \frac{1}{\beta^\alpha \Gamma(\alpha)} s^{\alpha-1} e^{-\frac{s}{\beta}} ds.$$

If α is an integer, it can be shown by successive integration by parts that

$$R(t) = e^{-\frac{t}{\beta}} \sum_{i=0}^{\alpha-1} \frac{\left(\frac{t}{\beta}\right)^i}{i!} \tag{2.16}$$

and

$$h(t) = \frac{f(t)}{R(t)}$$

$$= \frac{\frac{1}{\beta^\alpha \Gamma(\alpha)} t^{\alpha-1} e^{-\frac{t}{\beta}}}{e^{-\frac{t}{\beta}} \sum_{i=0}^{\alpha-1} \frac{\left(\frac{t}{\beta}\right)^i}{i!}}.$$

The gamma failure density function has shapes that are very similar to the Weibull distribution. At $\alpha = 1$, the gamma distribution becomes the exponential distribution with the constant failure rate $1/\beta$.

The gamma distribution can also be used to model the time to the nth failure of a system if the underlying failure distribution is exponential. Thus, if X_i is exponentially distributed with parameter $\theta = 1/\beta$, then $T = X_1 + X_2 + \cdots + X_n$ is gamma distributed with parameters β and n.

Example 2.7   The time to failure of a component has a gamma distribution with $\alpha = 3$ and $\beta = 5$. Determine the reliability of the component and the hazard rate at 10 time units.

Solution   Using Eq. (2.16), we compute

$$R(10) = e^{-\frac{10}{5}} \sum_{i=0}^{2} \frac{\left(\frac{10}{5}\right)^i}{i!} = 0.6767.$$

From Eq. (2.15), we obtain

$$h(10) = \frac{f(10)}{R(10)}$$

$$= \frac{0.054}{0.6767} = 0.798 \text{ failures/unit time.}$$

2.4.8 Beta Distribution

This is a two-parameter distribution that also has applications in reliability engineering. This probability density function, $f(t)$, is defined as

$$f(t) = \frac{\Gamma(\alpha + \beta)}{\Gamma(\alpha)\Gamma(\beta)} t^{\alpha}(1 - t)^{\beta} \quad 0 < t < 1, \alpha > 0, \beta > 0$$

where α and β are the distribution parameters.

2.4.9 Hazard Rate Model I

This is a two-parameter distribution that can have increasing and decreasing hazard rates. The hazard rate, $h(t)$, is defined as

$$h(t) = \frac{\lambda(b+1)[\ln(\lambda t + \alpha)]^{b}}{(\lambda t + \alpha)} \quad b \geq 0, \lambda > 0, \alpha \geq 1, t \geq 0$$

where
$b =$ shape parameter
$\lambda =$ scale parameter
$\alpha =$ third parameter.

The reliability function $R(t)$ for $\alpha = 1$ is

$$R(t) = e^{-[\ln(\lambda t + \alpha)]^{b+1}}.$$

The probability density function $f(t)$ is

$$f(t) = R(t) \cdot h(t)$$

$$= e^{-[\ln(\lambda t + \alpha)]^{b+1}} \frac{\lambda(b+1)[\ln(\lambda t + \alpha)]^{b}}{(\lambda t + \alpha)}.$$

2.4.10 Hazard Rate Model II

This is another two-parameter distribution that can have increasing and decreasing hazard rates. The hazard rate, $h(t)$, and the reliability function, $R(t)$, are defined as

$$h(t) = \frac{n\lambda t^{n-1}}{\lambda t^{n} + 1} \quad \text{for } n \geq 1, \lambda > 0, t \geq 0$$

where
$n =$ shape parameter
$\lambda =$ scale parameter.

The reliability function is given by

$$R(t) = e^{-\ln(\lambda t^N + 1)}.$$

The probability density function $f(t)$ is

$$f(t) = \frac{n\lambda t^{n-1}}{\lambda t^n + 1} e^{-\ln(\lambda t^n + 1)} \quad n \geq 1, \lambda > 0, t \geq 0.$$

2.5 Maintainability and Availability

2.5.1 Maintainability

When a system fails to perform satisfactorily, repair is normally carried out to locate and correct the fault. The system is restored to operational effectiveness by making an adjustment or by replacing a component.

Maintainability is defined as the probability that a failed system will be restored to specified conditions within a given period of time when maintenance is performed according to prescribed procedures and resources. In other words, maintainability is the probability of isolating and repairing a fault in a system within a given time. Maintainability engineers must work with system designers to ensure that the system product can be maintained by the customer efficiently and cost effectively. This function requires the analysis of part removal, replacement, tear-down, and build-up of the product in order to determine the required time to do the operation, the necessary skill, the type of support equipment and the documentation.

Let T denote the random variable of the time to repair or the total downtime. If the repair time T has a repair time density function $g(t)$, then the maintainability, $V(t)$, is defined as the probability that the failed system will be back in service by time t, i.e.,

$$V(t) = P(T \leq t) = \int_0^t g(s)ds.$$

For example, if $g(t) = \mu e^{-\mu t}$, where $\mu > 0$ is a constant repair rate, then

$$V(t) = 1 - e^{-\mu t}$$

which represents the exponential form of the maintainability function.

An important measure often used in maintenance studies is the mean time to repair (MTTR) or the mean downtime. MTTR is the expected value of the random variable repair time, not failure time, and is given by

$$\text{MTTR} = \int\limits_0^\infty tg(t)dt.$$

When the distribution has a repair time density given by $g(t) = \mu e^{-\mu t}$, then, from the above equation, MTTR $= 1/\mu$. When the repair time T has the log normal density function $g(t)$, and the density function is given by

$$g(t) = \frac{1}{\sqrt{2\pi}\sigma t} e^{-\frac{(\ln t - \mu)^2}{2\sigma^2}} \quad t > 0,$$

then it can be shown that

$$\text{MTTR} = m e^{\frac{\sigma^2}{2}}$$

where m denotes the median of the log normal distribution.

In order to design and manufacture a maintainable system, it is necessary to predict the MTTR for various fault conditions that could occur in the system. This is generally based on past experiences of designers and the expertise available to handle repair work.

The system repair time consists of two separate intervals: passive repair time and active repair time. Passive repair time is mainly determined by the time taken by service engineers to travel to the customer site. In many cases, the cost of travel time exceeds the cost of the actual repair. Active repair time is directly affected by the system design and is listed as follows:

1. The time between the occurrence of a failure and the system user becoming aware that it has occurred.
2. The time needed to detect a fault and isolate the replaceable component(s).
3. The time needed to replace the faulty component(s).
4. The time needed to verify that the fault has been corrected and the system is fully operational.

The active repair time can be improved significantly by designing the system in such a way that faults may be quickly detected and isolated. As more complex systems are designed, it becomes more difficult to isolate the faults.

2.5.2 Availability

Reliability is a measure that requires system success for an entire mission time. No failures or repairs are allowed. Space missions and aircraft flights are examples of systems where failures or repairs are not allowed. Availability is a measure that allows for a system to repair when failure occurs.

The availability of a system is defined as the probability that the system is successful at time t. Mathematically,

$$\text{Availability} = \frac{\text{System up time}}{\text{System up time} + \text{System down time}}$$
$$= \frac{\text{MTTF}}{\text{MTTF} + \text{MTTR}}.$$

Availability is a measure of success used primarily for repairable systems. For non-repairable systems, availability, $A(t)$, equals reliability, $R(t)$. In repairable systems, $A(t)$ will be equal to or greater than $R(t)$.

The mean time between failures (MTBF) is an important measure in repairable systems. This implies that the system has failed and has been repaired. Like MTTF and MTTR, MTBF is an expected value of the random variable time between failures. Mathematically,

$$\text{MTBF} = \text{MTTF} + \text{MTTR}$$

The term MTBF has been widely misused. In practice, MTTR is much smaller than MTTF, which is approximately equal to MTBF. MTBF is often incorrectly substituted for MTTF, which applies to both repairable systems and non-repairable systems.

If the MTTR can be reduced, availability will increase, and the system will be more economical. A system where faults are rapidly diagnosed is more desirable than a system that has a lower failure rate but where the cause of a failure takes longer to detect, resulting in a lengthy system downtime.

When the system being tested is renewed through maintenance and repairs, $E(T)$ is also known as MTBF.

Problems

2.1 Assume that the hazard rate, $h(t)$, has a positive derivative. Show that the hazard distribution

$$H(t) = \int_0^t h(x)dx$$

is strictly convex.

2.2 An operating unit is supported by $n-1$ identical units on cold standby. When it fails, a unit from standby takes its place. The system fails if all n units fail. Assume that units on standby cannot fail and the lifetime of each unit follows the exponential distribution with failure rate λ.

(a) What is the distribution of the system lifetime?

(b) Determine the reliability of the standby system for a mission of 100 hours when $\lambda = 0.0001$ per hour and $n = 5$.

2.3 Assume that there is some latent deterioration process occurring in the system. During the interval $[0, a\text{-}h]$ the deterioration is comparatively small so that the shocks do not cause system failure. During a relatively short time interval $[a\text{-}h, a]$, the deterioration progresses rapidly and makes the system susceptible to shocks. Assume that the appearance of each shock follows the exponential distribution with failure rate λ. What is the distribution of the system lifetime?

2.4 Consider a series system of n Weibull components. The corresponding lifetimes $T_1, T_2, \ldots, T_n$ are assumed to be independent with pdf

$$f(t) = \begin{cases} \lambda_i^\beta \beta t^{\beta-1} e^{-(\lambda_i t)^\beta} & \text{for } t \geq 0, \\ 0 & \text{otherwise,} \end{cases}$$

where $\lambda > 0$ and $\beta > 0$ are the scale and shape parameters, respectively.

(a) Show that the lifetime of a series system has the Weibull distribution with pdf

$$f_s(t) = \begin{cases} \left(\sum_{i=1}^n \lambda_i^\beta\right) \beta t^{\beta-1} e^{-[\sum_{i=1}^n \lambda_i^\beta) t^\beta]} & \text{for } t \geq 0 \\ 0 & \text{otherwise.} \end{cases}$$

(b) Find the reliability of this series system.

2.5 Consider the pdf of a random variable that is equally likely to take on any value *only* in the interval from a to b.

(a) Show that this pdf is given by

$$f(t) = \begin{cases} \dfrac{1}{b-a} & \text{for } a < t < b \\ 0 & \text{otherwise.} \end{cases}$$

(b) Derive the corresponding reliability function $R(t)$ and failure rate $h(t)$.

(c) Think of an example where such a distribution function would be of interest in reliability application.

2.6 The failure rate function, denoted by $h(t)$, is defined as

$$h(t) = \lim_{dt \to 0} \frac{F(t+dt) - F(t)}{dt\, R(t)}$$
$$= -\frac{1}{R(t)} \frac{dR(t)}{dt}$$
$$= -\frac{d}{dt} \ln[R(t)].$$

Show that the constant failure rate function implies an exponential distribution.

2.7 One thousand new streetlights are installed in Saigon. Assume that the lifetime of these streetlights follow the normal distribution. The average life of these lamps is estimated at 980 burning-hours with a standard deviation of 100 hours.

(a) What is the expected number of lights that will fail during the first 800 burning-hours?

(b) What is the expected number of lights that will fail between 900 and 1,100 burning-hours?

(c) After how many burning-hours would 10% of the lamps be expected to fail?

2.8 A fax machine with constant failure rate λ will survive for a period of 720 hours without failure, with probability 0.80.

(a) Determine the failure rate λ.

(b) Determine the probability that the machine, which is functioning after 600 hours, will still function after 800 hours.

(c)   Find the probability that the machine will fail within 900 hours, given that the machine was functioning at 720 hours.

2.9   The time to failure T of a unit is assumed to have a log normal distribution with pdf

$$f(t) = \frac{1}{\sqrt{2\pi}} \frac{1}{\sigma t} e^{-\frac{(\ln t - \mu)^2}{2\sigma^2}} \quad t > 0.$$

Show that the failure rate function is unimodal.

2.10   A diode may fail due to either open or short failure modes. Assume that the time to failure T_0 caused by open mode is exponentially distributed with pdf

$$f_0(t) = \lambda_0 e^{-\lambda_0 t} \quad t \geq 0$$

and the time to failure T_1 caused by short mode has the pdf

$$f_s(t) = \lambda_s e^{-\lambda_s t} \quad t \geq 0.$$

The pdf for the time to failure T of the diode is given by

$$f(t) = p f_0(t) + (1 - p) f_s(t) \quad t \geq 0.$$

(a)   Explain the meaning of p in the above pdf function.

(b)   Derive the reliability function $R(t)$ and failure rate function $h(t)$ for the time to failure T of the diode.

(c)   Show that the diode with pdf $f(t)$ has a decreasing failure rate.

2.11   A diesel is known to have an operating life (in hours) that fits the following pdf:

$$f(t) = \frac{2a}{(t + b)^2} \quad t \geq 0.$$

The average operating life of the diesel has been estimated to be 8,760 hours.

(a)   Determine a and b.

(b)   Determine the probability that the diesel will not fail during the first 6,000 operating-hours.

(c)   If the manufacturer wants no more than 10% of the diesels returned for warranty service, how long should the warranty be?

2.12   The failure rate for a hydraulic component is

$$h(t) = \frac{t}{t + 1} \quad t > 0$$

where t is in years.

(a)   Determine the reliability function $R(t)$.

(b)   Determine the MTTF of the component.

2.13   A 18-month guarantee is given based on the assumption that no more than 5% of new cars will be returned.

(a)   The time to failure T of a car has a constant failure rate. What is the maximum failure rate that can be tolerated?

(b)   Determine the probability that a new car will fail within three years assuming that the car was functioning at 18 months.

2.14   Show that if

$$R_1(t) \geq R_2(t) \quad \text{for all } t,$$

where $R_i(t)$ is the system reliability of the structure i, then MTTF of the system structure 1 is always $\geq$ MTTF of the system structure 2.

References

Birnbaum, Z.W. and S.C. Saunders, "A new family of life distributions," *J. Applied Probability,* Vol. 6(2), 1969, 319–327.

Dhillon, B.S., "A hazard rate model," *IEEE Trans. Reliability,* Vol. 28, June 1979, 150.

Dhillon, B.S., "Life distributions," *IEEE Trans. Reliability,* Vol. 30, December 1981, 457–460.

Dovich, R.A., *Reliability Statistics*, ASQC Quality Press, Milwaukee, 1990.

Drake, A., *Fundamentals of Applied Probability Theory*, McGraw-Hill, New York, 1967.

Feller, W., *An Introduction to Probability Theory and its Applications*, Vol. I, Wiley, New York, 1957.

Kapur, K.C. and L.R. Lamberson, *Reliability in Engineering Design*, Wiley, 1977.

Leemis, L., "Probabilistic properties of the exponential distribution," *Microelectronics and Reliability*, Vol. 28(2), 1988, 257–262.

Shooman, M.L., *Probabilistic Reliability: An Engineering Approach*, McGraw-Hill, New York, 1968.

Weibull, W., "A statistical distribution function of wide applicability," *J. Applied Mech.*, Vol. 18, 1951, 293–297.

3

Software Engineering Assessment

"You can't control what you can't measure."

3.1   Introduction

As software becomes increasingly important in systems that perform complex and critical functions, e.g., military defense, nuclear reactors, so too have the risks and impacts of software-caused failures. There is now general agreement on the need to increase software reliability and quality by eliminating errors created during software development. Industry and academic institutions have responded to this need by improving developmental methods in the technology known as software engineering and by employing systematic checks to detect software errors during and in parallel with the developmental process. Many organizations make the reduction of defects their first quality goal. The consumer electronics business, however, pursues a different goal: maintaining the number of defects in the field at zero. When electronic products leave the showroom, their destination is unknown. Therefore, detecting and correcting a serious software defect would entail recalling hundreds of thousands of products.

In the past 25 years, hundreds of research papers have been published in the areas of software quality, software engineering development process, software reliability modeling, software independent verification and validation (IV&V) and software fault tolerance. Software engineering is evolving from an art to a practical engineering discipline (Lyu, 1996). A large number of analytical models have been proposed and studied over the last two decades for assessing the quality of a software system. Each model must make some assumptions about the development process and test environment. The environment can change depending on the software application, the lifecycle development process as well as the capabilities of the engineering design team (Malaiya, 1990). Therefore, it is important for software users and practitioners to be familiar with all the relevant models in order to make informed decisions about the quality of any software product.

This chapter provides the basic concepts of software reliability and testing, the software engineering assessment including software lifecycle, software development process and its applications, software verification and validation, and data collection and analysis.

3.2   Software Versus Hardware Reliability

The development of hardware reliability theory has a long history and was established to improve hardware reliability greatly while the size and complexity of software applications have increased (Xie, 1991). Hardware reliability encompasses a wide spectrum of analyses that strive systematically to reduce or eliminate system failures which adversely affect product performance. Reliability also provides the basic approach for assessing safety and risk analysis.

Software reliability strives systematically to reduce or eliminate system failures which adversely affect performance of a software program. Software systems do not degrade over time unless modified. There are many differences between the reliability and testing concepts and techniques of hardware and software. Therefore, a comparison of software and hardware reliability would be useful in developing software reliability modeling. Table 3.1 shows the differences and similarities between the two. Figure 3.1 also shows the sequence of failure either by the hardware or software. The result is that software quality and reliability must be built into software during the developmental process.

Example 3.1: Chilled Water System   The chilled water system acts as a heat sink for an air-conditioning system and the electronics cooling system. It provides a supply of chilled water to the individual air-handling units (AHUs) and to the suction of the electronics cooling water system pumps. The system is a closed loop design that utilizes two large chiller units to cool the warm water returning from the cooling loads. The chillers in turn reject the waste heat to the cooling tower system. The chillers require 4,160 volt power for operation. The chilled water pumps provide circulation of the

TABLE **3.1.** Software reliability versus hardware reliability.

Software reliability	Hardware reliability
Without considering program evolution, failure rate is statistically non-increasing.	Failure rate has a bathtub curve. The burn-in state is similar to the software debugging state.
Failures never occur if the software is not used.	Material deterioration can cause failures even though the system is not used.
Most models are analytically derived from assumptions. Emphasis is on developing the model, the interpretation of the model assumptions, and the physical meaning of the parameters.	Failure data are fitted to some distributions. The selection of the underlying distribution is based on the analysis of failure data and experiences. Emphasis is placed on analyzing failure data.
Failures are caused by incorrect logic, incorrect statements, or incorrect input data. This is similar to design errors of a complex hardware system.	Failures are caused by material deterioration, random failures, design errors, misuse, and environment.
Software reliability can be improved by increasing the testing effort and by correcting detected faults. Reliability tends to change continuously during testing due to the addition of problems in new code or to the removal of problems by debugging errors.	Hardware reliability can be improved by better design, better material, applying redundancy and accelerated life testing.
Software repairs establish a new piece of software.	Hardware repairs restore the original condition.
Software failures are rarely preceded by warnings.	Hardware failures are usually preceded by warnings.
Software components have rarely been standardized.	Hardware components can be standardized.
Software essentially requires infinite testing.	Hardware can usually be tested exhaustively.

chilled water from the output of the chillers to the chilled water loads and then back to the chillers, requiring 480-volt power for operation. As it circulates through the AHU heat exchanger, it picks

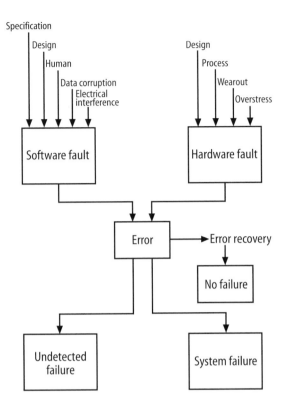

F<small>IG.</small> **3.1.** Sequence of failure.

up waste heat and is then returned to the chillers as warm water to be rechilled. The chillers, in turn, are cooled by emitting the waste heat into the atmosphere. Also, the system requires a supply of instrument air for proper operation of the system controllers. Figure 3.2(a) is a simplified diagram of the chilled water system.

The chilled water system consists of two chillers and two chilled water pumps that are used to circulate chilled water to the various chilled water loads. The system success criteria are based on supplying chilled water to the operations building AHUs. There are five AHUs in the operations building: two main units supply cool air to the entire building and three critical units supply cooling to three cooling zones in the critical operations area. Successful operation was designed as follows: one of the two chillers operating, and one of the two chilled water pumps running, supplying chilled water to the operations building. Additionally, one of the two main operations building AHUs must be operating along with two of the three critical area AHUs. Figure 3.2(b) is the chilled water system reliability block diagram. Its success criteria

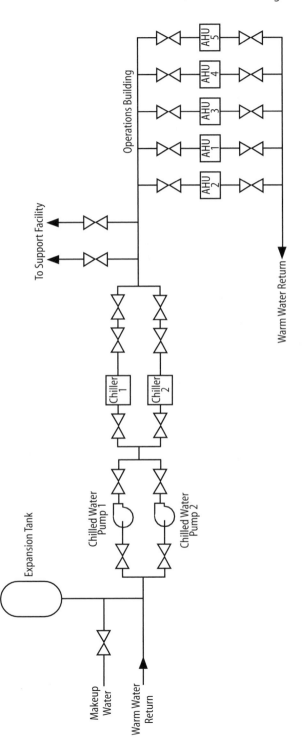

Fig. 3.2(a). Chilled water system simplified diagram.

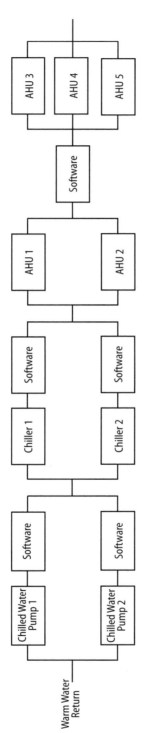

FIG. 3.2(b). Chilled water system reliability block diagram.

are based on the successful operation of one of the chillers and one of the chilled water pumps. It is also assumed that success requires one of the main operations building AHUs and two of the three critical area AHUs.

All the water pumps, the chillers, and AHUs can be represented by exponential distributions with failure rates λ_p, λ_c, and λ_a, respectively. Assume that

$$\lambda_p = 0.001 \text{ per hour}$$
$$\lambda_c = 0.0005 \text{ per hour}$$
$$\lambda_a = 0.0001 \text{ per hour.}$$

Assume a mission of $t = 72$ hours, then

(i) the reliability of the water pumps subsystem is

$$R_p = 1 - (1 - e^{-(0.001)(72)})^2$$
$$= 0.99517.$$

(ii) the reliability of the chillers' subsystem is

$$R_c = 1 - (1 - e^{-(0.0005)(72)})^2$$
$$= 0.99875.$$

(iii) the reliability of the AHU subsystem I is

$$R_{Ia} = 1 - (1 - e^{-(0.0001)(72)})^2$$
$$= 0.99995.$$

(iv) the reliability of the AHU subsystem II is

$$R_{IIa} = \sum_{i=2}^{3} \binom{3}{i} (e^{-(0.0001)(72)})^i (1 - e^{-(0.0001)(72)})^{3-i}$$
$$= 0.99985.$$

Assuming all the software systems do not fail, the overall chilled water system reliability is

$$R_s = R_p \times R_c \times R_{Ia} \times R_{IIa}$$
$$= 0.99373. \tag{3.1}$$

The software, however, does fail. The water pumps, chillers, and AHU software each contain approximately 250,000 lines of source code in ground control and processing to operate innumerable hardware units. One observes that the software reliabilities of the water pumps, chillers, and AHUs for a 72-hour mission are

$$p_p = 0.97, p_c = 0.99, \text{ and } p_a = 0.995,$$

respectively.

Therefore, the reliability of the overall chilled water hardware–software system becomes

$$R_s = (1 - [1 - (0.97)e^{-(0.001)(72)}]^2)$$
$$(1 - [1 - (0.99)e^{-(0.0005)(72)}]^2)(0.995)R_{Ia} \cdot R_{IIa}$$
$$= 0.983354$$

which is far less than the reliability result in Eq. (3.1).

3.3   Software Reliability and Testing Concepts

Software is essentially an instrument for transforming a discrete set of inputs into a discrete set of outputs. It comprises a set of coded statements or instructions whose functions may be to evaluate an expression and store the result in a temporary or permanent location, decide which statement to execute, or to perform input/output operations [Goel, 1985]. Hence, a software can be regarded as a function f, mapping the input space to the output space (f: input → output), where the input space is the set of all input states and the output space is the set of all output states (see Fig. 3.3).

Software reliability is the probability that a given software functions without failure in a given environmental condition during a specified time. Another deterministic model defines software reliability as the probability of successful execution(s) of an input state randomly selected from the input space under specified operating conditions. Another definition is the probability of failure-free execution of the software for a specified time in a specified environment. For example, an operating system with a reliability of 95% for 8 hours for an average user should work 95 out of 100 periods of 8 hours without any problems. Software failure means the inability to perform an intended task specified by a requirement. A software fault (or bug) is an error in the program source-text, which causes software failure when the program is executed under certain conditions. Hence, a software fault is generated when a mistake is made.

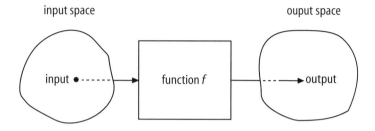

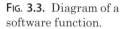

Fɪɢ. **3.3.** Diagram of a
software function.

In vehicular applications, for instance, errors can be divided
into four categories: critical, high, moderate and low.

Critical: It may affect a federally mandated item. (A change
            is required immediately, i.e., brake system.)

High: It may affect a necessary function of vehicle opera-
            tion or subsystem, i.e., inoperative engine, air-con-
            ditioning, locks, etc. (Potential customer satis-
            faction item, a change is required immediately.)

Moderate: It may affect a convenience feature, i.e., chimes,
            cruise control, compass, mini-trip computer, head-
            lamp delay, etc. (Change at the next opportunity.)

Low: It is unreasonable to expect that the minor nature of
            this item would cause any real affect on the vehicle
            or system performance. (No change required at this
            time.)

In general, the definition of what constitutes a software failure is
an area open for debate since it depends on the application. When a
program crashes, it has obviously failed due to an error in design,
specifications, coding, or testing. One can define a failure as not
meeting the user's requirements or expectations of the software
operation. This can be due to a number of criteria which are not
always well-defined. An example is that the speed of execution,
accuracy of the computations, etc., can be the criteria for failure
of the software.

Software testing is the process of executing a program to locate
an error. A good test case is one that has a high probability of finding undiscovered error(s). It is impossible to continue testing the
software until all faults are detected and removed as testing of all
possible inputs would require millions of years! Therefore, failure
probabilities must be inferred from testing a sample of all possible
input states called the *input space*. In other words, input space is

the set of all possible input states. Similarly, output space is the set of all possible output states for a given software and input space.

It is generally very difficult to exhaustively test a large computer program because of problems with dimensionality. If the input space consists of a single unbounded variable, then an infinite number of input cases will be needed to provide an exhaustive test of the program. If the input space is bounded, but contains a large number of independent variables, then the number of input cases needed for an exhaustive test will tend to be impossibly large, even if one accepts the use of discretization for each input variable. Similarly, it will probably be an impossible task to test each pathway through the program because of the very large number of paths involved. For instance, the flow graph of a small program in Fig. 3.4 shows the schematics of a fairly small program, with a number of

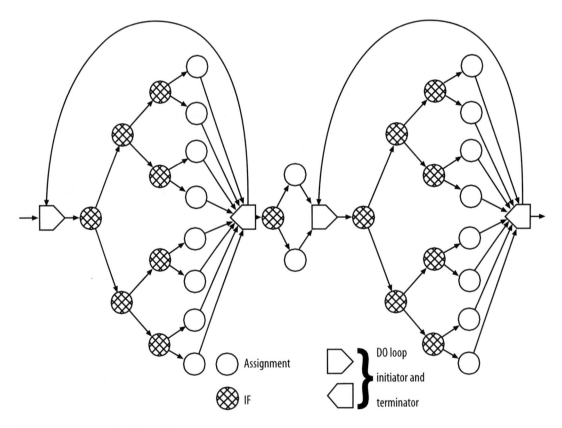

FIG. **3.4.** Flow graph of a small program.

DO loops and IF branches, with 10^{18} unique paths through it. Even though exhaustive software testing may not be feasible for a very large software system, it makes sense to carry out a series of tests on each functional area. In the context of the user relationship, the user should obviously stipulate the tests to be carried out and be actively involved in their execution (Churchley, 1991).

As we know, different inputs have different chances of being selected, and we can never be sure which inputs are selected in the operational phase of real world applications. During the operational phase, some input states are executed more frequently than others and a probability can be assigned to each input state to form the operational profile of the program. This operational profile can be used to construct the software reliability model. This type of model is also called an *input-domain model*. The interesting questions here are: (1) Is it possible to determine the sizes and locations of the fault regions in the input space? (2) How do we determine the reliability that a program will execute correctly for a particular length of time?

To select a good software model in order to make an accurate reliability prediction, the testing strategy should be incorporated into the software reliability model. It should be noted that the best models may vary from time to time and differ from application to application.

The evaluation of software reliability cannot be performed without software failure data. Therefore, the establishment of a software failure database is useful to both practitioners and researchers for predicting and estimating software reliability and to determine the total testing time needed to reach a desired reliability goal. Collected data are grouped into four categories: component data, management data, dynamic failure data, and fault removal data, each with a unique set of information.

The information in the component data category contains the number of executable source lines of code, the total number of comments and instructions, and the source language used for each system component.

The information in the management data category is the starting and ending date for each lifecycle phase (analysis, design, code, test, and operation), the definitions and requirements of each lifecycle phase, and the models used for estimating software reliability.

The information in the dynamic failure data category is the number of CPU hours since the last failure, the number of test cases executed since the last failure, the severity of the failure, the method of failure detection, and the unit complexity and size where the fault was detected.

The information in the fault removal data category is the date and time of fixing an error, the CPU hours required to fix an error, and the labor hours required to fix an error for each failure corrected.

The data collection addresses these questions:

- Are the defects discovered as a result of simply testing artifacts of prior modifications or are they previously undetected defects?
- What taxonomic categories are required for defense, aerospace, military, and commercial systems, and what defect percentages reside in each category?

3.4   Software Lifecycle

A software lifecycle provides a systematic approach to developing, using, operating, and maintaining a software system. The standard IEEE computer dictionary has defined the software lifecycle as: "That period of time in which the software is conceived, developed and used." There are many different definitions of software lifecycle (Boehm, 1981; Pressman, 1983).

A software lifecycle consists of the following five successive phases shown in Fig. 3.5:

1. Analysis (requirements and functional specifications)
2. Design
3. Coding
4. Testing
5. Operating

The phases and activities of the software lifecycle, in general, are given in Figs. 3.6–3.11. Table 3.2 shows the errors introduced and errors detected in the software lifecycle of a commercial application. In the early phases of the software lifecycle, a predictive model is needed because no failure data are available. This type of model predicts the number of initial faults in the software before testing. In the testing phase, the software reliability improves through perfect debugging.

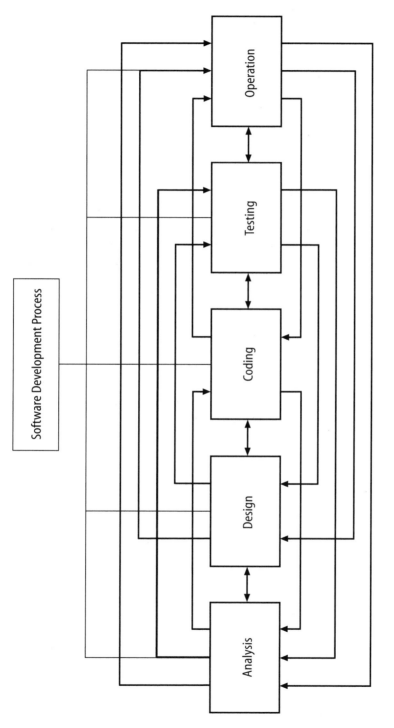

FIG. 3.5. Software development process lifecycle.

TABLE **3.2.** Software error introduction and discovery.

Lifecycle phase	Errors introduced (%)	Errors detected (%)
Analysis	55	18
Design	30	10
Coding and testing	10	50
Operations	5	22

By assuming perfect debugging, i.e., a fault is removed with certainty whenever a failure occurs, the number of remaining faults is a decreasing function of debugging time. With an imperfect debugging assumption, i.e., faults may or may not be removed, introduced, or changed at each debugging, the number of remaining faults may increase or decrease.

A reliability growth model is needed to estimate the current reliability level and the time and resources required to achieve the desired reliability goal. During this phase, reliability estimation is based on the analysis of failure data. After the release of a software program, the addition of new modules, removal of old ones, removal of detected errors, mixing of newly and previously written code, change of user environment, and change of hardware and management involvement have to be considered in the evaluation of software reliability. An evolution model is thus needed.

3.4.1 Analysis Phase

The analysis phase is the first step in the software development process. It is also the most important phase in the whole process and the foundation of building a successful software product. A survey at the North Jersey Software Process Improvement Network workshop in August 1995 showed that about 35% of the effort in software development projects should be concentrated in the analysis phase.

The purpose of the analysis phase is to define the requirements and provide specifications for the subsequent phases and activities. The analysis phase is composed of three major activities: problem definition, requirements, and specifications (see Fig. 3.6).

Problem definition develops the problem statement and the scope of the project. It is important to understand what the user's problem is and why the user needs a software product to solve the problem. This is determined by the frequent interactions with

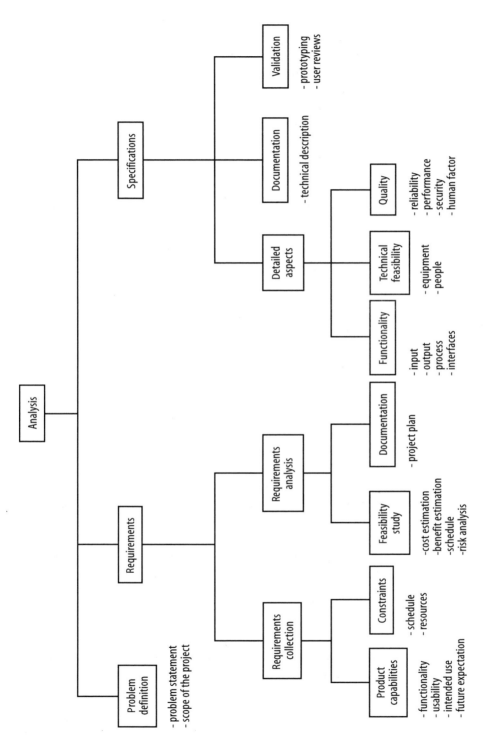

Fig. 3.6. Analysis phase.

customers. A well-defined problem and its scope can help focus further development activities.

The requirement activity consists of collecting and analyzing requirements. Requirement collection includes product capabilities and constraints. Product capabilities qualitatively describe how well the system will perform. To obtain the qualitative requirements, we need to collect information on product functionality, usability, intended use, and future expectations. Usability refers to system reliability, performance, security, and human factors. Intended use describes the generality of a solution and how and where the system will be operated and who will use it. Future expectation describes the ease of adapting the product for new uses and how easy it is to keep the software in operation when there is a need to modify the code. Constraints are another important part of requirements collection. Schedule and resources are the two major constraints in requirements. There is a user schedule requirement. Resources refer to the limitations on the user side during development and operation of the software. These limitations can be computer and peripheral equipment, staff availability to operate and maintain the software, management support, and the cost for development and operation.

Requirement analysis includes a feasibility study and documentation. Based on the collected user requirements, further analysis is needed to determine if the requirements are feasible. A feasibility study includes cost estimation, benefit estimation, schedule and risk analysis. The documentation for requirements is the project plan, which is the foundation document of the entire project. It contains a proposal for the product itself, a description of the environment in which it is to be used, and development plans. These plans indicate the schedule, budget, and procedures of the project.

The next activity in the analysis phase is specifications, which is transforming the user-oriented requirements into a precise form oriented to the needs of software engineers. According to the ANSI/IEEE standards: "The requirements specification shall clearly and precisely describe the essential functions, performances, design constraints, attributes, and external interfaces. Each requirement shall be defined such that its achievement is capable of being objectively verified..." (Jones, 1978). There are three major activities in the specification process: detailed aspects, documentation, and validation. The focus points of the detailed aspects are functionality, technical feasibility, and quality. Functionality refers to how to process the input information into expected results, and how the software interacts with other systems in the user environment. Technical feasibility is to examine the possibility of implementing the given functionalities, and the need of technical support, e.g., equipment and people. Quality measurement is needed to achieve the quality standard required by users. To check the quality standard, we need to focus on reliability, performance, security, and human factors.

Based on the project plan, the specification document provides technical details. It is written for the software team in a technical language. It is necessary to let the user review the specifications to ensure the proposed product is what the user wants. Prototyping can also be used for validation of the specifications. This allows the user and the software team to see the software in action and to find aspects that do not meet the requirements.

The importance of the analysis phase has been strongly reinforced in recent software development. A well-developed specification can reduce the incidence of faults in the software and minimize rework. Research indicates that increased effort and care during specification will generate significant rewards in terms of dependability, maintainability, productivity, and general software quality.

3.4.2 Design Phase

The design phase is concerned with building the system to perform as required. There are two stages of design: system architecture design and detailed design (see Fig. 3.7).

The system architecture design includes system structure and the system architecture document. System structure design is the process of partitioning a software system into smaller parts. Before subdividing the system, we need to do further specification analysis, examine the details of performance requirements, security requirements, assumptions and constraints, and the need for hardware and software. System decomposition includes subsystem process control and interface relationship. Besides determining how to control the process of each subsystem by identifying major modules, the internal and external interfaces need to be defined. Internal interface refers to how the subsystems interact with each other. External interface defines how the software interacts with its environment, e.g., user, operation, other software and hardware. The last activity in system architecture design is to initiate a system architecture document, which is part of the design

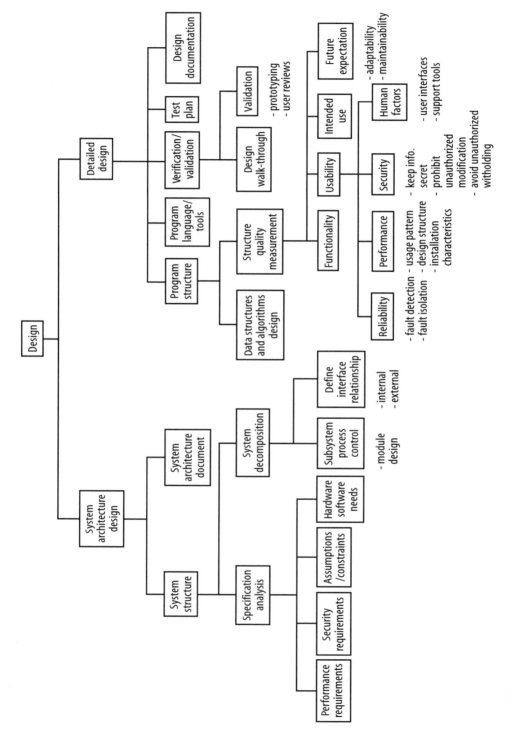

FIG. **3.7.** Design phase.

document in the design phase. The system architecture document describes system components, subsystems and their interfaces.

Detailed design is about designing the program and algorithmic details. The activities within detailed design are program structure, program language and tools, validation and verification, test planning, and design documentation. Program structure optimizes the design of selecting data structures and algorithms to achieve the goal of the project. Structure quality measurement checks if the selected program structure meets the quality requirements. The four major measurements are functionality, usability, intended use, and future expectations. To comply with the functionality requirements means that the designed algorithm should implement the functional characteristics of the proposed system. Usability consists of reliability, performance, security, and human factors requirements. There are two important parts to provide reliability: fault detection and fault isolation. The design has to consider both aspects. Since performance requirements influence the selection of data structures and algorithms, it is important to check performance factors at the design phase. To estimate the performance of the design, the information on usage pattern, design structure, and installation characteristics are needed. The specifications describe the level and what security looks like while design considers its implementation. Different types of systems may have their own particular security needs. There is a series of issues that need to be evaluated for system security: information must be kept confidential, unauthorized modification of information must be prohibited, and unauthorized withholding of information must be avoided. To check if human factors meet the requirements, the designed user interfaces and support tools are two issues to be considered. Intended use is to measure if the designed algorithms meet the stated requirements . Future expectations focus on adaptability and maintainability of the designed system. During detailed design, the selected data structures and algorithms are implemented in a particular programming language on a particular machine. Thus, choosing the appropriate program language and tools is essential.

Test plans should be initiated at design phase. These include identifying items to be tested, creating test case specifications, and generalizing the test approach. A design document is the deliverable of the design phase. It describes system architecture, module design, data object design, how quality expectations will be achieved, and the test plan.

3.4.3 Coding Phase

Coding involves translating the design into the code of a programming language, beginning when the design document is baselined. Coding comprises of the following activities: identifying reusable modules, code editing, code inspection, and final test planning (see Fig. 3.8).

Identifying reusable modules is an effective way to save time and effort. Before writing the code, there may be existing code for modules of other systems or projects which is similar to the current system. These models can be reused with modification. When writing the code, developers should adopt good program styles. A good program style is characterized by simplicity, readability, good documentation, changeability, and module independence. Generally, programming standards should be followed to ensure that the written programs are easily understood by all project team members. Writing structured code also helps to make the program easy to read and maintain. When modifying reusable code, the impact of reusable modules and the interfaces with other modules need to be considered.

Code inspection includes code reviews, quality, and maintainability. Code reviews is to check program logic and readability. This is normally conducted by other developers on the same project team and not the author of the program. Quality verification ensures that all the modules perform the functionality as described in the detailed design. Quality check focuses on reliability, performance, and security. Maintainability is also checked to ensure the programs are easy to maintain.

The final test plan should be ready at the coding phase. Based on the test plan initiated at the design phase, with the feedback of coding activities, the final test plan should provide details of what needs to be tested, testing strategies and methods, testing schedules, and all necessary resources.

3.4.4 Testing Phase

Testing is the verification and validation activity for the software product. The goals of the testing phase are (1) to affirm the quality of the product by finding and eliminating faults in the program, (2) to demonstrate the presence of all specified functionality in the product, and (3) to estimate the operational reliability of the software.

During the testing phase, program components are combined into the overall software code and testing is performed according to a developed test (Software Verification and Validation) plan.

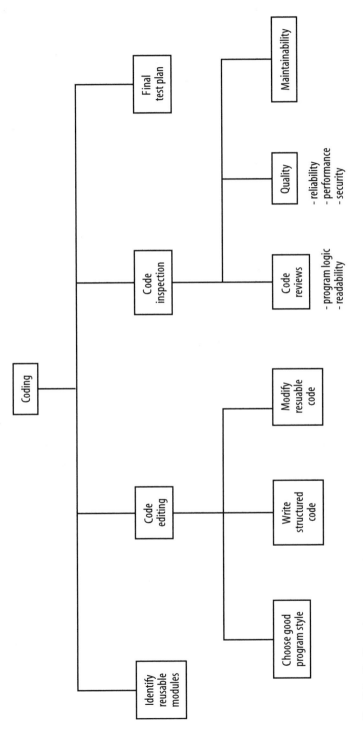

FIG. 3.8. Coding phase.

System integration of the software components and system accept-ance tests are performed against the requirements. In other words, testing during this phase determines whether all requirements have been satisfied and is performed in accordance with the reviewed software verification and validation plan. Test results are evaluated and test and verification reports are prepared to describe the outcome of the process. The testing phase consists of unit test, integration test, and acceptance test (see Fig. 3.9).

The unit test is the process of taking a program module and run-ning it in isolation from the rest of the software product by using prepared inputs and comparing the actual results with the results predicted by the specifications and design of the module. The unit test is the responsibility of programmers while the later stages of testing may be done by an independent testing group.

The integration test includes subsystem and system tests. The subsystem test focuses on testing the interfaces and interdependen-cies of subsystems or modules. The system test tests all the subsys-tems as a whole to determine whether specified functionality is performed correctly as the results of the software. The integration test also includes the system integration testing process which brings together all system components, hardware and software, and humanware. This testing is conducted to ensure that system requirements in real or simulated system environments are satisfied.

The acceptance test acts as a validation of the testing phase, con-sisting of internal test and field test. The internal test includes cap-ability test and guest test, both performed in-house. The capability test tests the system in an environment configured similar to the customer environment. The guest test is conducted by the users in their software organization sites. The field test is to install the pro-duct in a user environment and allows the user to test the product where customers often lead the test and define and develop the test cases. The field test is also called "beta test". The acceptance test is defined as formal testing conducted to determine whether a soft-ware system satisfies its acceptance criteria and to enable the cus-tomer to determine whether the system is acceptable. When the developer's testing and system installation have been completed, acceptance testing that leads to ultimate certification begins. It is recommended that acceptance testing be performed by an indepen-dent group to ensure that the software meets all requirements. Testing by an independent group (without the developers' precon-ceptions about the functioning of the system) provides assurance

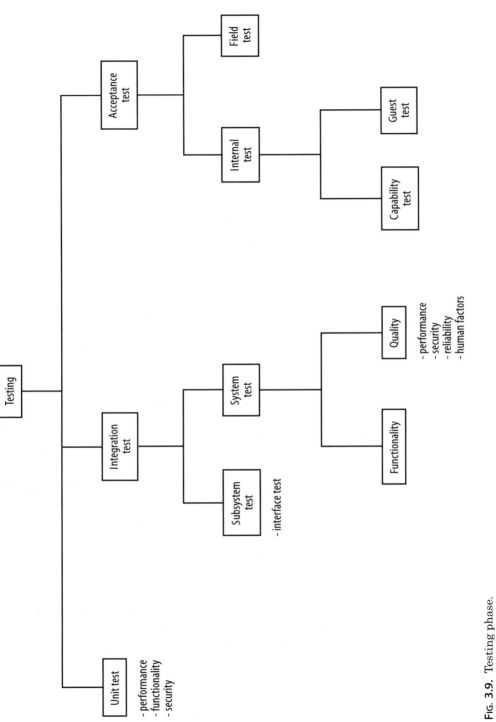

FIG. **3.9.** Testing phase.

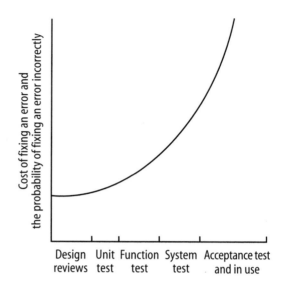

FIG. **3.10.** Relationships between
error corrections and time.

Design   Unit Function System Acceptance test
reviews test   test   test   and in use

that the system satisfies the intent of the original requirements. The acceptance test group usually consists of analysts who will use the system and members of the requirements definition group. Concurrent with all of the previous phases is the preparation of the user and maintenance manuals.

Figure 3.10 shows two relationships concerning error corrections versus schedule time during testing phase. The first relationship showing that the cost of correcting an error increases rapidly during the latter parts of the development cycle should be obvious to most practitioners and software designers. However, a second and lesser known relationship also follows the same general curve. The probability of fixing a known error incorrectly also increases rapidly during the latter stages. This phenomenon is of interest because an incorrect fix to a problem often causes more harm than the original problem. This may lead to an important question: Is there a good reason to deliberately not correct a software error?

3.4.5 Operating Phase

The final phase in the software lifecycle is operation. The operating phase usually contains activities such as installation, training, support, and maintenance (see Fig. 3.11).

After completion of the testing phase, the turnover of the software product plays a very small but important role of the lifecycle. It transfers responsibility for software maintenance from the developer to the user by installing the software product. The user is

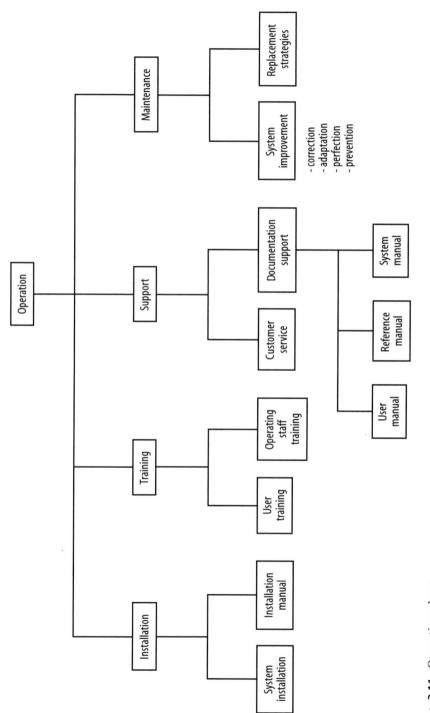

Fɪɢ. **3.11.** Operating phase.

then responsible for establishing a program to control and manage the software. Installation may include system installation and providing the installation manual to the users. The training includes user training and operating staff training. User training is based primarily on major system functions and the user's need for access to them so that they understand what the functions are and how to perform them. Documentation support provides the customer with the user, reference, and system manuals. The software product is thus accepted for operational use.

Maintenance is defined as any change made to the software, either to correct a deficiency in performance (as required by the original software requirements specification), to compensate for environmental changes, or to enhance its operation. Maintenance activities include system improvement and replacement strategies. There are four types of activities in system improvement: correction of errors, adaptation to other changes, perfection of acceptable functions, and prevention of future errors. System improvement activities are similar to those of previous phases of development: analysis, design, coding, and testing.

3.5   Software Development Process and Its Applications

In this section, we will discuss the details of applying the generalized analytic hierarchy process (AHP) (Lee, 1999) to the software development process.

3.5.1 Analytic Hierarchy Process

The AHP is a comprehensive mathematical framework for priority setting in a complex system. The AHP has been applied in a variety of areas since it was first developed by Saaty (1980) in the 1970s. According to Saaty, a complex system is decomposed into subsystems and represented in the hierarchical form. The element at the highest level is called the goal. The elements at each level are the criteria of the elements at the level below. The elements at the bottom level are called the alternatives. In this way, the AHP organizes the basic rationality of the priority setting process by breaking down a multi-element complex system into its smaller constituent parts called components (or levels). The process setting can be divided into three phases: system structuring, pairwise comparison and priorities synthesis. The generalization of the AHP to systems with feedback, i.e., systems with both inter- and intra-component dependence, is given in (Lee, 1999).

3.5.2 Evaluation of Software Development Process

Software has a lifecycle, which goes through the periods of initiation, growth, maturity, and phase-out. The development of software goes through a number of phases or stages. There are several ways to structure software lifecycle into phases according to different activities during software development. After analyzing the software develoment process, we construct the software lifecycle into five phases: analysis, design, coding, testing, and operations. The development phases overlap and feed information to each other. Each phase can be decomposed further to show the detailed activities under that phase (see Section 3.4). The results of a recent study on software development and its environmental factors show that analysis, design, coding, and testing take about 25%, 18%, 36%, and 21% of the whole software development efforts, respectively. In this application, the hierarchy structure for the design phase is discussed in this section. The same procedure can be applied to analyze the impacts among the activities to decompose the system into components and to construct the weight diagrams. Figure 3.12 shows details of the hierarchy structure of the design phase. For convenience, we label the activities in design phase by e_0 to e_{27}. As we can see from Fig. 3.12, the structure is primarily a hierarchical structure. Figure 3.13 shows the impact diagram of the elements in the design phase. The diagram shows that there is no bi-directional impact path between any two components. The techniques to handle the priority setting in this section can be obtained in (Lee, 1999).

The overall priorities of the activities in the design phase are summarized in Fig. 3.14. At the top-most level, we see that the weight of detailed design is approximately two to three times that of system architecture design. This is consistent with our intuition because there are more activities under detailed design and more resources are needed for these activities. Here we want to emphasize that the generalized AHP (Lee, 1999) not only provides qualitative information consistent with intuition, but also quantitative information which is very instrumental in resource allocation. In the next level, we see that the results obtained are consistent with the intuition. In system architecture design, the resource allocated to system structure is about four times that of system architecture document. This should be evident from the amount of activities involved in system structure. Similarly, in detailed design, almost

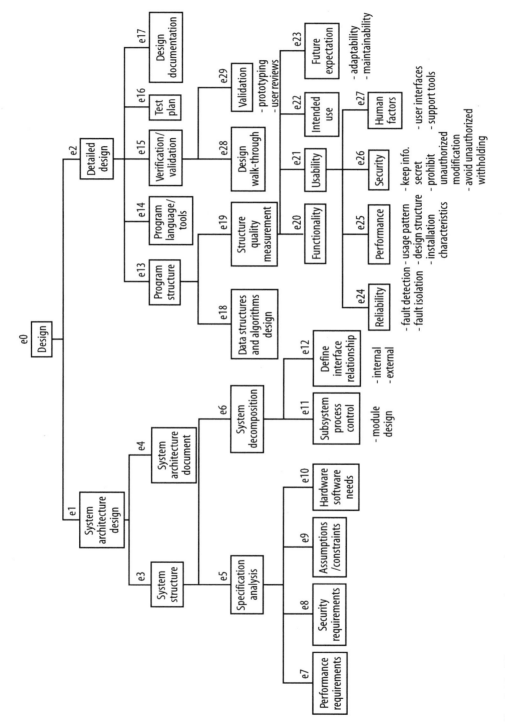

FIG. 3.12. Element number assignment.

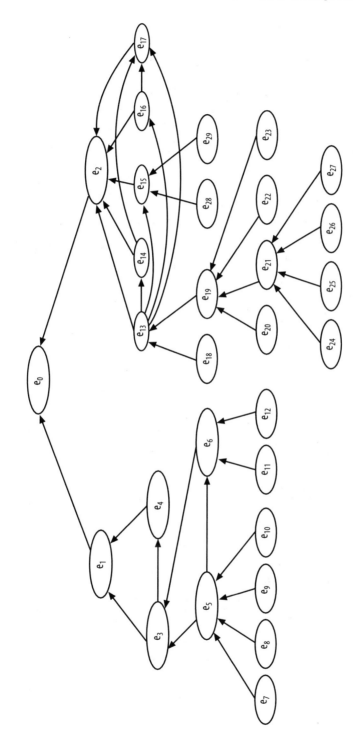

FIG. 3.13. Impact diagram of the design phase of software development.

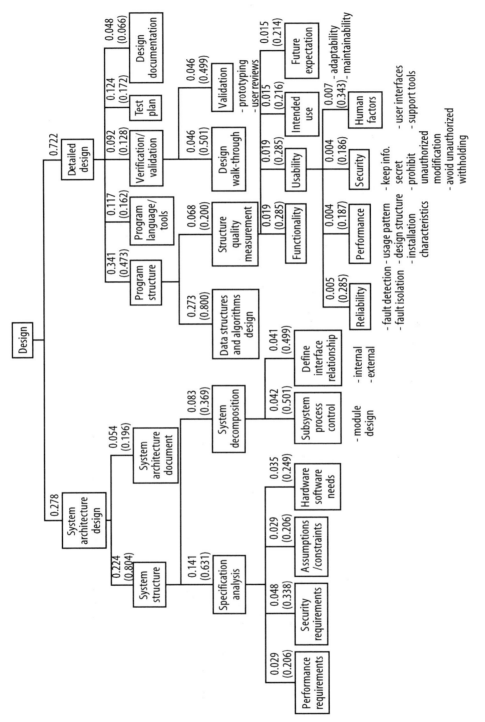

FIG. 3.14. Final weight diagram. The numbers on the top are the final weights, the numbers at the bottom (in the parentheses) are the relative weights.

half of the resources should be allocated to program structure since it is the major activity at this level. To improve the design progress, we need to focus more on detailed design, especially on program structure. Other activities are also important, and resources allocated to these activities should also be proportional to their final weights.

Based on the results of this application using the generalized AHP, software developers can plan their schedules according to the relative weights within the design time frame. The weight diagram is useful in identifying the most important activities while conducting multiple tasks. For instance, suffcent time and effort should be spent on specification analysis before conducting system decomposition. Since the relative weight of system decomposition is about 37% and that of specification analysis is about 63%, software developers should spend roughly twice as much time and effort on the latter. If the right amount of time and effort is spent on each of the activities, the development process can be made smoother, more efficient, and a better outcome can be achieved. This application presents a methodology which provides an effective way in quantifying resource allocation to enhance the software development process.

3.6   Software Verification and Validation

Verification and validation (V&V) are two ways to check whether the design satisfies the user's requirements. According to the IEEE Standard Glossary of Software Engineering Terminology:

> Software verification is the process of evaluating a system or component to determine whether the products of a given development phase satisfy the conditions imposed at the start of that phase.

> Software validation is the process of evaluating a system or component during or at the end of the development process to determine whether it satisfies specified requirements.

In short, Boehem (1981) expressed the difference between software verification and software validation as follows:

> Verification: "Are we building the product right?"
> Validation: "Are we building the right product?"

In other words, verification checks whether the product under construction meets the requirements definition. Validation checks whether the product's functions are in accordance to the customer's needs. Recently, an "eagle-eyed math lover" high school student, Colin Rizzio, who took the Scholastic Assessment Tests (SAT) with about 350,000 high school students on 12 October 1996, recognized the flaw in the multiple-choice answers for a question dealing with an algebraic equation. The problem was that the question-writer had used an alphabet, in this case "a," to represent any number, a standard practice in an algebra equation. The original "correct" answer assumed that "a" was positive and did not account for the possibility that it was a negative number such as −4. Students who assumed the number could be negative, had a different answer. It was the aim of the SAT to "check your work," which apparently, the testers did not (The Home News & Tribune, 1997). As a result, the scores of up to 45,000 high-school students, who took the SAT last fall, were boosted by as much as 30 points. The math portion of the SAT test was worth 800 points.

Often, programming is done primarily by scientists or engineers, who have little training in software development or programming. They are, however, highly motivated to get a program running in the shortest time possible. The consequence of expedited results is that the users find bugs in the software program after the software product is put into operation. Although it costs the developer very little to fix faults during the development phase, i.e., the testing phase, it would definitely cost orders of magnitude more to fix faults during the operating and maintenance phases. The cost of fixing an error increases dramatically as the software lifecycle progresses. Figure 3.15 illustrates this case.

Verification should be integrated not only in the testing phase but in all phases of the software development lifecycle. It should be noted that testing is one aspect of verification but it cannot replace the task of the verification process. In fact, verification is most effective and efficient when applied from the beginning of the development process. Verification should be performed independently by a group other than the software developer whose interests lie in showing that the software program works and not in finding bugs or faults in the software. Therefore, an independent group of the development process is more likely to do a thorough testing and execute the software verification, coming up with a series of complex tests that may create blind spots in the evaluation process. A

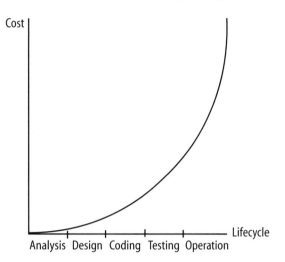

FIG. **3.15.** Software cost versus the development phase.

good independent verification and validation (IV&V) should: (1) have significant experience, (2) have a base of personnel skilled in IV&V, (3) hold a repertoire of qualified and transferable tools, (4) be a development team participant, and (5) be adaptable to any project.

In general, validation determines end product accuracy, e.g., code, with respect to the software requirements. It determines if the output conform with what was required? Verification is performed at each phase and between each phase of the development lifecycle. It determines that each phase and subphase product is correct, complete, and consistent with itself and with its predecessor product.

3.7 Data Collection and Analysis

Traditionally, there are two common types of failure data: time-domain data and interval-domain data. These data are usually used by practitioners when analyzing and predicting reliability applications. Some software reliability models can handle both types of data. The time-domain approach involves recording the individual times at which failure occurred, as illustrated in Table 3.3. The first failure occurred 25 minutes into the test, the second at 55 minutes, the third at 70 minutes, the fourth at 95 minutes, the fifth at 112 minutes, the sixth at 119 minutes, the seventh at 143 minutes, and the eighth failure at 187. Some models may require obtaining

TABLE **3.3.** Data recording for the time-domain approach.

Failure records	Actual failure time (minutes)	Time between failures (minutes)
1	25	25
2	55	30
3	70	15
4	95	15
5	112	17
6	119	7
7	143	26
8	187	44

the time between failures in lieu of the actual failure time. In this example, the values 25, 30, 15, 15, 17, 7, 26, and 44 should be used as the time-domain data set.

The interval-domain approach is characterized by counting the number of failures occurring during a fixed period (e.g., test session, hour, week, day). Using this method, the collected data are a count of the number of failures in the interval. This approach is illustrated in Table 3.4. Using the same failures as in the time-domain example, we would record two failures in the first 1-hour interval, four failures in the second interval, one failure in the third interval, and one in the fourth. Intervals, however, do not need to be equally spaced for data collection. For example, if the interval for data collection is a test session, one session may last 4 hours and the next may last 8 hours. Models with assumptions that handle this situation should be considered for higher fidelity forecasts for systems with interval-domain data.

The time-domain approach always provides higher accuracy in the parameter estimates with current tools but involves more data collection efforts than the interval-domain approach. The

TABLE **3.4.** Data recording for the interval-domain approach.

Time (hours)	Observed number of failures
1	2
2	4
3	1
4	1

practitioners must trade off the cost of data collection with the accuracy reliability level required by the model predictions.

Further Reading

Some interesting research papers and books are:

Anderson, T., *Software Requirements Specification and Testing*, Blackwell Scientific Publications, London, 1995.

Briand, L.C., V.R. Basili, and C.J. Hetmanski, "Providing an Empirical Basis for Optimizing the Verification and Testing Phases of Software Development," in *Third International Symposium on Software Reliability Engineering*, 1992, IEEE Computer Society Press, Los Angeles, 1992.

Smith, D.J., *Achieving Quality Software*, 3rd edition, Chapman & Hall, New York, 1995.

Voas, J.M. and K.W. Miller, "Improving the Software Development Process Using Testability Research," in *Third International Symposium on Software Reliability Engineering*, 1992, IEEE Computer Society Press, Los Angeles, 1992.

Problems

3.1   List several practical system failures (since 1990) caused by software.

3.2   What are the differences between verification and validation? Provide several real world applications and examples.

3.3   Identify from articles and books at least two methodologies and/or models to support software V&V processes. Prepare a summary of these methodologies and/or models.

3.4   Read three papers on software lifecycle published in the IEEE Transactions on Software Engineering. Summarize the material and relate it to the software development process.

3.5   Why it is important to integrate analysis, design, coding and testing in the software development process? Who should be involved in this integration and which techniques and/or methodologies are available to support this integrated approach?

3.6   What are the main difficulties involved in the data collection of time-domain approach and interval-domain approach?

3.7   Using the generalized AHP in Section 3.5, analyze the impacts among the activities and construct the weight diagrams for the analysis phase (see Section 3.4) in the development process.

3.8   Using the generalized AHP in Section 3.5, analyze the impacts among the activities and construct the weight diagrams for the testing phase (see Section 3.4) in the development process.

References

Boehm, B.W., *Software Engineering Economics*, Prentice-Hall, Englewood Cliffs, 1981.

Churchley, A. (ed.), *Microprocessor Based Protection Systems*, Elsevier Applied Science, Amsterdam, 1991

Goel, A.L., "Software reliability models: Assumptions, limitations, and applicability," *IEEE Trans. Software Engineering*, Vol. SE-2 (12), 1985.

Jones, T.C., "Measuring programming quality and productivity," *IBM Systems J.*, Vol. 17(1), 1978.

Lee, M., H. Pham, and X. Zhang, "A methodology for priority setting with application to software development process," *European J. Operational Research*, Vol. 118(2), October 1999.

Lyu, M.R., *Handbook of Software Reliability Engineering*, McGraw-Hill, New York, 1996.

Malaiya, Y.K. and P.K. Srimani (eds.), *Software Reliability Models: Theoretical Developments, Evaluation and Applications*, IEEE Computer Society, 1990.

Pham, H., "Study of environmental factors and software reliability," IE Working Paper, Rutgers University, Piscataway, 1998.

Pressman, R.S., *Software Engineering, A Practitioner's Approach*, Addison-Wesley, 1983.

Saaty, T.L., *The Analytic Hierarchy Process*, McGraw-Hill, New York, 1980.

The Home News & Tribune, "An eagle-eyed math lover," 7 February 1997.

Xie, M., *Software Reliability Modelling*, World Scientific, Singapore, 1991.

4

Software Reliability Modeling

"If you don't test your software,
how do you know it works."

4.1   Introduction

For software qualification, it is highly desirable to have an estimate of the remaining errors in a software system. It is difficult to determine such an important finding without knowing what the initial errors are. Research activities in software reliability engineering have been studied over the past 25 years and more than 50 statistical models and various techniques have been developed for estimating software reliability and numbers of residual errors in software. From historical data on programming errors, there are likely to be about eight errors per 1,000 program statements after the unit test. This, of course, is just an average and does not take into account any tests on the program.

There are two main types of software reliability models: the deterministic and the probabilistic. The deterministic model is used to study the number of distinct operators and operands in a program as well as the number of errors and the number of machine instructions in the program. Performance measures of the deterministic type are obtained by analyzing the program texture and do not involve any random event. Two well-known models are: Halstead's software metric and McCabe's cyclomatic complexity metric. Halstead's software metric is used to estimate the number of errors in the program, whereas McCabe's cyclomatic complexity metric (McCabe, 1976) is used to determine an upper bound on the number of tests in a program. Recently, Cai (1998) proposed a new model for estimating the number of remaining software defects. In general, these models represent a growing quantitative approach to the measurement of computer software.

The probabilistic model represents the failure occurrences and the fault removals as probabilistic events. The probabilistic software reliability models can be classified into different groups:

- error seeding
- failure rate
- curve fitting
- reliability growth
- nonhomogeneous Poisson process
- Markov structure.

In this chapter, we discuss various types of software reliability models and methods for estimating software reliability and other performance measures such as software complexity, software safety, and the number of remaining errors.

4.2   Halstead's Software Metric

Halstead's theory of software metric is probably the best-known technique to measure the complexity in a software program and the amount of difficulty involved in testing and debugging the software. Halstead (1977) uses the number of distinct operators and the number of distinct operands in a program to develop expressions for the overall program length, volume and the number of remaining defects in a program.

The following notations are used:

n_1 = number of unique or distinct operators appearing in a program;

n_2 = number of unique or distinct operands appearing in a program;

N_1 = total number of operators occurring in a program;

N_2 = total number of operands occurring in a program;

N = length of the program;

V = volume of the program;

E = number of errors in the program;

I = number of machine instructions.

Halstead (1977) showed that the length and the volume measure of the program can be estimated:

$$N = N_1 + N_2$$

and

$$V = N \log_2(n_1 + n_2),$$

respectively, where

$$N_1 = n_1 \log_2 n_1$$
$$N_2 = n_2 \log_2 n_2.$$

Halstead proposed two empirical formulae to estimate the number of remaining defects in the program, E, from program volume. The two formulae, namely, Halstead empirical model 1 and Halstead empirical model 2, respectively, are

$$\hat{E} = \frac{V}{3,000}$$

and

$$\hat{E} = \frac{A}{3,000}$$

where

$$A = \left(\frac{V}{\frac{2n_2}{n_1 N_2}} \right)^{\frac{2}{3}}.$$

To examine whether Halstead's software science can offer reasonable estimates for the number of remaining defects, we will discuss the following two examples.

Example 4.1 Consider the Interchange Sort Program (Fitzsimmons, 1978) for the Fortran version in Table 4.1. The length and the volume for this program can be calculated as follows:

$$N = N_1 + N_2 = 50$$

and

$$V = N \log_2(n_1 + n_2)$$
$$= 50 \log_2(10 + 7)$$
$$= 204.$$

From Halstead empirical model 1, the number of remaining defects in this program can be expected to be:

$$\hat{E} = \frac{204}{3,000} = 0.068.$$

TABLE **4.1.** Operators and operands for an interchange sort program. (Fitzsimmons, 1978)

Interchange sort program

```
SUBROUTINE SORT (X,N)
DIMENSION X(N)
IF (N.LT.2) RETURN
DO 20 I = 2,N
    DO 10 J = 1,I
    IF (X(I).GE.X(J)) GO TO 10
        SAVE = X(I)
        X(I) = X(J)
        X(J) = SAVE
10 CONTINUE
20 CONTINUE
    RETURN
    END
```

Operators of the interchange sort program

Operator	Count
1 End of statement	7
2 Array subscript	6
3 =	5
4 IF ( )	2
5 DO	2
6 ,	2
7 End of program	1
8 .LT.	1
9 .GE.	1
$n_1 = 10$ GO TO 10	1
	$N_1 = 28$

Operands of the interchange sort program

Operand	Count
1 X	6
2 I	5
3 J	4
4 N	2
5 2	2
6 SAVE	2
$n_2 = 7$ 1	1
	$N_2 = 22$

It should be noted that the volume for this program under the assembly language version would be 328 since it needs more effort to specify a program in assembly programming language.

Example 4.2 Let us consider Akiyama's software data (Halstead, 1977) in Table 4.2, which will be used to validate Halstead's empirical model 2. The software system consists of nine modules and is written in assembly language.

TABLE **4.2.** Akiyama's published data. (Halstead, 1977)

Program module	Program (S)	Decisions (D)	Subroutine calls (J)	Number of defects observed (O)
MA	4,032	372	283	102
MB	1,329	215	49	18
MC	5,453	552	362	146
MD	1,674	111	130	26
ME	2,051	315	197	71
MF	2,513	217	186	37
MG	699	104	32	16
MH	3,792	233	110	50
MX	3,412	416	230	80

TABLE **4.3.** Computational results of Akiyama's software data.

Program module	N	n_1	n_2	$A^{1.5}(\times 10^6)$	E	$V/3{,}000$	Number of defects observed (O)
MA	8,064	471	442	170.3	102	26.4	102
MB	2,658	180	176	15.3	21	7.5	18
MC	10,906	610	574	322.6	157	37.1	146
MD	3,348	231	201	28.2	31	9.7	26
ME	4,102	336	138	100.2	72	12.3	71
MF	5,026	322	287	65.5	54	15.5	37
MG	1,398	131	76	6.5	12	3.6	16
MH	7,584	252	603	58.5	50	24.6	50
MX	6,824	433	357	135.9	88	21.9	80

Assuming that each of the S machine language steps includes one operator and one operand, we obtain

$$N_1 = S, \quad N_2 = S, \quad \text{and} \quad N = 2S.$$

Halstead assumed that the number of distinct operators appearing in a program, n_1, was equal to the sum of the number of machine language instruction types, program calls, and unique program decisions. He further assumed that there were 64 types of machine language instructions and that only one-third of the decisions were unique. Therefore, Halstead proposed n_1 as follows:

$$n_1 = \frac{D}{3} + J + 64.$$

Now we use the following formula:

$$N = n_1 \log_2 n_1 + n_2 \log_2 n_2$$

to obtain n_2 when n_1 and N are known. Thus, A can be also obtained. The computational results are shown in Table 4.3. From the results in Table 4.3, Halstead's empirical model 2 is close to the number of observed defects than empirical model 1.

4.3 McCabe's Cyclomatic Complexity Metric

A cyclomatic complexity metric measure of software proposed by McCabe (1976) is a complexity measure of the digraph based on the control flow representation of a program. Cyclomatic complexity is a software metric that provides a quantitative measure of the logical complexity of a program by counting the decision points. For example, one should start with 1 for the straight path through the module or subroutine (McConnel, 1993). Then 1 should be added each time one of the following keywords appear: IF, REPEAT, WHILE, FOR, OR, AND. Also, 1 should be added for each case in a case statement and also if the case statement lacks a default case. If the total score is less than 10, then by using the McCabe's measure, the code is considered to be of high quality software. McCabe has suggested that a program with a high level of the metric is very difficult to produce and maintain. He recommended a total score of 10 as an upper limit for the Fortran environment (Rook, 1990).

In a strongly connected graph (with a path joining any pair of nodes), the cyclomatic number is equal to the maximum number of linearly independent circuits. The linearly independent circuits form a basis for the set of all circuits in G, and any path passing through G can be expressed as a linear combination of the circuits.

The cyclomatic number $V(G)$ of a graph G can be determined from the following formula:

$$V(G) = e - n + 2p$$

where

$e =$ number of edges in the control graph in which an edge is equivalent to a branching point in the program;

$n =$ number of vertices in the control graph and a vertex is equivalent to a sequential block of code in the program; and

$p =$ number of connected elements (usually 1).

The cyclomatic number of a graph with multiple connected components is equal to the sum of the cyclomatic numbers of the connected components.

In any program that can be represented by a state diagram, the McCabe cyclomatic complexity metric can also be used to calculate the number of control flow paths (FP), i.e.,

$$FP = e - n + 2$$

where e is the number of edges and n is the number of nodes. Here, edges are represented on the diagram by arrows and nodes are shown on the diagram with circles.

Example 4.3 A control flow path is a path that considers only a single loop iteration. The program state diagram, in Fig. 4.1, measures the time between two switch inputs. It calculates the speed of a Mini-Van in which the number of edges equals 8 and the number of nodes equals 7. Therefore, the number of control flow paths is:

$$FP = 8 - 7 + 2 = 3.$$

The program in Fig. 4.1 has three control flow paths which are: 1234567, 121234567, 123454567.

Another simple way of computing the cyclomatic number is as follows:

$$V(G) = \pi + 1$$

where π is the number of predicate nodes (decisions or branches) in the program.

In other words, the cyclomatic number is a measure of the number of branches in a program. A branch occurs in IF, WHILE, REPEAT, and CASE statements (a GO TO statement is normally excluded from the structured program). The cyclomatic number has been widely used in predicting the number of errors and in measuring software quality.

McCabe (1976) notes that, when used in the context of the basis path testing method, the cyclomatic complexity, $V(G)$, provides an upper bound for the number of independent paths in the basis set of a program. It also provides an upper bound on the number of tests that must be conducted to ensure that all program statements have been executed at least once.

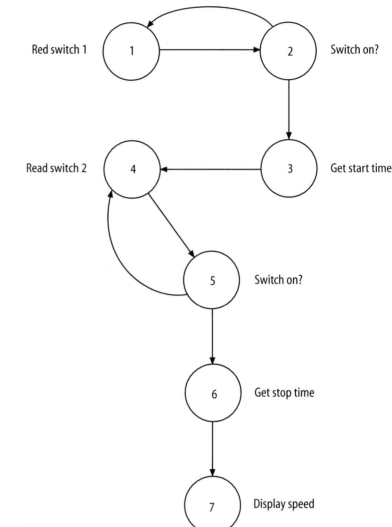

FIG. 4.1. Program state diagram.

It should be noted that the cyclomatic complexity is an interesting measure of whether the software designer has followed good structured programming and software standard practices. The cyclomatic complexity also estimates how difficult it will be to test all the paths in the program and thus, provide useful information of how difficult it will be to satisfy the software specifications and requirements (Basili, 1984). Also, McCabe's measure will be less useful if the software developers and testing debuggers are interested in detecting all the faults (Friedman, 1995).

4.4   Error Seeding Models

The error seeding group of models estimates the number of errors in a program by using the multistage sampling technique. Errors are divided into indigenous errors and induced errors (seeded errors). The unknown number of indigenous errors is estimated from the number of induced errors and the ratio of the two types of errors obtained from the debugging data. Models included in this group are:

- Mills' error seeding model (Mills, 1970);
- Cai's model (Cai, 1998); and
- Hypergeometric distribution model (Tohma *et al.*, 1991).

4.4.1   Mills' Error Seeding Model

Mills' error seeding model (Mills, 1970) proposed an error seeding method to estimate the number of errors in a program by introducing seeded errors into the program. From the debugging data, which consist of inherent errors and induced errors, the unknown number of inherent errors could be estimated. If both inherent errors and induced errors are equally likely to be detected, then the probability of k induced errors in r removed errors follows a hypergeometric distribution which is given by

$$P(k; N, n_1, r) = \frac{\binom{n_1}{k}\binom{N}{r-k}}{\binom{N+n_1}{r}}, \quad k = 1, 2, \ldots, r \qquad (4.1)$$

where

N = total number of inherent errors
n_1 = total number of induced errors
r = total number of errors removed during debugging
k = total number of induced errors in r removed errors
$r - k$ = total number of inherent errors in r removed errors.

Since n_1, r, and k are known, the MLE of N can be shown to be (Huang, 1984):

$$\hat{N} = \lfloor N_0 \rfloor + 1$$

where

$$N_0 = \frac{n_1(r-k)}{k} - 1. \tag{4.2}$$

If N_0 is an integer, then N_0 and $N_0 + 1$ are both the MLEs of N.

Example 4.4   Assume $n_1 = 10$, $r = 15$, and $k = 6$. The number of inherent errors can be estimated using Eq. (4.2), i.e.,

$$\begin{aligned}
N_0 &= \frac{n_1(r-k)}{k} - 1 \\
&= \frac{10(15-6)}{6} - 1 \\
&= 14.
\end{aligned}$$

$N_0 = 14$ is an integer, therefore, both values of 14 and 15 are the MLEs of the total number of inherent errors.

However, here are some drawbacks with this model. It is expensive to conduct testing of the software and at the same time, it increases the testing effort. This method was also criticized for its inability to determine the type, location, and difficulty level of the induced errors such that they would be detected equally likely as the inherent errors.

Another realistic method for estimating the residual errors in a program is based on two independent groups of programmers, testing the program for errors using independent sets of test cases.

Suppose that out of a total number of N initial errors, the first programmer detects n_1 errors (and does not remove them at all) and the second independently detects r errors from the same program. Assume that k common errors are found by both programmers (see Fig. 4.2). If all errors have the equal chance of being detected, then the fraction detected by the first programmer (k) of a randomly selected subset of errors (e.g., r) should equal the fraction that the first programmer detects (n_1) of the total number of initial errors N. In other words,

$$\frac{k}{r} = \frac{n_1}{N}$$

so that an estimate of the total number of initial errors, N, is

$$\hat{N} = \frac{n_1 r}{k}.$$

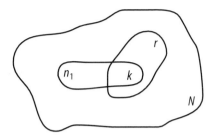

FIG. 4.2. k common errors versus N initial errors.

The probability of exactly N initial errors with k common errors in r detected errors by the second programmer can be obtained using a hypergeometric distribution as follows:

$$P(k; N, n_1, r) = \frac{\binom{n_1}{k}\binom{N-n_1}{r-k}}{\binom{N}{r}} \tag{4.3}$$

and the MLE of N is

$$\hat{N} = \frac{n_1 r}{k}$$

which is the same as the above.

Example 4.5   Given $n_1 = 10$, $r = 15$, and $k = 6$, the number of initial errors (or inherent errors) is given by

$$\hat{N} = \frac{n_1 r}{k}$$
$$= \frac{10(15)}{6} = 25.$$

4.4.2   Cai's Model

Cai (1998) recently modified Mills' model by dividing the software into two parts: Part 0 and Part 1. This model is used to estimate the number of defects remaining in the software. The following assumptions are applied to this model:

1. There are N defects remaining in the software where Part 0 contains N_0 and Part 1 has the remaining defects, i.e, $N = N_0 + N_1$.
2. Each of the remaining defects has the same probability of being detected.
3. The defect is removed when detected.
4. Only one remaining defect is removed each time and no new defects are introduced.
5. There are n remaining defects removed.

Let t_i represent the time instant of the ith remaining defect being removed, and let Y_i be random variables such that

$$Y_i = \begin{cases} 0 \text{ if the } i\text{th detected defect is contained in Part 0} \\ 1 \text{ if the } i\text{th detected defect is contained in Part 1.} \end{cases}$$

Let $N_j(i)$ be the number of defects remaining in Part j in the time interval $(t_i, t_{i+1}]$ for $i = 0, 1, 2, \ldots, n$ and $j = 0$ or 1. Assume y_i are the observed values of Y_i and $y_0 = 0$. Therefore,

$$N_0(i) = N_0 - i + \sum_{j=0}^{i} y_i$$

$$N_1(i) = N_1 - \sum_{j=0}^{i} y_j.$$

Let $p_j(i)$ be the probability of having a defect remaining in Part j detected during the time interval $(t_i, t_{i+1}]$ where $i = 0, 1, 2, \ldots, n$ and $j = 0$ or 1. Then we obtain

$$p_0(i) = \frac{N_0(i)}{N_0(i) + N_1(i)}$$

$$= \frac{N_0 - i + \sum_{j=0}^{i} y_i}{N_0 + N_1 - i} \tag{4.4}$$

and

$$p_1(i) = \frac{N_1(i)}{N_0(i) + N_1(i)}$$

$$= \frac{N_1 - \sum_{j=0}^{i} y_i}{N_0 + N_i - i}. \tag{4.5}$$

Now we wish to estimate N_0 and N_1 using the MLE method. Then the likelihood function can be determined as follows (Cai, 1998):

$$L(y_1, \ldots, y_n) = P\{Y_1 = y_1, \ldots, Y_n = y_n\}$$

$$= \prod_{i=1}^{n} P\{Y_i = y_i | Y_1 = y_1, \ldots, Y_{i-1} = y_{i-1}\}.$$

Note that

$$P\{Y_i = y_i | Y_1 = y_1, \ldots, Y_{i-1} = y_{i-1}\} = \begin{cases} p_0(i-1) & \text{if } y_i = 0 \\ p_1(i-1) & \text{if } y_i = 1, \end{cases}$$

or, equivalently, that

$$P\{Y_i = y_i | Y_1 = y_1, \ldots, Y_{i-1} = y_{i-1}\} = [p_0(i-1)]^{1-y_i}[p_1(i-1)]^{y_i}.$$

Thus,

$$L(y_1, \ldots, y_n) = \prod_{i=1}^{n}[p_0(i-1)]^{1-y_i}[p_1(i-1)]^{y_i}.$$

Taking the ln of the likelihood function, we obtain

$$\ln L(y_1, \ldots, y_n) = \sum_{i=1}^{n}(1-y_i)\ln[p_0(i-1)] + \sum_{i=1}^{n}y_i[p_1(i-1)].$$

Substituting p_0 and p_1 of Eqs. (4.4) and (4.5), respectively, into the above equation and taking the first derivatives with respect to N_0 and N_1, the estimates of N_0 and N_1 are determined by the following equations:

$$\sum_{i=1}^{n} \frac{1-y_i}{N_0 - i + 1 + \sum\limits_{j=0}^{i-1} y_j} = \frac{1}{N_0 + N_1 - i + 1}$$

and

$$\sum_{i=1}^{n} \frac{y_i}{N_1 - \sum\limits_{j=0}^{i-1} y_j} = \frac{1}{N_0 + N_1 - i + 1}.$$

4.4.3 Hypergeometric Distribution Model

We now present a model, proposed by Tohma *et al.* (1991), for estimating the number of faults initially resident in a program at the beginning of the test or debugging process based on the hypergeometric distribution. Let C_{i-1} be the cumulative number of errors already detected so far by $t_1, t_2, \ldots, t_{i-1}$, and let N_i be the number of newly detected errors by time t_i. Assume:

1. A program initially contains m faults when the test phase starts.
2. A test is defined as a number of test instances which are couples of input data and output data. In other words, the collection of test operations performed in a day or a week is called a *test instance*. The test instances are denoted by t_i for $i = 1, 2, \ldots, n$.
3. Detected faults are not removed between test instances.

Therefore, from the latter assumption, same faults can be experienced at several test instances. Let W_i be the number of faults experienced by test instance t_i. It should be noted that some of the W_i faults may be those that are already counted in C_{i-1}, and the remaining W_i faults account for the newly detected faults.

If n_i is an observed instance of N_i, then we can see that $n_i \leq W_i$. Each fault can be classified into one of two categories:

– newly discovered faults
– rediscovered faults.

If we assume that the number of newly detected faults N_i follows a hypergeometric distribution, then the probability of obtaining exactly n_i newly detected faults among W_i faults is (Tohma, 1991):

$$P(N_i = n_i) = \frac{\binom{m-C_{i-1}}{n_i}\binom{C_{i-1}}{W_i-n_i}}{\binom{m}{W_i}} \tag{4.6}$$

where

$$C_{i-1} = \sum_{k=1}^{i-1} n_k$$

$$C_0 = 0, \qquad n_0 = 0$$

and

$$\max\{0, W_i - C_{i-1}\} \leq n_i \leq \{W_i, m - C_{i-1}\}$$

for all i. Since N_i is assumed to be hypergeometrically distributed, the expected number of newly detected faults during the interval $[t_{i-1}, t_i]$ is

$$E(N_i) = \frac{m - C_{i-1}}{m} W_i$$

and the expected value of C_i is given by

$$E(C_i) = m \left[1 - \prod_{j=1}^{i}(1 - p_j) \right]$$

where

$$p_i = \frac{W_i}{m} \quad i = 1, 2, \dots .$$

4.5   Failure Rate Models

The failure rate group of models is used to study the program failure rate per fault at the failure intervals. Models included in this group are:

- Jelinski and Moranda (Jelinski, 1972)
- Schick and Wolverton (Schick, 1978)
- Jelinski–Moranda geometric (Moranda, 1979)
- Moranda geometric Poisson (Littlewood, 1979)
- Negative-binomial Poisson
- Modified Schick and Wolverton (Sukert, 1977)
- Goel and Okumoto imperfect debugging (Goel, 1979).

This group of models studies how failure rates change at the failure time during the failure intervals. As the number of remaining faults changes, the failure rate of the program changes accordingly. Since the number of faults in the program is a discrete function, the failure rate of the program is also a discrete function with discontinuities at the failure times.

4.5.1 Jelinski–Moranda Model

The Jelinski–Moranda (J–M) model (Jelinski, 1972) is one of the earliest software reliability models. Many existing software reliability models are variants or extensions of this basic model. The assumptions in this model include the following:

- The program contains N initial faults which is an unknown but fixed constant.
- Each fault in the program is independent and equally likely to cause a failure during a test.
- Time intervals between occurrences of failure are independent of each other.
- Whenever a failure occurs, a corresponding fault is removed with certainty.
- The fault that causes a failure is assumed to be instantaneously removed, and no new faults are inserted during the removal of the detected fault.
- The software failure rate during a failure interval is constant and is proportional to the number of faults remaining in the program.

The program failure rate at the ith failure interval is given by

$$\lambda(t_i) = \phi[N - (i - 1)], \quad i = 1, 2, \ldots, N \qquad (4.7)$$

where

ϕ = a proportional constant, the contribution any one fault makes to the overall program;

N = the number of initial faults in the program;

t_i = the time between the $(i - 1)$th and the ith failures.

For example, the initial failure intensity is

$$\lambda(t_1) = \phi N$$

and after the first failure, the failure intensity decreases to

$$\lambda(t_2) = \phi(N - 1)$$

and so on. The pdf and cdf of t_i are

$$f(t_i) = \lambda(t_i) e^{-\int_0^{t_i} \lambda(x_i)dx_i}$$

$$= \phi[N - (i - 1)] e^{-\int_0^{t_i} \phi[N-(i-1)]dx_i}$$

$$= \phi[N - (i - 1)] e^{-\phi(N-(i-1))t_i}$$

and

$$F(t_i) = \int_0^{t_i} f(x_i)dx_i$$

$$= \int_0^{t_i} \phi[N - (i - 1)] e^{-\phi[N-(i-1)]x_i} dx_i$$

$$= 1 - e^{-\phi(N-i+1)t_i},$$

respectively. The software reliability function is, therefore,

$$R(t_i) = e^{-\phi(N-i+1)t_i}. \qquad (4.8)$$

The property of this model is that the failure rate is constant and the software during the testing stage is unchanged or frozen.

Suppose that the failure data set $\{t_1, t_2, \ldots, t_n\}$ is given and assume that ϕ is known. Using the MLE method, we obtain the likelihood function as follows:

$$L(N) = \prod_{i=1}^{n} f(t_i)$$

$$= \prod_{i=1}^{n} [\phi(N - (i - 1))e^{-\phi(N-(i-1))t_i}]$$

$$= \phi^n \prod_{i=1}^{n} [N - (i - 1)]e^{-\phi \sum_{i=1}^{n}[N-(i-1)]t_i}$$

and the log of the likelihood function is

$$\ln L(N) = n\ln\phi + \sum_{i=1}^{n} \ln[N - (i - 1)] - \phi \sum_{i=1}^{n}[N - (i - 1)]t_i.$$

Taking the first partial derivative of the above function with respect to N, we obtain

$$\frac{\partial}{\partial N} \ln L = \sum_{i=1}^{n} \frac{1}{N - (i - 1)} - \phi \sum_{i=1}^{n} t_i.$$

Set

$$\frac{\partial}{\partial N} \ln L(N) = 0,$$

then

$$\sum_{i=1}^{n} \frac{1}{N - (i - 1)} = \phi \sum_{i=1}^{n} t_i.$$

Thus, the MLE of N can be obtained by solving the following equation:

$$\sum_{i=1}^{n} \frac{1}{N - (i + 1)} = \phi \sum_{i=1}^{n} t_i.$$

In many applications, the parameter ϕ is not known. In this case, we wish to estimate both the parameters N and ϕ which are unknown. Again, the log likelihood function is

$$L(N, \phi) = n\ln\phi + \sum_{i=1}^{n} \ln[N - (i - 1)] - \phi \sum_{i=1}^{n}[N - (i - 1)]t_i.$$

Taking the derivatives of $\ln L(N, \phi)$ with respect to N and ϕ, we obtain

$$\frac{\partial}{\partial N}[\ln L(N, \phi)] = \sum_{i=1}^{n} \frac{1}{N - (i-1)} - \phi \sum_{i=1}^{n} t_i \equiv 0$$

and

$$\frac{\partial}{\partial \phi}[\ln L(N, \phi)] = \frac{n}{\phi} - \sum_{i=1}^{n}[N - (i-1)]t_i \equiv 0.$$

From the two equations above, we obtain

$$\phi = \frac{\sum\limits_{i=1}^{n} \frac{1}{N-(i-1)}}{\sum\limits_{i=1}^{n} t_i}$$

and

$$n \sum_{i=1}^{n} t_i = \left[\sum_{i=1}^{n}[N - (i-1)]t_i\right]\left[\sum_{i=1}^{n} \frac{1}{N - (i-1)}\right].$$

4.5.2 Schick–Wolverton Model

The Schick–Wolverton (S–W) model (Schick, 1978) is a modification to the J–M model. It is similar to the J–M model except that it further assumes that the failure rate at the ith time interval increases with time t_i since the last debugging. In the model, the program failure rate function between the $(i-1)$th and the ith failure can be expressed as

$$\lambda(t_i) = \phi[N - (i-1)]t_i \qquad (4.9)$$

where ϕ and N are the same as that defined in the J–M model and t_i is the test time since the $(i-1)$st failure.

The pdf of t_i can be obtained as follows:

$$f(t_i) = \phi(N - i + 1)t_i e^{-\frac{\phi(N-i+1)t_i^2}{2}} \quad \text{for } i = 1, 2, \ldots, N.$$

Hence, the software reliability function is

$$R(t_i) = e^{-\int_0^{t_i} \lambda(t_i)dt_i} \qquad (4.10)$$

$$= e^{-\frac{\phi(N-i+1)t_i^2}{2}}.$$

We now wish to estimate N assuming that ϕ is given. Using the MLE method, the log likelihood function is given by

$$
\begin{aligned}
\ln L(N) &= \ln \left\{ \prod_{i=1}^{n} f(t_i) \right\} \\
&= \ln \left\{ \prod_{i=1}^{n} \left(\phi(N - i + 1)t_i e^{-\frac{\phi(N-i+1)t_i^2}{2}} \right) \right\} \\
&= n\ln\phi + \sum_{i=1}^{n} \ln[N - (i-1)] + \sum_{i=1}^{n} \ln t_i \\
&\quad - \sum_{i=1}^{n} \phi[N - (i-1)]\frac{t_i^2}{2}.
\end{aligned}
\tag{4.11}
$$

Taking the first derivative with respect to N, we have

$$
\frac{\partial}{\partial N}[\ln L(N)] = \sum_{i=1}^{n} \frac{1}{N - (i-1)} - \phi \sum_{i=1}^{n} \frac{t_i^2}{2} \equiv 0.
$$

Therefore, the MLE of N can be obtained by solving the following equation:

$$
\sum_{i=1}^{n} \frac{1}{N - (i-1)} = \phi \sum_{i=1}^{n} \frac{t_i^2}{2}.
$$

Next, we assume that both N and ϕ are unknown. From Eq. (4.11), we obtain

$$
\frac{\partial}{\partial N}[\ln L(N, \phi)] = \sum_{i=1}^{n} \frac{1}{N - (i-1)} - \phi \sum_{i=1}^{n} \frac{t_i^2}{2} \equiv 0
$$

and

$$
\frac{\partial}{\partial \phi}[\ln L(N, \phi)] = \frac{n}{\phi} - \sum_{i=1}^{n}[N - (i-1)]\frac{t_i^2}{2} \equiv 0.
$$

Therefore, the MLEs of N and ϕ can be found by solving the two equations simultaneously as follows:

$$
\begin{aligned}
\phi &= 2 \sum_{i=1}^{n} \frac{1}{(N - i + 1)T} \\
N &= \frac{2n}{\phi T} + \frac{\sum_{i=1}^{n}(i - 1)t_i^2}{T}
\end{aligned}
\tag{4.12}
$$

where

$$T = \sum_{i=1}^{n} t_i^2.$$

4.5.3 Jelinski–Moranda Geometric Model

The J–M geometric model (Moranda, 1979) assumes that the program failure rate function is initially a constant D and decreases geometrically at failure times. The program failure rate and reliability function of time-between-failures at the ith failure interval can be expressed, respectively, as

$$\lambda(t_i) = Dk^{i-1}$$

and

$$R(t_i) = e^{-\int_0^{t_i} \lambda(x_i)dx_i} \qquad (4.13)$$
$$= e^{-Dk^{i-1}t_i}$$

where

$D =$ initial program failure rate;
$k =$ parameter of geometric function $(0 < k < 1)$.

If we allow multiple error removal in a time interval, then the failure rate function becomes

$$\lambda(t_i) = Dk^{n_{i-1}}$$

where n_{i-1} is the cumulative number of errors found up to the $(i-1)$st time interval. The software reliability function can be written as

$$R(t_i) = e^{-Dk^{n_{i-1}}t_i}. \qquad (4.14)$$

4.5.4 Moranda Geometric Poisson Model

The Moranda geometric Poisson model (Moranda, 1975) assumes fixed times $T, 2T, \ldots$ of equal length intervals, and that the number of failures occurring at interval i, N_i, follow a Poisson distribution with intensity rate Dk^{i-1}. The probability of getting m failures at the ith interval is

$$Pr\{N_i = m\} = \frac{e^{-Dk^{i-1}}(Dk^{i-1})^m}{m!}. \qquad (4.15)$$

The reliability and other performance measures can be easily derived in the same manner as in the J–M model.

4.5.5 Negative-Binomial Poisson Model

Assume that the intensity λ is a random variable with the gamma density function having parameters k and m, that is,

$$f(\lambda) = \frac{1}{\Gamma(m)} k^m \lambda^{m-1} e^{-k\lambda}, \quad \lambda \geq 0,$$

then the probability that there are exactly n software failures occuring during the time interval $(0, t)$ is given by

$$P\{N(t) = n\} = \binom{n+m-1}{n} p^m q^n, \quad n = 0, 1, 2, \ldots, \qquad (4.16)$$

where

$$p = \frac{k}{t+k} \quad \text{and} \quad q = \frac{t}{t+k} = 1 - p.$$

This probability is also called a negative binomial density function.

4.5.6 Modified Schick–Wolverton Model

Sukert (1977) modifies the S–W model to allow more than one failure at each time interval. The software failure rate function is given by

$$\lambda(t_i) = \phi[N - n_{i-1}] t_i, \qquad (4.17)$$

where n_{i-1} is the cumulative number of failures at the $(i-1)$th failure interval. Thus, the software reliability function is

$$R(t_i) = e^{-\phi[N-n_{i-1}]\frac{t_i^2}{2}} \qquad (4.18)$$

4.5.7 Goel–Okumoto Imperfect Debugging Model

Goel and Okumoto (1979) extend the J–M model by assuming that a fault is removed with probability p whenever a failure occurs. The failure rate function of the J–M model with imperfect debugging at the ith failure interval becomes

$$\lambda(t_i) = \phi[N - p(i-1)]. \qquad (4.19)$$

The reliability function is

$$R(t_i) = e^{-\phi[N-p(i-1)]t_i}. \qquad (4.20)$$

It should be noted that

$$\lambda(t_i) = \phi[N - p(i-1)]$$
$$= p\phi\left[\frac{N}{p} - (i-1)\right]$$
$$= \phi'[N' - (i-1)]$$

is the same as in the J–M model where

$$\phi' = p\phi \text{ and } N' = \frac{N}{p}.$$

4.6   Curve Fitting Models

The curve fitting group of models uses statistical regression analysis to study the relationship between software complexity and the number of faults in a program, the number of changes, or failure rate. This group of models finds a functional relationship between dependent and independent variables by using the methods of linear regression, nonlinear regression, or time series analysis. The dependent variables, for example, are the number of errors in a program. The independent variables are the number of modules changed in the maintenance phase, time between failures, programmers skill, program size, etc. Models included in this group are

• Estimation of errors
• Estimation of complexity
• Estimation of failure rate.

4.6.1   Estimation of Errors Model

The number of errors in a program can be estimated by using a linear or nonlinear regression model. A simple nonlinear regression model to estimate the total number of initial errors in the program, N, can be presented as follows:

$$N = \sum_i a_i X_i + \sum_i b_i X_i^2 + \sum_i c_i X_i^3$$

where X_i is the ith error factor and a_i, b_i, c_i are the coefficients of the model.

Typical error factors are software complexity metrics and environmental factors. Most curve fitting models involve only one error factor. Several reliability models with environmental factors will be further discussed in Chapter 8.

4.6.2 Estimation of Complexity Model

Belady and Lehman (1976) proposed a model to estimate the software complexity, C_R, using the time series approach. The software complexity model is summarized as follows:

$$C_R = a_0 + a_1 R + a_2 E_R + a_3 M_R + a_4 I_R + a_5 D + \epsilon,$$

where

R = release sequence number
E_R = environmental factor(s) at release R
M_R = number of modules at release R
I_R = inter-release interval R
D = number of days since first release
ϵ = error.

This model is applicable for software having multiple release versions and evolving over a long period of time.

4.6.3 Estimation of Failure Rate Model

Miller and Sofer (1985) proposed a model to estimate the failure rate of software. Given failure times $t_1, t_2, \ldots, t_n$, a rough estimate of the failure rate at the ith failure interval is

$$\hat{\lambda}_i = \frac{1}{t_{i+1} - t_i}.$$

Assuming that the failure rate is monotonically non-increasing, an estimate of this function $\lambda_i^*, i = 1, 2, \ldots, n$ can be obtained by using the least squared method (see (Miller, 1985)).

4.7   Reliability Growth Models

The reliability growth group of models measures and predicts the improvement of reliability programs through the testing process. The growth model represents the reliability or failure rate of a system as a function of time or the number of test cases. Models included in this group are

- Coutinho model (Coutinho, 1973)
- Wall and Ferguson model (Wall, 1977)

4.7.1 Coutinho Model

Coutinho (1973) adapted the Duane growth model to represent the software testing process. Coutinho plotted the cumulative number of deficiencies discovered and the number of correction actions made versus the cumulative testing weeks on log-log paper. Let $N(t)$ denote the cumulative number of failures and let t be the total testing time. The failure rate, $\lambda(t)$, model can be expressed as

$$\lambda(t) = \frac{N(t)}{t}$$
$$= \beta_0 t^{-\beta_1},$$

where β_0 and β_1 are the model parameters. The least squares method can be used to estimate the parameters of this model.

4.7.2 Wall and Ferguson Model

Wall and Ferguson (1977) proposed a model similar to the Weibull growth model for predicting the failure rate of software during testing. The cumulative number of failures at time t, $m(t)$, can be expressed as

$$m(t) = a_0[b(t)]^{\beta},$$

where a_0 and β are the unknown parameters. The function $b(t)$ can be obtained as the number of test cases or total testing time. Similarly, the failure rate function at time t is given by

$$\lambda(t) = m'(t) = a_0 \beta b'(t)[b(t)]^{\beta-1}.$$

Wall and Ferguson (1977) tested this model using several software failure data and observed that failure data correlate well with the model.

4.8 Non-Homogeneous Poisson Process Models

The non-homogeneous Poisson Process (NHPP) group of models provides an analytical framework for describing the software failure phenomenon during testing. The main issue in the NHPP model is to estimate the mean value function of the cumulative number of failures experienced up to a certain point in time. Models included in this group are

- Musa exponential (Musa, 1987)
- Goel and Okumoto NHPP (Goel, 1979)
- S-shaped growth (Ohba, 1984; Yamada, 1983, 1984a)
- Hyperexponential growth (Huang, 1984; Ohba, 1984, Yamada, 1984b)

- Discrete reliability growth (Yamada, 1985a)
- Testing-effort dependent reliability growth (Yamada, 1986, 1993)
- Generalized NHPP (Pham, 1997a, 1997b, 1999)

In Appendix B, the NHPP represents the number of failures experienced up to time t as an NHPP, $\{N(t), t \geq 0\}$. The main issue in the NHPP model is to determine an appropriate mean value function to denote the expected number of failures experienced up to a certain point in time. With different assumptions, the model will result with different functional forms of the mean value function.

Based on the NHPP assumptions in Appendix B, it can be shown that $N(t)$ has a Poisson distribution with mean $m(t)$, i.e.,

$$P\{N(t) = k\} = \frac{[m(t)]^k}{k!} e^{-m(t)} \quad k = 0, 1, 2, \ldots .$$

By definition, the mean value function of the cumulative number of failures, $m(t)$, can be expressed in terms of the failure intensity function of the software, i.e.,

$$m(t) = \int_0^t \lambda(s)ds. \tag{4.21}$$

The reliability function of the software is

$$R(t) = e^{-m(t)} = e^{-\int_0^t \lambda(s)ds}. \tag{4.22}$$

Goel and Okumoto's NHPP model (Goel, 1979, 1980) belongs to this class. Other types of mean value functions suggested by Ohba (1984), Yamada and Osaki (1985), Pham and Zhang (1997), and Pham (1999) are the delayed S-shaped growth model, inflection S-shaped growth model, and hyperexponential growth model. We will further discuss NHPP models in Chapter 5.

4.9   Markov Structure Models

A Markov process has the property that the future behavior of the process depends only on the current state and is independent of its past history (see Appendix B). The Markov structure group of models is a general way of representing the failure process of software. This group of models can also be used to study the reliability and interrelationship of the modules. It is assumed that failures of the modules are independent of each other. This assumption seems reasonable at the module level since they can be designed, coded and tested independently, but may not be true at the system level. Models included in this group are:

- Markov model with imperfect debugging (Goel, 1979)
- Littlewood Markov (Littlewood, 1979)
- Software safety (Tokuno, 1997; Yamada, 1998)

4.9.1 Markov Model with Imperfect Debugging

Goel and Okumoto (1979) proposed a linear Markov model with imperfect debugging and the transition probabilities of the model can be expressed as

$$
p_{ij} = \begin{cases} p & \text{for } j = i - 1 \\ q & \text{for } j = i \\ 1 & \text{for } i = j = 0 \\ 0 & \text{otherwise,} \end{cases}
$$

where p is the probability of successful debugging and $q = 1 - p$ for $i, j = 0, 1, \dots, N$. In other word, q is the probability of unsuccessfully debugging the fault whenever a failure occurs. The reliability function of the kth failure interval is given by Goel (1979):

$$
R_k(t) = \sum_{j=0}^{k-1} \binom{k-1}{j} p^{k-j-1} q^j e^{-[N-(k-j-1)]\phi t}.
$$

4.9.2 Littlewood Markov Model

Littlewood's model (Littlewood, 1979) represents the transitions between program modules during execution as the Markov process. Two types of failures are considered in the model. The first type of failure comes from a Poisson failure process at each module. It is recognized that new errors will be introduced as modules are integrated. The second type of failure is the interface between modules. Let

n = number of modules
a_{ij} = transition process from module i to module j
λ_i = Poisson failure rate of module i
q_{ij} = probability that transition from module i to module j fails
π_i = limiting distribution of the process.

Assuming that failures at modules and interfaces are independent of each other, Littlewood (1979) has shown that the program failure process is asymptotically a Poisson process with failure rate

$$\sum_{i=1}^{n} \pi_i(\lambda_i + \sum_{i \neq j} a_{ij}q_{ij})$$

as λ_i and q_{ij} approach zero.

4.9.3 Software Safety Model

Yamada *et al.* (1998) proposed a software safety model to describe the time-dependent behavior of the software using Markov processes. The assumptions in this safety model include:

- When the software system is operating, the holding times of the safety and the unsafe state follow exponential distributions with means $1/\theta$ and $1/\eta$, respectively.
- A debugging activity is performed when a software failure occurs. Debugging activities are perfect with probability a, while they are imperfect with probability b.
- Software reliability growth occurs in cases of perfect debugging. The time interval between software failure occurrences follows an exponential distribution with mean $1/\lambda_n$ where $n = 0, 1, 2, \ldots$ denotes the cumulative number of corrected faults.
- The probability that two or more software failures occur simultaneously is negligible.

Consider a stochastic process $\{X(t), t \geq 0\}$ representing the state of the software system at time t. The state space of $\{X(t), t \geq 0\}$ is defined as

$W_n =$ the system is operating safety
$U_n =$ the system falls into the unsafe state.

Yamada *et al.* (1998) use Moranda's model to describe the software reliability growth process. When n faults have been corrected, the failure intensity for the next software failure occurrence λ_n is

$$\lambda_n = Dk^n,$$

where $D > 0$ and $0 < k < 1$ are the initial failure rate and the decreasing ratio of the failure rate, respectively.

Software safety is defined as the probability that the system does not fall into any unsafe states at time point t and is given as follows (Yamada, 1998):

$$S(t) = \sum_{n=0}^{\infty} P_n(t)$$

where

$$P_n(t) = P\{X(t) = W_n\}$$

$$= A_n e^{-(\lambda_n + \theta + \eta)t} + \sum_{i=0}^{n} B_{ni} e^{-a\lambda_i t}$$

and the constant coefficients A_n and B_{ni} are given by

$$A_n = \frac{-\theta \prod\limits_{j=0}^{n-1} a\lambda_j}{\prod\limits_{j=0}^{n} (a\lambda_j - \lambda_n - \theta - \eta)}$$

$$B_{ni} = \frac{(\lambda_n + \eta - a\lambda_i) \prod\limits_{j=0}^{n-1} \lambda_j}{(\lambda_n + \theta + \eta - a\lambda_i) \prod\limits_{j=0 \& j \neq i}^{n} (\lambda_j - \lambda_i)}.$$

Problems

4.1 Assume that Fig. P.4.1 illustrates a program used in a toaster. The program scans the lever switch until a user pushes down the bread to be toasted. The heater is energized. The heat time is determined from the light-dark switch on the side of the toaster. After the time is input in Step 5, it is checked against valid times. If the number is less than 1 or greater than 10, the heat is turned off and the toast pops up because the input is incorrect. This time number should be an integer between one and ten. A value of one dictates a short heating time of ten seconds, resulting in light toast. Higher values dictate longer heating times. The program decreases the time number until the remaining time equals zero. The heater is then turned off and the toast pops up. The number of program control flow paths equals three. Determine the number of control flow paths in the program in Fig. P.4.1.

4.2 Suppose that the failure data $\{t_1, t_2, \ldots, t_n\}$ is given. Find the maximum likelihood estimates for the parameters N and ϕ of the modified Schick–Wolverton model.

4.3 Show that the mean time to the next failure of the S–W model is given by

$$\text{MTTF}_i = \sqrt{\frac{\pi}{2\phi(N - i + 1)}}$$

4.4 Derive Eqs. (4.1) and (4.2).

4.5 Assume that a new fault is introduced during the removal of a detected fault with a probability q. Determine the probability function of removing k induced errors and $r-k$ indigenous errors in m tests.

4.6 Using the Naval Tactical Data Systems (NTDS) failure data:
   (a) Calculate the maximum likelihood estimates for the parameters of the Jelinski–Moranda (J–M) model based on all available NTDS data.
   (b) Choose another software reliability model, other than NHPP models, and repeat question 4.6(a). Is the new model better or worse? Explain and justify your results.

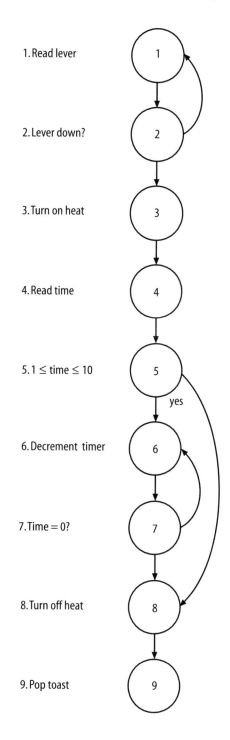

1. Read lever

2. Lever down?

3. Turn on heat

4. Read time

5. $1 \le time \le 10$

yes

6. Decrement timer

7. Time = 0?

8. Turn off heat

9. Pop toast

FIG. P.4.1. Program state diagram used in a toaster.

References

Basili, V.R. and B.T. Perricone, "Software errors and complexity: An empirical investigation," *Commun. ACM*, Vol. 27(1), 1984.

Belady, L.A. and M.M. Lehman, "A model of large program development," *IBM System J., Vol. 3, 1976.*

Cai, K.-Y., "On estimating the number of defects remaining in software," *Journal of Systems and Software*, Vol. 40(1), 1998.

Coutinho, J.S., "Software reliability growth," in *Proceedings International Conference on Reliable Software*, IEEE Computer Society Press, Los Angeles, 1973.

Fitzsimmons, A. and T. Love, "A review and evaluation of software science," *ACM Computing Surveys*, Vol. 10(1), March 1978.

Friedman, M.A. and J.M. Voas, *Software Assessment — Reliability, Safety, Testability*, John Wiley & Sons, New York, 1995.

Goel, A.L., "A summary of the discussion on an analysis of computing software reliability models," *IEEE Trans. Software Engineering*, Vol. SE-6(5), 1980.

Goel, A.L. and K. Okumoto, "A Markovian model for reliability and other performance measures of software systems," in *Proc. COMPCON*, IEEE Computer Society Press, Los Angeles, 1979.

Goel, A.L. and K. Okumoto, "Time-dependent error-detection rate model for software and other performance measures," *IEEE Trans. Reliability*, Vol. R-28(3), 1979.

Halstead, M.H, *Elements of Software Science*, Elsevier, New York, 1977.

Huang, X.Z., "The hypergeometric distribution model for predicting the reliability of software," *Microelectronics and Reliability*, Vol. 24(1), 1984.

Jelinski, Z. and P.B. Moranda, "Software reliability research," in *Statistical Computer Performance Evaluation*, W. Freiberger (ed.), Academic Press, New York, 1972.

Littlewood, B., "How to measure software reliability and how not to," *IEEE Trans. Reliability*, Vol. R-28(2), 1979.

Littlewood, B., "Software reliability model for modular program structure," *IEEE Trans. Reliability*, Vol. R-28(3), 1979.

McCabe, T.J., "A complexity measure," *IEEE Trans. Software Engineering*, Vol. SE-2(4), 1976.

McConnell, S.C., *Code Complete*, Microsoft Press, Richmond, 1993, p. 395

Miller, D.R. and A. Sofer, "Completely monotone regression estimation of software failure rate," in *Proc. International Conference on Software Engineering*, IEEE Computer Society Press, Los Angeles, 1985.

Mills, H.D. "On the statistical validation of computer programs," *IBM FSD*, July 1970 (unpublished).

Moranda, P.B., "A comparison of software error-rate models," in *Proc. Texas Conference on Computing Systems*, IEEE Computer Society Press, Los Angeles, 1975.

Moranda, P.B., "An error detection model for application during software development," *IEEE Trans. Reliability*, Vol. R-28(5), 1979.

Musa, J.D., A. Iannino, and K. Okumoto, *Software Reliability: Measurement, Prediction, and Application*, McGraw-Hill, New York, 1987.

Ohba, M. "Software reliability analysis models," *IBM J. Research Development,* Vol. 21(4), 1984.

Pham, H., "A software cost model with imperfect debugging, random life-cycle and penalty cost," *Int. J. Systems Science,* Vol. 27(5), 1996, 455–463.

Pham, H. and L. Nordmann, "A generalized NHPP software reliability model," in *Third International Conference on Reliability and Quality in Design,* Anaheim, CA, March 1997, ISSAT Press, Anaheim, 1997.

Pham, H. and X. Zhang, "An NHPP software reliability model and its comparison," *Int. J. Reliability, Quality and Safety Engineering,* Vol. 4(3), 1997.

Pham H., "Software reliability", in *Encyclopedia Electrical and Electronics Engineering,* John Wiley & Sons, New York, 1999.

Rook, P., *Software Reliability Handbook,* Elsevier Applied Science, Amsterdam, 1990.

Schick, G.J. and R.W. Wolverton, "An analysis of competing software reliability Models," *IEEE Trans. Software Engineering,* Vol. SE-4(2), 1978.

Sukert, A.N., "An investigation of software reliability models," in *Proceedings Annual Reliability & Maintainability Symposium,* IEEE Reliability Society, Piscataway, 1977.

Tohma, Y., H. Yamano, M. Ohba, and R. Jacoby, "The estimation of parameters of the hypergeometric distribution and its application to the software reliability growth model," *IEEE Trans. Software Engineering,* Vol. SE-17(5), 1991.

Tokuno, K. and S. Yamada, "Markovian availability measurement and assessment for hardware-software systems," *Int. J. Reliability, Quality and Safety Engineering,* Vol. 4(3), 1997.

Wall, J.K. and P.A. Ferguson, "Pragmatic software reliability prediction," in *Proceedings Annual Reliability & Maintainability Symposium,* IEEE Reliability Society, Piscataway, 1977.

Yamada, S., M. Ohba, and S. Osaki, "S-shaped reliability growth modeling for software error detection," *IEEE Trans. Reliability,* Vol. R-32(5), 1983.

Yamada, S. and S. Osaki, "Nonhomogeneous error detection rate models for software reliability growth," in *Stochastic Models in Reliability Theory,* S. Osaki and Y. Hatoyama (eds.), Springer-Verlag, Berlin, 1984.

Yamada, S., M. Ohba, and S. Osaki, "S-shaped software reliability growth models and their applications," *IEEE Trans. Reliability,* Vol. R-33, October 1984.

Yamada, S. and S. Osaki, "Discrete software reliability growth models," *J. Applied Stochastic Models and Data Analysis,* Vol. 1(4), 1985.

Yamada, S. and S. Osaki, "Software reliability growth modeling: Models and applications," *IEEE Trans. Software Engineering,* Vol. SE-11(12), 1985.

Yamada, S., H. Ohtera, and H. Narihisa, "Software reliability growth models with testing-effort," *IEEE Trans. Reliability,* Vol. R-35(1), 1986.

Yamada, S., J. Hishitani, and S. Osaki, "Software reliability growth models with a Weibull test-effort function," *IEEE Trans. Reliability,* Vol. R-42(1), 1993.

Yamada, S., K. Tokuno, and Y. Kasano, "Quantitative assessment models for software safety/reliability," *Electronics and Communications in Japan,* Part 2, Vol. 81(5), 1998.

5

NHPP Software Reliability Models

"Every problem has in it the seeds of its own solution.
If you don't have any problems, you don't get any seeds."

5.1 Introduction

Software reliability assessment is increasingly important in developing and testing new software products. Before newly developed software is released to the user, it is extensively tested for errors that may have been introduced during development. Although detected errors are removed immediately, new errors may be introduced during debugging. Software that contains errors and released to the market incurs high failure costs. Debugging and testing, on the other hand, reduces the error content but increases development costs. Thus, there is a need to determine the optimal time to stop software testing. During system testing, reliability measure is an important criterion in deciding when to release the software. Several other criteria, such as the number of remaining errors, failure rate, reliability requirements, or total system cost, may be used to determine optimal testing time.

In this chapter, we present stochastic reliability models for the software failure phenomenon based on a Non-homogeneous Poisson Process (NHPP). Allowing both the error content function and the error detection rate to be time-dependent, a generalized software reliability model and an analytical expression for the mean value function are presented. Numerous existing models based on NHPP are also summarized. Several applications and numerical examples are included to illustrate the results. A general function for calculating the mean time between failures (MTBF) of software systems based on the NHPP is also presented. An NHPP is a realistic model for predicting software reliability and has a very interesting and useful interpretation in debugging and testing the software.

Notation

$m(t)$	expected number of errors detected by time t ("mean value function")
$a(t)$	error content function, i.e., total number of errors in the software including the initial and introduced errors at time t
$b(t)$	error detection rate per error at time t
$N(t)$	random variable representing the cumulative number of software errors detected by time t
$y(t)$	actual values of $N(t)$ ($y_i := y(t_i)$)
S_j	actual time at which the jth error is detected
$R(s/t)$	reliability during $(t, t + s]$ given that the last error occured at time t

5.2   Parameter Estimation

Parameter estimation is of primary importance in software reliability prediction. Once the analytical solution for $m(t)$ is known for a given model, the parameters in the solution need to be determined. Parameter estimation is achieved by applying a technique of maximum likelihood estimation (MLE), the most important and widely used estimation technique. In many cases, the maximum likelihood estimators are consistent and asymptotically normally distributed as the sample size increases (Zhao, 1996). In this chapter, we only discuss the MLE technique to estimate the unknown parameters for the NHPP models. Depending on the format in which test data are available, two different approaches are frequently used. A set of failure data is usually collected in one of two common ways (see Chapter 3) and is discussed next.

Type 1 Data: Interval Domain Data
Assuming that the data are given for the cumulative number of detected errors y_i in a given time-interval $(0, t_i)$ where $i = 1, 2, \ldots,$ n and $0 < t_1 < t_2 < \ldots < t_n$, then the log likelihood function (LLF) takes on the following form:

$$\text{LLF} = \sum_{i=1}^{n}(y_i - y_{i-1}) \cdot \log[m(t_i) - m(t_{i-1})] - m(t_n). \qquad (5.1)$$

Thus the maximum of the LLF is determined by the following system of equations:

$$0 = \sum_{i=1}^{n} \frac{\frac{\partial}{\partial \theta} m(t_i) - \frac{\partial}{\partial \theta} m(t_{i-1})}{m(t_i) - m(t_{i-1})} (y_i - y_{i-1}) - \frac{\partial}{\partial \theta} m(t_n), \qquad (5.2)$$

where for θ each of the unknown parameters is to be substituted.

Using the observed failure data (t_i, y_i) for $i = 1, 2, \ldots, n$, we can use the mean value function $m(t_i)$ to determine the expected number of errors to be detected by time t_i for $i = n+1, n+2$, etc.

Type 2 Data: Time Domain Data
Assuming that the data are given for the occurrence times of the failures or the times of successive failures, i.e., the realization of random variables S_j for $j = 1, 2, \ldots, n$. Given that the data provide n successive times of observed failures s_j for $0 \le s_1 \le s_2 \ldots \le s_n$, we can convert these data into the time between failures x_i where $x_i = s_i - s_{i-1}$ for $i = 1, 2, \ldots, n$. Given the recorded data on the time of failures, the log likelihood function takes on the following form:

$$\text{LLF} = \sum_{i=1}^{n} \log[\lambda(s_i)] - m(s_n). \qquad (5.3)$$

The MLE of unknown parameters $\theta = (\theta_1, \theta_2, \ldots, \theta_n)$ can be obtained by solving the following equations:

$$0 = \sum_{i=1}^{n} \frac{\frac{\partial}{\partial \theta} \lambda(S_i)}{\lambda(S_i)} - \frac{\partial}{\partial \theta} m(S_n) \qquad (5.4)$$

where

$$\lambda(t) = \frac{\partial}{\partial t} m(t)$$

and for θ each of the unknown parameters is to be substituted.

The equations to be solved for the MLE of the system parameters are nonlinear. To make use of the iterative Newton method, we may need the first and second partial derivatives of the functions $m(t)$ and $\lambda(t)$. The "Solver" add-on package for Microsoft Excel provides a convenient tool for maximizing objective functions LLF. The author developed an Excel macro that computes the MLE of free parameters for an arbitrary mean value function of a given set of test data, using the iterative Newton method. This user friendly software program, called NHPP software, computes the MLEs, estimates the parameters of the function $m(t)$, charts the results at a single click of the mouse, and can be easily run from the diskette included in this book.

5.3 NHPP Models

In this section, we will describe the following types of NHPP models:

1. NHPP exponential model
2. NHPP S-shaped model
3. NHPP imperfect debugging model
4. NHPP S-shaped imperfect debugging model.

5.3.1 NHPP Exponential Model

The exponential NHPP model is based on the following assumptions:

1. All faults in a program are mutually independent from the failure detection point of view.
2. The number of failures detected at any time is proportional to the current number of faults in a program. This means that the probability of the failures for faults actually occurring, i.e., detected, is constant.
3. The isolated faults are removed prior to future test occasions.
4. Each time a software failure occurs, the software error which caused it is immediately removed, and no new errors are introduced.

This is shown in the following differential equation:

$$\frac{\partial m(t)}{\partial t} = b[a - m(t)], \tag{5.5}$$

where a is the expected total number of faults that exist in the software before testing and b is the failure detection rate or the failure intensity of a fault.

The solution of the above differential equation is given by

$$m(t) = a[1 - e^{-bt}]. \tag{5.6}$$

This model is known as Goel–Okumoto model (Goel, 1979).

For type 1 data, the estimate of parameters a and b of the Goel–Okumoto model using the MLE method discussed in Section 5.2, can be obtained by solving the following equations simultaneously:

$$a = \frac{y_n}{(1 - e^{-bt_n})}$$

$$\frac{y_n t_n e^{-bt_n}}{1 - e^{-bt_n}} = \sum_{k=1}^{n} \frac{(y_k - y_{k-1})(t_k e^{-bt_k} - t_{k-1} e^{-bt_{k-1}})}{(e^{-bt_{k-1}} - e^{-bt_k})}. \tag{5.7}$$

Similarly, for type 2 data, the estimate of parameters a and b using the MLE method can be obtained by solving the following equations:

$$a = \frac{n}{(1 - e^{-bs_n})}$$

$$\frac{n}{b} = \sum_{i=1}^{n} s_i + \frac{ns_n e^{-bs_n}}{(1 - e^{-bs_n})}. \tag{5.8}$$

Let $\hat{a}$ and $\hat{b}$ be the MLE of parameters a and b, respectively. We can then obtain the MLE of the mean value function (MVF) and the reliability function as follows:

$$\hat{m}(t) = \hat{a}[1 - e^{-\hat{b}t}]$$

$$\hat{R}(x|t) = e^{-\hat{a}[e^{-\hat{b}t} - e^{-\hat{b}(t+x)}]}.$$

It is of interest to determine the variability of the number of failures at time t, $N(t)$. One can approximately obtain the confidence intervals for $N(t)$ based on the Poisson distribution as

$$\hat{m}(t) - z_\alpha \sqrt{\hat{m}(t)} \leq N(t) \leq \hat{m}(t) + z_\alpha \sqrt{\hat{m}(t)},$$

where z_α is $100(1+ \alpha)/2$ percentile of the standard normal distribution, i.e., $N(0, 1)$.

Example 5.1 The data set in Table 5.1 was reported by Musa (1987) based on failure data from a real-time command and control system, which represents the failures observed during system testing for 25 hours of CPU time. The delivered number of object instructions for this system was 21,700 and was developed by Bell Laboratories.

It should be noted that this data set belongs to the concave class, therefore, it seems reasonable to use the Goel–Okumoto NHPP model to describe the failure process of the software system. From the failure data, the two unknown parameters, a and b, can be obtained from Eq. (5.7) and the estimated values for the two parameters are

$$\hat{a} = 142.3153$$

$$\hat{b} = 0.1246.$$

Recall that $\hat{a}$ is an estimate of the expected total number of failures to be eventually detected and b represents the number of faults detected per fault per unit time (hour). The estimated mean value function and software reliability function are

$$\hat{m}(t) = 142.3153(1 - e^{-0.1246t})$$

and

$$\hat{R}(x|t) = e^{-(142.3153)[e^{-(0.1246)t} - e^{-(0.1246)(t+x)}]},$$

respectively.

The above two functions can be used to determine when to release the software system or the additional testing effort required when the system is ready for release. Let us assume that failure data from only 16 hours of testing are available and from Table 5.1, a total of 122 failures have been observed. Based on these data and

TABLE 5.1. Failure data in a one-hour interval and number of failures.

Hour	Number of failures	Cummulative failures
1	27	27
2	16	43
3	11	54
4	10	64
5	11	75
6	7	82
7	2	84
8	5	89
9	3	92
10	1	93
11	4	97
12	7	104
13	2	106
14	5	111
15	5	116
16	6	122
17	0	122
18	5	127
19	1	128
20	1	129
21	2	131
22	1	132
23	2	134
24	1	135
25	1	136

using the MLE method, the estimated values for the two parameters are

$$\hat{a} = 138.3779 \text{ and } \hat{b} = 0.1334$$

and the estimated mean value function becomes

$$\hat{m}(t) = 138.3779(1 - e^{-0.1334t}).$$

The reliability of the software system is

$$\hat{R}(x|t) = e^{-(138.3779)[e^{-(0.1334)t} - e^{-(0.1334)(t+x)}]}.$$

An estimate number of remaining errors after 16 hours of testing is
16.38 with a 90% confidence interval of (4.64, 28.11). Similarly, the
estimated current software reliability for the next hour is 0.129
and the corresponding 90% confidence interval is (0.019, 0.31).

Next, suppose the problem of interest is to know how much additional testing is needed in order to achieve an acceptable number of
remaining errors so that the software can be released for operational use. For example, we would want to release the software if
the expected number of remaining errors is less than or equal to
10. In the above analysis, we learned that the best estimate of the
remaining errors in the software after 16 hours of testing is about
17. Therefore, testing has to continue in the hope that additional
faults can be detected. If we were to carry on a similar task after
each additional hour of testing, we can expect to obtain another
seven additional errors during the next four hours (see Table 5.2).

TABLE 5.2. Software reliability performance measures.

Test time (T)	a	b	Remaining errors	Reliability R(0.1/T)	R(1/T)
16	138.3779	0.1333	16.3780	0.8049	0.1294
17	133.7050	0.1432	11.7050	0.8466	0.2096
18	141.2543	0.1274	14.2544	0.8349	0.1817
19	139.7190	0.1304	11.7190	0.8591	0.2386
20	138.8495	0.1323	9.8495	0.8786	0.2951
21	140.3408	0.1290	9.3408	0.8871	0.3228
22	140.1002	0.1296	8.1002	0.9010	0.3737
23	141.9104	0.1255	7.9104	0.9060	0.3933
24	142.0264	0.1252	7.0265	0.9162	0.4372
25	142.3153	0.1246	6.3154	0.9248	0.4772

In other words, the expected number of remaining errors after 20 hours would be 9.8 so that the above objective would be met.

Musa (1985) proposed a similar model to the Goel–Okumoto model by considering the relationship between execution time and calendar time. Let $m(t)$ be the number of failures discovered as a result of test case runs up to the time of observation. Musa obtained the differential equation as follows:

$$\frac{\partial m(t)}{\partial t} = \frac{c}{nT}[a - m(t)], \tag{5.9}$$

where

a = number of failures in the program
c = the testing compression factor
T = mean time to failure at the beginning of the test
n = total number of failures possible during the maintained life of the program
t = execution time or the total CPU time utilized to complete the test case runs up to a time of observation.

The above differential equation can be easily solved as follows:

$$m(t) = a\left(1 - e^{-\frac{ct}{nT}}\right). \tag{5.10}$$

This model is also called the Musa exponential model.

The failure intensity function is

$$\begin{aligned}\lambda(t) &= \frac{\partial}{\partial t}[m(t)] \\ &= \frac{c}{nT}[a - m(t)].\end{aligned} \tag{5.11}$$

The reliability function and pdf are given by

$$R(t) = e^{-a(1-e^{-\frac{c}{nT}t})} \tag{5.12}$$

and

$$f(t) = \frac{c}{nT}ae^{-\frac{c}{nT}t}e^{-a(1-e^{-\frac{c}{nT}t})} \tag{5.13}$$

respectively. Suppose we have observed k failures of the software and suppose that the failure data set $\{t_1, t_2, \ldots, t_k\}$ is given where t_i is the observed time between the $(i-1)$th and the ith failure. Here, we want to estimate the unknown parameters a and c. Using the MLE method, the likelihood function is obtained as:

$$L(a, c) = \prod_{i=1}^{k} f(t_i)$$

$$= \left(\frac{ac}{nT}\right)^k \left(e^{-\frac{c}{nT}\sum_{i=1}^{k} t_i}\right) e^{-a\sum_{i=1}^{k}\left[1-e^{-\frac{c}{nT}t_i}\right]}$$

The LLF is given by

$$\ln L = k \ln\left(\frac{ac}{nT}\right) - \frac{c}{nT}\sum_{i=1}^{k} t_i - a\sum_{i=1}^{k}[1 - e^{-\frac{c}{nT}t_i}].$$

Then we have

$$\frac{\partial}{\partial c}\ln L = kc - \frac{1}{nT}\sum_{i=1}^{k} t_i - \frac{a}{nT}\sum_{i=1}^{k} t_i e^{-\frac{c}{nT}t_i} \equiv 0$$

and

$$\frac{\partial}{\partial a}\ln L = \frac{k}{a} - \sum_{i=1}^{k}(1 - e^{-\frac{c}{nT}t_i}) \equiv 0.$$

Thus, a and c can be obtained by solving the following two equations:

$$c = \frac{1}{knT}\sum_{i=1}^{k} t_i + \frac{a}{knT}\sum_{i=1}^{k} t_i e^{-\frac{c}{nT}t_i}$$

and

$$a = \frac{k}{\sum_{i=1}^{k}[1 - e^{-\frac{c}{nT}t_i}]}.$$

Hyperexponential Growth Model

The hyperexponential growth model (Ohba, 1984) is based on the assumption that a program has a number of clusters of modules, each having a different initial number of errors and a different failure rate. Examples are new modules versus reused modules, simple modules versus complex modules, and modules which interact with hardware versus modules which do not. Since the sum of exponential distributions becomes a hyperexponential distribution, the mean value function of the hyperexponential class NHPP model is

$$m(t) = \sum_{i=1}^{n} a_i[1 - e^{-b_i t}],$$

where

n = number of clusters of modules
a_i = number of initial faults in cluster i
b_i = failure rate of each fault in cluster i.

The failure intensity function can be obtained as follows:

$$\lambda(t) = \sum_{i=1}^{n} a_i b_i e^{-b_i t}.$$

A similar extension of the exponential growth model has been suggested by Yamada and Osaki (1985) by dividing software into k modules. The failure intensity of faults within different modules are assumed to be different, while the failure intensity of faults within the same module are assumed to be the same. The expected number of faults detected for each module are exponential. Thus, the expected number of faults detected for the entire software can be obtained as

$$m(t) = a \sum_{i=1}^{k} p_i[1 - e^{-b_i t}],$$

where

k = number of modules in the software
b_i = error detection rate of one fault within the ith module
p_i = probability of faults for the ith module
a = expected number of software errors to be detected eventually or total number of faults existing in the software before testing.

For type 1 data, the MLEs of the parameters a and b_i for $i = 1, 2, \ldots, k$ can be obtained by solving the following equations simultaneously:

$$a = \frac{y_n}{\sum\limits_{i=1}^{k} p_i(1 - e^{-b_i t_n})}$$

$$\frac{y_n t_n e^{-b_i t_n}}{\sum\limits_{i=1}^{k} p_i(1 - e^{-b_i t_n})} = \sum_{j=1}^{n} \frac{(y_j - y_{j-1})(t_j e^{-b_i t_j} - t_{j-1} e^{-b_i t_{j-1}})}{\sum\limits_{i=1}^{k} p_i(e^{-b_i t_{j-1}} - e^{-b_i t_j})}.$$

Similarly, for type 2 data, the MLEs of the parameters a and b_i for $i = 1, 2, \ldots, k$ can be obtained by solving the following equations:

$$a = \frac{n}{\sum\limits_{i=1}^{k} p_i(1 - e^{-b_i s_n})}$$

$$\frac{n s_n e^{-b_i s_n}}{\sum\limits_{i=1}^{k} p_i(1 - e^{-b_i s_n})} = \sum_{j=1}^{n} \frac{(e^{-b_i s_j} - b_i s_j e^{-b_i s_j})}{\sum\limits_{i=1}^{k} p_i b_i e^{-b_i s_j}}.$$

Connective NHPP Model

Nakagawa (1994) proposed a model, called connective NHPP model, where the basic shape of the growth curve is exponential and that an S-curve forms due to the test. In the connective NHPP model, a group of modules called "main route modules" are tested first, followed by the rest of the modules. Even if the failure intensity of the faults in the main route module and the other modules are similar, the growth curve becomes an S-curve since the search for their detection starts at different points in time. The expected number of faults detected for the software as a whole can be expressed as follows:

$$m(t) = a_1(1 - e^{-bI_{[0,t_0)}(t)})$$
$$+ a_2(1 - e^{-bI_{[t_0,\infty)}(t)}),$$

where $a_2 > a_1 > 0$, and

a_1 = number of faults that are expected to be detected in the main route modules

a_2 = number of faults that are expected to be detected in modules other than the main route modules

b = failure intensity

t_0 = starting time for testing modules other than the main route modules

$I_{[.]}$ = indicator function.

5.3.2 NHPP S-Shaped Model

In the NHPP S-shaped model, the software reliability growth curve is an S-shaped curve which means that the curve crosses the exponential curve from below and the crossing occurs once and only once. The detection rate of faults, where the error detection rate changes with time, become the greatest at a certain time after testing begins, after which it decreases exponentially. In other words, some faults are covered by other faults at the beginning of the testing phase, and before these faults are actually removed, the covered faults remain undetected. Yamada (1984) also determined that the software testing process usually involves a learning process where testers become familiar with the software products, environments, and software specifications. Several S-shaped models (Yamada, 1984; Pham, 1997) such as delayed S-shaped, inflection S-shaped, etc., will also be discussed in this section.

The NHPP S-shaped model is based on the following assumptions:

1. The error detection rate differs among faults.
2. Each time a software failure occurs, the software error which caused it is immediately removed, and no new errors are introduced.

This can be shown as the following differential equation:

$$\frac{\partial m(t)}{\partial t} = b(t)[a - m(t)], \tag{5.14}$$

where

a = expected total number of faults that exist in the software before testing

$b(t)$ = failure detection rate, also called the failure intensity of a fault

$m(t)$ = expected number of failures detected at time t.

The above differential equation can be easily solved and is given by

$$m(t) = a[1 - e^{-\int_0^t b(u)du}]. \tag{5.15}$$

NHPP Inflection S-Shaped Model

The inflection S-shaped model (Ohba, 1984) is based on the dependency of faults by postulating the following assumptions:

1. Some of the faults are not detectable before some other faults are removed.
2. The probability of failure detection at any time is proportional to the current number of detectable faults in the software.
3. Failure rate of each detectable fault is constant and identical.
4. The isolated faults can be entirely removed.

Assume

$$b(t) = \frac{b}{1 + \beta e^{-bt}}, \tag{5.16}$$

where the parameters b and β represent the failure-detection rate and the inflection factor, respectively. From Eq. (5.15), the mean value function is given by

$$m(t) = \frac{a}{1 + \beta e^{-bt}}(1 - e^{-bt}). \tag{5.17}$$

This model is called the inflection S-shaped NHPP model (Ohba, 1984). The failure intensity function is given by

$$\lambda(t) = \frac{ab(1 + \beta)e^{-bt}}{(1 + \beta e^{-bt})^2}.$$

We then obtain the expected number of remaining errors at time t

$$m(\infty) - m(t) = \frac{a(1 + \beta)e^{-bt}}{(1 + \beta e^{-bt})}.$$

For type 1 data, the estimate of parameters a and b for specified β using the MLE method can be obtained by solving the following equations simultaneously:

$$a = \frac{y_n(1 + \beta e^{-bt_n})}{(1 - e^{-bt_n})}$$

and

$$\sum_{i=1}^{n}(y_i - y_{i-1})\left(\frac{(t_i e^{-bt_i} - t_{i-1}e^{-bt_{i-1}})}{(e^{-bt_{i-1}} - e^{-bt_i})} + \frac{\beta t_i e^{-bt_i}}{(1 + \beta e^{-bt_i})} + \frac{\beta t_{i-1}e^{-bt_{i-1}}}{(1 + \beta e^{-bt_{i-1}})}\right)$$
$$= \frac{y_n t_n e^{-bt_n}(1 - \beta + 2\beta e^{-bt_n})}{(1 - e^{-bt_n})(1 + \beta e^{-bt_n})}.$$

Similarly, for type 2 data, the estimate of parameters a and b for specified β using the MLE method can be obtained by solving the following equations:

$$a = \frac{n(1 + \beta e^{-bs_n})}{(1 - e^{-bs_n})}$$

and

$$\frac{ns_n e^{-bs_n}(1 + \beta)}{(1 - e^{-bs_n})(1 + \beta e^{-bs_n})} = \frac{n}{\beta} - \sum_{i=1}^{n} s_i + 2\sum_{i=1}^{n}\frac{\beta s_i e^{-bs_i}}{(1 + \beta e^{-bs_i})}$$

NHPP Delayed S-Shaped Model

We now discuss a stochastic model for a software error detection process based on NHPP in which the growth curve of the number of detected software errors for the observed failure data is S-shaped, called delayed S-shaped NHPP model (Yamada, 1984). The software error detection process described by an S-shaped curve can be characterized as a learning process in which test-team members become familiar with the test environment, testing tools, or project requirements, i.e., their test skills gradually improve. The delayed S-shaped model is based on the following assumptions:

1. All faults in a program are mutually independent from the failure detection point of view.
2. The probability of failure detection at any time is proportional to the current number of faults in a software.
3. The proportionality of failure detection is constant.
4. The initial error content of the software is a random variable.
5. A software system is subject to failures at random times caused by errors present in the system.
6. The time between failures $(i-1)$th and ith depends on the time to the $(i-1)$th failure.
7. Each time a failure occurs, the error which caused it is immediately removed and no other errors are introduced.

Assume

$$b(t) = \frac{b^2 t}{bt + 1},\qquad (5.18)$$

where b is the error detection rate per error in the steady-state. From Eq. (5.15), the mean value function can be obtained as

$$m(t) = a[1 - (1 + bt)e^{-bt}],\qquad (5.19)$$

which shows an S-shaped curve. This model is called the delayed S-shaped NHPP model for such an error detection process, in which the observed growth curve of the cumulative number of detected errors is S-shaped (Yamada, 1983). The corresponding failure intensity function is

$$\lambda(t) = ab^2 t\, e^{-bt}.\qquad (5.20)$$

The reliability of the software system is

$$R(s|t) = e^{-[m(t+s)-m(t)]}$$
$$= e^{-a[(1+bt)e^{-bt}-(1+b(t+s))e^{-b(t+s)}]}.$$

(5.21)

The expected number of errors remaining in the system at time t is given by

$$n(t) = m(\infty) - m(t)$$
$$= a(1 + bt)e^{-bt}.$$

For type 1 data, the estimate of parameters a and b using the MLE method can be obtained by solving the following equations simultaneously:

$$a = \frac{y_n}{[1 - (1 + bt_n e^{-bt_n})]}$$

and

$$\frac{y_n t_n^2 e^{-bt_n}}{[1 - (1 + bt_n e^{-bt_n})]} = \sum_{i=1}^{n} \frac{(y_i - y_{i-1})(t_i^2 e^{-bt_i} - t_{i-1}^2 e^{-bt_{i-1}})}{[(1 + bt_{i-1})e^{-bt_{i-1}} - (1 + bt_i)e^{-bt_i}]}.$$

Similarly, for type 2 data, the estimate of parameters a and b for specified β using the MLE method can be obtained by solving the following equations:

$$a = \frac{n}{[1 - (1 + bs_n e^{-bs_n})]}$$

and

$$\frac{2n}{b} = \sum_{i=1}^{n} s_i + \frac{nbs_n^2 e^{-bs_n}}{[1 - (1 + bs_n)e^{-bs_n})]}.$$

Example 5.2   The small on-line data entry software package test data, available since 1980 in Japan (Ohba, 1984), is shown in Table 5.3 (data set #5). The size of the software has approximately 40,000 LOC. The testing time was measured on the basis of the number of shifts spent running test cases and analyzing the results. The pairs of the observation time and the cumulative number of faults detected are presented in Table 5.3.

TABLE **5.3.** On-line data entry software package test data
(data set #5).

Time of observation (days)	Cumulative number of failures
1	2
2	3
3	4
4	5
5	7
6	9
7	11
8	12
9	19
10	21
11	22
12	24
13	26
14	30
15	31
16	37
17	38
18	41
19	42
20	45
21	46

The MLEs of the unknown parameters a and b for the delayed S-shaped NHPP model are

$$a = 71.725$$
$$b = 0.104.$$

The estimated mean value function $m(t)$ is

$$m(t) = (71.725)[1 - (1 + 0.104t)e^{-0.104t}].$$

Figure 5.1 shows the analysis results using the delayed S-shaped NHPP model. We can see that the model fits the observed failure data well in this data set.

5.3.3 NHPP Imperfect Debugging Model

In this section, the development of a software reliability model (Pham, 1996) that addresses the problems of multiple failure types and imperfect debugging based on an NHPP for predicting software performance measures is discussed. The model allows for

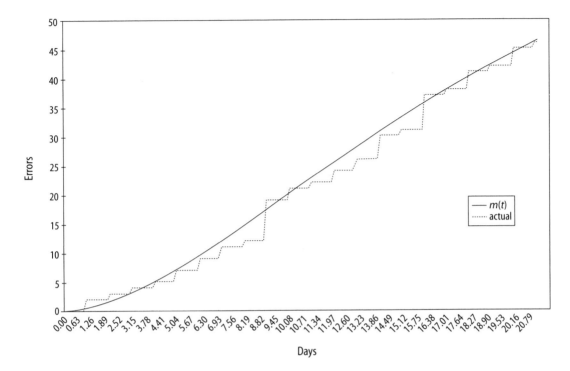

FIG. 5.1. Mean value function versus actual error data.

three different error types, categorized by the difficulty of removal and detection. Critical errors (type 1) are very difficult to detect and remove, major errors (type 2) are difficult to detect and remove, and minor errors (type 3) are easy to detect and remove.

Notation

a	expected number of software errors to be eventually detected
b_i	error detection rate per type i error, $i = 1, 2, 3$; $0 < b_1 < b_2 < b_3 < 1$
p_i	content proportion of type i errors
$\lambda(t)$	intensity function or error detection rate
$N_i(t)$	cumulative number of type i errors
$n(t)$	number of errors to be eventually detected plus the number of errors introduced to the program by time t

β_i type i error introduction rate that satisfies, $0 \le \beta_i < 1$

$m_i(t)$ expected number of software type i detected errors by time t

The NHPP imperfect debugging model is based on the following assumptions:

1. When detected errors are removed, it is possible to introduce new errors.
2. The probability of finding an error in a program is proportional to the number of remaining errors in the program.
3. The probability of introducing a new error is constant.
4. Three types of errors exist:

   type 1 errors (critical): very difficult to detect

   type 2 errors (major): difficult to detect

   type 3 errors (minor): easy to detect.
5. The parameters a and b_i for $i = 1, 2, 3$ are unknown constants.
6. The error detection phenomenon in the software is modeled by an NHPP.

The function $m(t)$ is given as the solution of the following system of differential equations:

$$\frac{\partial}{\partial t}[m_i(t)] = b_i[n_i(t) - m_i(t)]$$

$$\frac{\partial}{\partial t}[n_i(t)] = \beta_i \frac{\partial}{\partial t}[m_i(t)]$$

$$m(t) = \sum_{i=1}^{3} m_i(t)$$

$$n_i(0) = ap_i$$

$$m_i(0) = 0.$$

Solving the above system of differential equations simultaneously, we obtain the following results (Pham, 1996):

$$m_i(t) = \frac{ap_i}{(1 - \beta_i)}[1 - e^{-(1-\beta_i)b_i t}] \qquad (5.22)$$

$$\lambda_i(t) = ap_i b_i e^{-(1-\beta_i)b_i t} \qquad (5.23)$$

and

$$n_i(t) = \frac{ap_i}{1 - \beta_i}[1 - \beta_i e^{-(1-\beta_i)b_i t}]. \qquad (5.24)$$

The software reliability function $R(x|t)$ is given by:

$$R(x|t) = e^{-\left[\sum_{i=1}^{3} \frac{ap_i}{1-\beta_i}(e^{-(1-\beta_i)b_i t})[1-e^{-(1-\beta_i)b_i x}]\right]}. \tag{5.25}$$

Parameter Estimation

The model parameters a, b_1, b_2, and b_3 are estimated using the MLE method. For type 1 data (the data on the cumulative number of detected errors), suppose that the data are available in the form of $(t_i, y_{i,j})$, where $y_{i,j}$ are the cumulative number of failures type j detected up to time t_i for $i = 1, 2, \ldots, n$, and $j = 1, 2, 3$. Assuming the fault detection process is NHPP, the likelihood function $L(a, b_1, b_2, b_3)$ for given data $(t_i, y_{i,j})$, $i = 1, 2, \ldots, n$, and $j = 1, 2, 3$ is as follows:

$$L(a, b_1, b_2, b_3) = Pr\{\prod_{j=1}^{3}(m_j(0) = 0, m_j(t_1) = y_{1,j}, m_j(t_2)$$

$$= y_{2,j}, \ldots, m_j(t_n) = y_{n,j})\}$$

$$= \prod_{j=1}^{3}\prod_{i=1}^{n} \frac{[m_j(t_i) - m_j(t_{i-1})]^{y_{i,j}-y_{i-1,j}}}{(y_{i,j} - y_{i-1,j})!} e^{-[m_j(t_i)-m_j(t_{i-1})]},$$

where

$$m_j(t_i) = \frac{ap_j}{(1-\beta_j)}[1 - e^{-(1-\beta_j)b_j t_i}].$$

Taking the log likelihood function, we obtain

$$\ln[L(a, b_1, b_2, b_3)] = \sum_{j=1}^{3}\sum_{i=1}^{n}[(y_{i,j} - y_{i-1,j})\ln[m_j(t_i) - m_j(t_{i-1})]$$

$$- \ln[(y_{i,j} - y_{i-1,j})!] - [m_j(t_i) - m_j(t_{i-1})]].$$

Taking the partial derivatives of the log likelihood function, $\ln[L(a, b_1, b_2, b_3)]$, with respect to the unknown parameters, a, b_1, b_2, and b_3, and setting them equal to zero, we obtain the following system of equations:

$$a = \frac{\sum_{j=1}^{3} y_{n,j}}{\sum_{j=1}^{3} \frac{p_j}{(1-\beta_j)}[1 - e^{-(1-\beta_j)b_j t_n}]} \tag{5.26}$$

$$\sum_{i=1}^{n}(y_{i,j} - y_{i-1,j})\frac{(1 - \beta_j)[t_i e^{-(1-\beta_j)b_j t_i} - t_{i-1}e^{-(1-\beta_j)b_j t_{i-1}}]}{[e^{-(1-\beta_j)b_j t_{i-1}} - e^{-(1-\beta_j)b_j t_i}]}$$
$$= ap_j t_n e^{-(1-\beta_j)b_j t_n} \qquad (5.27)$$

for $j = 1$, 2, and 3. Solving the above system of equations simultaneously gives the maximum likelihood estimates of parameters a, b_1, b_2, and b_3.

For type 2 data (the data on failure occurrence times), assume that the data set is available in the form of n_1 type 1 errors, n_2 type 2 errors, and n_3 type 3 errors, and $S_{1,1} \leq S_{1,2} \leq \cdots \leq S_{1,n1}$, $S_{2,1} \leq S_{2,2} \leq \cdots \leq S_{2,n2}$, and $S_{3,1} \leq S_{3,2} \leq \cdots \leq S_{3,n3}$, where $S_{i,j}$ is the actual time that the jth failure of type i error occurs. Again, using the MLE method, the likelihood function for the NHPP model in a given data set is as follows:

$$L(a, b_1, b_2, b_3) = \prod_{j=1}^{3}\prod_{i=1}^{n_j} e^{-m_j(S_r)}\lambda_j(S_{j,i}),$$

where

$$S_r = \max\{S_{1,n_1}, S_{2,n_2}, S_{3,n_3}\}.$$

Taking the partial derivatives with respect to the unknown parameters and setting them equal to zero, we obtain the following results:

$$a = \frac{\sum_{j=1}^{3} n_j}{\sum_{j=1}^{3}\frac{p_j}{(1-\beta_j)}[1 - e^{-(1-\beta_j)b_j S_r}]} \qquad (5.28)$$

$$\sum_{i=1}^{n_j} S_{j,i} = \frac{n_j - ap_j b_j S_r e^{-(1-\beta_j)b_j S_r}}{b_j(1 - \beta_j)} \qquad (5.29)$$

for $j = 1, 2$, and 3. Solving Eqs. (5.28) and (5.29) simultaneously gives the maximum likelihood estimates of parameters a, b_1, b_2, and b_3.

Example 5.3   The failure data in Table 5.4 (Misra, 1983) consists of three types of errors: critical, major, and minor. The observation time (week, hour) and the number of failures detected per week are presented in Table 5.4. Given

$$p_1 = 0.0173; p_2 = 0.3420; p_3 = 0.6407$$
$$\beta_1 = 0.5; \beta_2 = 0.2; \beta_3 = 0.05.$$

TABLE 5.4. Failure data (data set #6).

Test		Failures		
Week	Hours	Critical	Major	Minor
1	62.5	0	6	9
2	44	0	2	4
3	40	0	1	7
4	68	1	1	6
5	62	0	3	5
6	66	0	1	3
7	73	0	2	2
8	73.5	0	3	5
9	92	0	2	4
10	71.4	0	0	2
11	64.5	0	3	4
12	64.7	0	1	7
13	36	0	3	0
14	54	0	0	5
15	39.5	0	2	3
16	68	0	5	3
17	61	0	5	3
18	62.6	0	2	4
19	98.7	0	2	10
20	25	0	2	3
21	12	0	1	1
22	55	0	3	2
23	49	0	2	4
24	64	0	4	5
25	26	0	1	0
26	66	0	2	2
27	49	0	2	0
28	52	0	2	2
29	70	0	1	3
30	84.5	1	2	6
31	83	1	2	3
32	60	0	0	1
33	72.5	0	2	1
34	90	0	2	4
35	58	0	3	3
36	60	0	1	2
37	168	1	2	11
38	111.5	0	1	9

Using the MLE method, the parameters for the reliability model are obtained as follows:

$$a = 428$$
$$b_1 = 0.00024275$$
$$b_2 = 0.00029322$$
$$b_3 = 0.00030495.$$

Substituting the known and estimated parameters into the reliability equation, we obtain

$$R(x|t) = e^{-A}$$

where

$$A = 14.81(e^{-0.00012138t})(1 - e^{0.00012138x})$$
$$+ 182.97(e^{-0.00023458t})(1 - e^{-0.00023458x})$$
$$+ 288.65(e^{-0.0002897t})(1 - e^{-0.0002897x}).$$

Other reliability performance measures are given by

$$m_1(T) = 14.81(1 - e^{-0.00012138T}), \quad n_1(T) = 14.81(1 - 0.5e^{-0.00012138T})$$
$$m_2(T) = 182.97(1 - e^{-0.00023458T}), \quad n_2(T) = 182.97(1 - 0.2e^{-0.00023458T})$$
$$m_3(T) = 288.65(1 - e^{-0.0002897T}), \quad n_3(T) = 288.65(1 - 0.05e^{-0.0002897T}).$$

Other Imperfect Debugging Models

Many existing models describe perfect debugging in Subsections 5.3.1 and 5.3.2, that is, $a(t) = a$, with a time-dependent error detection rate $b(t)$. In this section, we discuss several software reliability models with imperfect debugging processes and a constant error-detection rate $b(t) = b$.

The NHPP imperfect debugging model is based on the following assumptions:

1. When detected errors are removed, it is possible to introduce new errors.
2. The probability of finding an error in a program is proportional to the number of remaining errors in the program.

This imperfect debugging model can be formulated as follows:

$$\frac{\partial m(t)}{\partial t} = b[a(t) - m(t)] \tag{5.30}$$

with the initial condition $m(0) = 0$, and $a(t)$ as the error content function of time t during software testing.

The solution for the mean value function $m(t)$ of Eq. (5.30) is given by

$$m(t) = be^{-bt} \int_0^t a(s)e^{bs}ds. \tag{5.31}$$

If we let

$$a(t) = ae^{\alpha t},$$

then the mean value function, as given in Eq. (5.31), can be obtained as

$$m(t) = \frac{ab}{b + \alpha}(e^{\alpha t} - e^{-bt}). \tag{5.32}$$

This model has been studied by Yamada (1984). Next, we show how to estimate the parameters of a, b, and α. Using the MLE method, the log of the likelihood function is

$$\ln L = \sum_{i=1}^{n} \left[(y_i - y_{i-1})\ln[m(t_i) - m(t_{i-1})] - [m(t_i) - m(t_{i-1})] \right.$$
$$\left. - \ln[(y_i - y_{i-1})!] \right],$$

where

$$m(t_i) - m(t_{i-1}) = \frac{ab}{b + \alpha}[(e^{\alpha t_i} - e^{\alpha t_{i-1}}) - (e^{-bt_i} - e^{-bt_{i-1}})].$$

Taking the partial derivatives of the above function with respect to the unknown parameters a, b, and, α, set

$$\frac{\partial}{\partial a} \ln L = 0$$

$$\frac{\partial}{\partial b} \ln L = 0$$

$$\frac{\partial}{\partial \alpha} \ln L = 0.$$

We obtain the results by solving the following equations simultaneously:

$$a = \frac{b + \alpha}{b} \frac{y_n}{(e^{\alpha t_n} - e^{-bt_n})}$$

$$\sum_{i=1}^{n}\left[(y_i - y_{i-1})\frac{a}{B}\left(\frac{\alpha A}{(b+\alpha)^2} + \frac{bC}{b+\alpha}\right) - a\left(\frac{\alpha A}{(b+\alpha)^2} + \frac{bC}{b+\alpha}\right)\right] = 0$$

and

$$\sum_{i=1}^{n}\left[(y_i - y_{i-1})\frac{a}{B}\left(\frac{-bA}{(b+\alpha)^2} + \frac{bC}{b+\alpha}\right) - a\left(\frac{-bA}{(b+\alpha)^2} + \frac{bC}{b+\alpha}\right)\right] = 0,$$

where

$$A = (e^{\alpha t_i} - e^{\alpha t_{t-1}}) - (e^{-bt_i} - e^{-bt_{t-1}})$$

$$B = \frac{abA}{b+\alpha}$$

and

$$C = t_i e^{\alpha t_i} - t_{i-1} e^{\alpha t_{i-1}}.$$

Similarly, if we consider the function

$$a(t) = a(1 + \alpha t),$$

then, from Eq. (5.31), the mean value function can be obtained as

$$m(t) = a(1 - e^{-bt})\left(1 - \frac{\alpha}{\beta}\right) + a\alpha t. \tag{5.33}$$

This model is also called the Yamada imperfect debugging model (Yamada, 1984).

5.3.4 NHPP Imperfect Debugging S-Shaped Model

The NHPP S-shaped imperfect debugging model (Pham, 1997) is formulated based on the following assumptions:

1. The error detection rate differs among faults.
2. Each time a software failure occurs, the software error which caused it is immediately removed, and new faults can be introduced.

The model can be formulated as the following differential equation:

$$\frac{\partial m(t)}{\partial t} = b(t)[a(t) - m(t)], \tag{5.34}$$

where

$a(t)$ = total errors content function that exist in a software at time t

$b(t)$ = failure detection rate function.

Pham–Nordmann NHPP Model

This model (Pham, 1997) assumes that

1. the introduction rate is a linear function of testing time, and
2. the error detection rate function is non-decreasing with an inflection S-shaped model.

Mathematically,

$$a(t) = a(1 + \alpha t)$$
$$b(t) = \frac{b}{1 + \beta e^{-bt}} \,.$$

(5.35)

Substituting both the functions $a(t)$ and $b(t)$ into Eq. (5.34), and assuming the initial condition $m(0) = 0$, we obtain the mean value function $m(t)$ as follows:

$$m(t) = \frac{a}{1 + \beta e^{-bt}} \left([1 - e^{-bt}] \left[1 - \frac{\alpha}{\beta} \right] + \alpha t \right).$$

(5.36)

In other words, the model assumes a constant error introduction rate α, and therewith an error content function that is linear in time. It assumes furthermore an error detection rate $b(t)$ for software reliability growth models with an inflection S-shaped mean value function.

Pham Exponential Imperfect Debugging Model

This model (Pham, 1998) assumes that

1. the introduction rate is an exponential function of testing time, and
2. the error detection rate function is non-decreasing with an inflection S-shaped model.

Mathematically, assuming that

$$a(t) = \alpha_1 e^{\beta t}$$
$$b(t) = \frac{b}{1 + ce^{-bt}}$$

(5.37)

and solving Eq. (5.34) by substituting the two functions $a(t)$ and $b(t)$ above, we obtain

$$m(t) = \frac{\alpha_1 b}{b + \beta} \left(\frac{e^{(\beta + b)t} - 1}{e^{bt} + c} \right).$$

(5.38)

Parameter Estimation

We are interested in estimating the parameters α_1, β, b, and c of function $m(t)$ in Eq. (5.38). If data are given on the cumulative number of errors at discrete times ($y_i := y(t_i)$ for $i = 1, 2, \ldots, n$), then we need to obtain the first partial derivative of $m(t)$ with respect to α_1, β, b, and c, respectively,

$$\frac{\partial}{\partial \alpha_1} m(t) = \frac{1}{\alpha_1} m(t)$$

$$\frac{\partial}{\partial \beta} m(t) = \left[\frac{-1}{(b+\beta)} + \frac{te^{(b+\beta)t}}{e^{(b+\beta)t} - 1} \right] m(t)$$

$$\frac{\partial}{\partial b} m(t) = \left[\frac{\beta}{(b+\beta)b} + \frac{te^{(b+\beta)t}}{e^{(b+\beta)t} - 1} + \frac{-te^{bt}}{e^{bt} + c} \right] m(t)$$

$$\frac{\partial}{\partial c} m(t) = -\frac{1}{e^{bt} + c} m(t).$$

The second derivative of $m(t)$ with respect to α_1, β, b, and c is

$$\frac{\partial^2}{\partial \alpha_1^2} [m(t)] = 0$$

$$\frac{\partial^2}{\partial \alpha_1 \partial \beta} m(t) = \frac{1}{\alpha_1} \frac{\partial}{\partial \beta} m(t)$$

$$\frac{\partial^2}{\partial \alpha_1 \partial b} m(t) = \frac{1}{\alpha_1} \frac{\partial}{\partial b} m(t)$$

$$\frac{\partial^2}{\partial \alpha_1 \partial c} m(t) = \frac{1}{\alpha_1} \frac{\partial}{\partial c} m(t)$$

$$\frac{\partial^2}{\partial \beta^2} m(t) = \left[\left(\frac{-1}{(b+\beta)} + \frac{te^{(b+\beta)t}}{e^{(b+\beta)t} - 1} \right)^2 \right.$$
$$\left. + \left(\frac{1}{(b+\beta)^2} - \frac{t^2 e^{(b+\beta)t}}{(e^{(b+\beta)t} - 1)^2} \right)^2 \right] m(t)$$

$$\frac{\partial^2}{\partial \beta \partial b} m(t) = m(t) \left[\left(\frac{-1}{(b+\beta)} + \frac{te^{(b+\beta)t}}{e^{(b+\beta)t} - 1} \right) \right.$$
$$\left(\frac{\beta}{(b+\beta)b} + \frac{te^{(b+\beta)t}}{e^{(b+\beta)t} - 1} - \frac{te^{bt}}{e^{bt} + c} \right)$$
$$\left. + \left(\frac{1}{(b+\beta)^2} - \frac{t^2 e^{(b+\beta)t}}{(e^{(b+\beta)t} - 1)^2} \right)^2 \right] m(t)$$

$$\frac{\partial^2}{\partial\beta\partial c}m(t) = -\left(\frac{1}{e^{bt}+c}\right)\frac{\partial}{\partial\beta}m(t)$$

$$\frac{\partial^2}{\partial b\partial b}m(t) = m(t)\left(\frac{\beta}{(b+\beta)b} + \frac{te^{(b+\beta)t}}{e^{(b+\beta)t}-1} - \frac{te^{bt}}{e^{bt}+c}\right)^2$$

$$+ \left(-\frac{\beta(2b+\beta)}{(b+\beta)^2b^2} + \frac{t^2e^{(b+\beta)t}}{(e^{(b+\beta)t}-1)^2} - \frac{ct^2e^{bt}}{(e^{bt}+c)^2}\right)m(t)$$

$$\frac{\partial^2}{\partial b\partial c}m(t) = -\left(\frac{1}{e^{bt}+c}\right)\frac{\partial}{\partial b}m(t) + \frac{te^{bt}}{(e^{bt}+c)^2}m(t)$$

$$\frac{\partial^2}{\partial c^2}m(t) = \frac{2}{(e^{bt}+c)^2}m(t)$$

$$\frac{\partial^2}{\partial\beta\partial\alpha_1}m(t) = \frac{\partial^2}{\partial\alpha_1\partial\beta}m(t)$$

$$\frac{\partial^2}{\partial b\partial\alpha_1}m(t) = \frac{\partial^2}{\partial\alpha_1\partial b}m(t)$$

Note that

$$\lim_{t\to 0}\frac{te^{(b+\beta)t}}{e^{(b+\beta)t}-1} = \frac{1}{b+\beta}$$

$$\lim_{t\to 0}\frac{t^2e^{(b+\beta)t}}{(e^{(b+\beta)t}-1)^2} = \frac{1}{(b+\beta)^2}$$

and $m(0) = 0$. Consequently,

$$\lim_{t\to 0}\frac{\partial^2}{\partial x\partial y}m(t) = \lim_{t\to 0}\frac{\partial}{\partial x}m(t) = m(0) = 0.$$

Therefore, one can obtain the model parameters α_1, β, b, and c by solving the following system of equations simultaneously,

$$m(t) = 0$$

$$\left[\frac{-1}{(b+\beta)} + \frac{te^{(b+\beta)t}}{e^{(b+\beta)t}-1}\right]m(t) = 0$$

$$\left[\frac{\beta}{(b+\beta)b} + \frac{te^{(b+\beta)t}}{e^{(b+\beta)t}-1} + \frac{-te^{bt}}{e^{bt}+c}\right]m(t) = 0.$$

If the given data represent the occurrence times of the errors (S_j for $j = 1, 2, \ldots, n$), then we need to obtain the first partial derivative of $\lambda(t)$ with respect to α_1, β, b, and c, where

$$\lambda(t) = \frac{\partial}{\partial t} m(t)$$

$$= \frac{\alpha_1 b}{b + \beta} \left[\frac{(b + \beta)e^{(b+\beta)t}}{e^{bt} + c} - \frac{e^{(b+\beta)} - 1}{(e^{bt} + c)^2} be^{bt} \right].$$

Using Eq. (5.4), we can easily obtain the estimate of α_1, β, b, and c.

Pham–Zhang NHPP Model

This model (Pham, 1997) assumes that:

1. The error introduction rate is an exponential function of the testing time. In other words, the number of errors increases quicker at the beginning of the testing process than at the end. This reflects the fact that more errors are introduced into the software at the beginning, while at the end, testers possess more knowledge and therefore introduce fewer errors into the program.
2. The error detection rate function is non-decreasing with an inflexion S-shaped model.

Mathematically, assuming that

$$a(t) = c + a(1 - e^{-\alpha t})$$
$$b(t) = \frac{b}{1 + \beta e^{-bt}} \tag{5.39}$$

and solving Eq. (5.34) by substituting the above two functions $a(t)$ and $b(t)$ and assuming $m(0) = 0$, we obtain the mean value function as follows:

$$m(t) = \frac{1}{(1 + \beta e^{-bt})} \left((c + a)(1 - e^{-bt}) - \frac{ab}{b - \alpha}(e^{-\alpha t} - e^{-bt}) \right). \tag{5.40}$$

In general, NHPP software reliability models can be used to estimate the expected number of errors. Obviously, different models use different assumptions and therefore provide different mathematical forms for the mean value function $m(t)$. Table 5.5 shows a summary of many existing NHPP software reliability models appearing in the software reliability engineering literature.

TABLE 5.5. Summary of NHPP software reliability models.

Model name	Model type	MVF ($m(t)$)	Comments
Goel–Okumoto (G–O)	Concave	$m(t) = a(1 - e^{-bt})$ $a(t) = a$ $b(t) = b$	Also called exponential model.
Delayed S-shaped	S-shaped	$m(t) = a(1 - (1 + bt)e^{-bt})$ $a(t) = a$ $b(t) = \dfrac{b^2 t}{1 + bt}$	Modification of G–O model to make it S-shaped.
Inflection S-shaped	Concave	$m(t) = \dfrac{a(1 - e^{-bt})}{1 + \beta e^{-bt}}$ $a(t) = a$ $b(t) = \dfrac{b}{1 + \beta e^{-bt}}$	Solves a technical condition with the G–O model. Becomes the same as G–O if $\beta = 0$.
Yamada exponential	Concave	$m(t) = a(1 - e^{-r\alpha(1 - e^{(-\beta t)})})$ $a(t) = a$ $b(t) = r\alpha\beta e^{-\beta t}$	Attempt to account for testing-effort.
Yamada Rayleigh	S-shaped	$m(t) = a(1 - e^{-r\alpha(1 - e^{(-\beta t^2/2)})})$ $a(t) = a$ $b(t) = r\alpha\beta t e^{-\beta t^2/2}$	Attempt to account for testing-effort.
Yamada imperfect debugging model (1)	S-shaped	$m(t) = \dfrac{ab}{\alpha + b}(e^{\alpha t} - e^{-bt})$ $a(t) = ae^{\alpha t}$ $b(t) = b$	Assume exponential fault content function and constant error detection rate.
Yamada imperfect debugging model (2)	S-shaped	$m(t) = a[1 - e^{-bt}][1 - \dfrac{\alpha}{b}] + \alpha a t$ $a(t) = a(1 + \alpha t)$ $b(t) = b$	Assume constant introduction rate α and the error detection rate.
Pham–Nordmann	S-shaped and concave	$m(t) = \dfrac{a[1 - e^{-bt}][1 - \dfrac{\alpha}{b}] + \alpha a t}{1 + \beta^{-bt}}$ $a(t) = a(1 + \alpha t)$ $b(t) = \dfrac{b}{1 + \beta e^{-bt}}$	Assume introduction rate is a linear function of testing time, and the error detection rate function is non-decreasing with an inflection S-shaped model.
Pham–Zhang	S-shaped and concave	$m(t) = \dfrac{1}{(1 + \beta e^{-bt})}[(c + a)(1 - e^{-bt})$ $\quad - \dfrac{a}{b - \alpha}(e^{-\alpha t} - e^{-bt})]$ $a(t) = c + a(1 - e^{-\alpha t})$ $b(t) = \dfrac{b}{1 + \beta e^{-bt}}$	Assume introduction rate is exponential function of the testing time, and the error detection rate is non-decreasing with an inflection S-shaped model.

5.4 Applications

In this section, we apply most, if not all, the models mentioned in Section 5.4 to several failure data sets in Section 5.3. The estimators and the curves of the mean value functions and actual failure data are obtained for each data set. First, parameters of these models are estimated and the mean value functions are provided, followed by comparison of all the models.

The criterion to be dicussed is the sum of squared error (SSE) to judge the performance of the models, which sum up the square of the residuals of the actual failure data and the mean value function of each model in terms of the number of actual failure errors at any point in time. Therefore, the lower the SSE value, the better the model is. The SSE of a model can be expressed as follows:

$$\text{SSE} = \sum_{i=1}^{n} [y_i - \hat{m}(t_i)]^2,$$

where

y_i = total number of failures observed at time t_i according to the failure data

$\hat{m}(t_i)$ = estimated cumulative number of failures at time t_i obtained from the fitted mean value function, $i = 1, 2, \ldots, n$.

Case Study 1 In this subsection, the software data set in Table 5.6, was extracted from information about failures in the development of software for the real-time, multi-computer complex which forms the core of the US Naval Fleet Computer Programming Center of the US Naval Tactical Data Systems (NTDS). The software consists of 38 different project modules. The time horizon is divided into four phases: production phase, test phase, user phase, and subsequent test phase. The 26 software failures were found during the production phase, five during the test phase and; the last failure was found on 4 January 1971. One failure was observed during the user phase, in September 1971, and two failures during the test phase in 1971.

The fact that the last three of the first 26 errors in Table 5.6 occur almost in a cluster, while there is a relatively long interval between errors before and after that cluster, leads to the conclusion that error number 26 is an unfortunate cut-off point if one wishes to fit a software reliability model. Instead, it seems more reasonable to use either the first 25 or the first 27 error data to fit models. Let us chose the second alternative and fit models 1 through 5 to the first

TABLE 5.6. Naval Tactical Data System (NTDS) software error data (data set #1) (Goel, 1979).

Error no. n	Time between errors x_k (days)	Cumulative time $S_n := \sum x_k$ (days)
Production (checkout) phase		
1	9	9
2	12	21
3	11	32
4	4	36
5	7	43
6	2	45
7	5	50
8	8	58
9	5	63
10	7	70
11	1	71
12	6	77
13	1	78
14	9	87
15	4	91
16	1	92
17	3	95
18	3	98
19	6	104
20	1	105
21	11	116
22	33	149
23	7	156
24	91	247
25	2	249
26	1	250
Test phase		
27	87	337
28	47	384
29	12	396
30	9	405
31	135	540
User phase		
32	258	798
Test phase		
33	16	814
34	35	849

27 error data. Although not presented here, we also fit the models to the first 25 error data, which results in only slide deviations with the same overall conclusions (see Problem 5.5).

It is worthwhile to note that parameter estimates derived from a set of test data are representative and meaningful only for working conditions similar to those under which the test data were obtained. In the NTDS data in Table 5.6, for example, the production and test phases can be considered similar, while the user phase is very different from the former two in two ways. Firstly, during the user phase, no new errors due to testing and debugging are introduced. Secondly, during the user phase, the software is not subject to such "hard" and extensive testing as it is during the production and testing phase. Both differences add towards the same effect — a reduced error occurrence rate. This can be verified by looking at the test data, which indicates a considerably longer time span between errors during the user phase. In general, changes in the working environment go along with shifts in model parameters, and consequently, parameter estimates interfering the production and first test phases are only applicable during the first two phases and meaningless beyond that point. For the ease of discussion, let us name the following five selected models to be analyzed in this example.

Model 1 (Goel–Okumoto):

$$m(t) = a(1 - e^{-bt})$$

Model 2 (Inflection S-shaped):

$$m(t) = \frac{a}{1 + \beta e^{-bt}}(1 - e^{-bt})$$

Model 3 (Yamada imperfect debugging 1):

$$m(t) = \frac{ab}{b + \alpha}[e^{\alpha t} - e^{-bt}]$$

Model 4 (Yamada imperfect debugging 2):

$$m(t) = a(1 - e^{-bt})\left(1 - \frac{\alpha}{b}\right) + a\alpha t$$

Model 5 (Pham–Nordmann):

$$m(t) = \frac{a}{1 + \beta e^{-bt}}\left[(1 - e^{-bt})\left(1 - \frac{\alpha}{b}\right) + \alpha t\right]$$

and the corresponding reliability of model i is R_i.

The maximum likelihood estimators of the parameters for several software reliability models based on the first 27 error data of the NTDS are obtained as follows:

Model 1: $a = 29.42827$ $b = 0.007402$

Model 2: $a = 27.44246$ $b = 0.015517$ $\beta = 2.042596$

Model 3: $a = 29.42827$ $b = 0.015517$ $\alpha = 0$

Model 4: $a = 29.42827$ $b = 0.007402$ $\alpha = 0$

Model 5: $a = 19.32872$ $b = 0.047526$ $\alpha = 0.001256$

$$\beta = 24.33569$$

Figure 5.2 shows the plot for the mean value function of Models 1 through 4 versus Model 5 and the actual error data, and Fig. 5.3 shows the corresponding conditional reliability functions after the 27th error. From Fig. 5.3, we observe that Models 1–4 overestimate the reliability as compared to Model 5. For example, at mission time $s = 20$, Model 5 estimates a reliability of $R_5 = 0.615$, whereas Models 1, 3, and 4 estimate a reliability of $R_1 = 0.716$ and Model 2 estimates a reliability of $R_2 = 0.889$.

After the maximum likelihood estimation, Models 1, 3, and 4 yield exactly the same mean value functions. Despite allowing for an imperfect debugging model, the perfect debugging model remains the one that provides the best fit to the NTDS data. The additional freedom introduced in Models 3 and 4 through an additional parameter takes no effect on the fitted mean value function, since the MLE of that additional parameter is found to be $\alpha = 0$, transforming Models 3 and 4 into Model 1. Model 2 allows for an inflection S-shaped mean value function. However, the fitted mean value function is only vaguely S-shaped if at all. On the upper end of the time horizon, the mean value function underestimates the actual failure numbers.

The mean value function of Model 5 provides a good fit to the S-shape of the actual NTDS data. It overestimates the actual failure numbers at the upper end of the time horizon. However, it provides an excellent fit to the 27 test data points it is fitted to, and a good overall fit to all data points of the production and first test phases, reducing fitting and forecast errors of Models 1–4. Further studies show that the parameter in Model 5 appears especially sensitive to data at further progressed times, indicating an increasingly better fit with a widening time span of the test data (Pham, 1997).

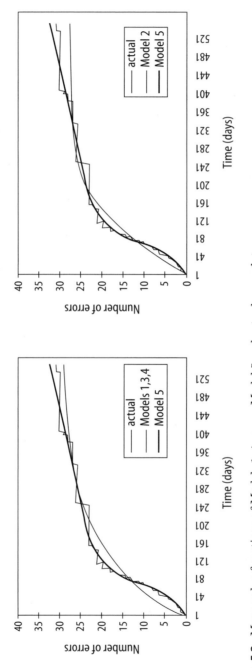

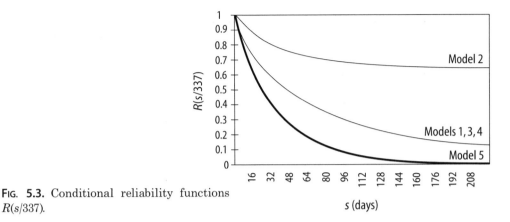

FIG. **5.3.** Conditional reliability functions $R(s/337)$.

Case Study 2: A Real-Time Control System   The program for monitoring a real-time control system consists of about 200 modules having, on average, 1,000 lines of a high-level language such as Fortran (Tohma, 1991). Since the test data (Table 5.7) are recorded daily, the test operations performed in a day are regarded to be a test instance. In Table 5.7, data marked with an asterisk (∗) are interpolated data.

Let us look at the goodness of fit test to most NHPP software models based on the data given in Table 5.7. The results of the estimated parameters of the models and their SSEs are given in Table 5.8. It is observed that the Pham–Zhang model, having an SSE equal to 59,549, fits significantly better than the other NHPP models for this failure data set. It is worthwhile to note that the inflection S-shaped model also performs well in this case. Figure 5.4 shows the actual failure data as well as the curves of the mean value functions of the models included in Table 5.8.

Case Study 3: An Application in Software Industry   The software failure data, given in Table 5.9, is presented from four major releases of the software products at Tandem Computers Company, Los Angeles, CA (Wood, 1996). It should be noted that the real test data in Table 5.9 has been transformed so that the same conclusions will be reached from models of the transformed data and that of the original data (Wood, 1996). In this application, two NHPP models will be analyzed: the Pham–Zhang (P–Z) and the Goel–Okumoto (G–O) models. Table 5.9 shows the results predicted based on the CPU execution hours for the P–Z and G–O models. From this application and Fig. 5.5, we can observe that the P–Z model predicts much more accurately than the G–O model.

TABLE 5.7. Real-time software failure data.

Days	Faults	Cumulative faults	Days	Faults	Cumulative faults	Days	Faults	Cumulative faults	Days	Faults	Cumulative faults
1	5*	5*	29	2	254	57	2	448	85	0	473
2	5*	10*	30	5	259	58	3	451	86	0	473
3	5*	15*	31	4	263	59	2	453	87	2	475
4	5*	20*	32	1	264	60	7	460	88	0	475
5	6*	26*	33	4	268	61	3	463	89	0	475
6	8	34	34	3	271	62	0	463	90	0	475
7	2	36	35	6	277	63	1	464	91	0	475
8	7	43	36	13	293	64	0	464	92	0	475
9	4	47	37	19	309	65	1	465	93	0	475
10	2	49	38	15	324	66	0	465	94	0	475
11	31	80	39	7	331	67	0	465	95	0	475
12	4	84	40	15	346	68	1	466	96	1	476
13	24	108	41	21	367	69	1	467	97	0	476
14	49	157	42	8	375	70	0	467	98	0	476
15	14	171	43	6	381	71	0	467	99	0	476
16	12	183	44	20	401	72	1	468	100	1	477
17	8	191	45	10	411	73	1	469	101	0	477
18	9	200	46	3	414	74	0	469	102	0	477
19	4	204	47	3	417	75	0	469	103	1	478
20	7	211	48	8	425	76	0	469	104	0	478
21	6	217	49	5	430	77	1	470	105	0	478
22	9	226	50	1	431	78	2	472	106	1	479
23	4	230	51	2	433	79	0	472	107	0	479
24	4	234	52	2	435	80	1	473	108	0	479
25	2	236	53	2	437	81	0	473	109	1	480
26	4	240	54	7	444	82	0	473	110	0	480
27	3	243	55	2	446	83	0	473	111	1	481
28	9	252	56	0	446	84	0	473			

Consider the calendar testing weeks as the timeframe and compare the accuracy of predictions using execution time versus calendar time for Release 4. The results are shown in Table 5.10. For both the execution time and the calendar time, the number of failures predicted by the P–Z model are significantly more accurate than those predicted by the G–O model. Table 5.11 also summarizes the results of many existing NHPP models for Release 4.

TABLE 5.8. MLEs and SSEs for NHPP models.

Model name	MVF ($m(t)$)	MLEs	SSE
Goel–Okumoto (G–O)	$m(t) = a(1 - e^{-bt})$ $a(t) = a$ $b(t) = b$	$\hat{a} = 497.282$ $\hat{b} = 0.0308$	216872
Delayed S-shaped	$m(t) = a(1 - (1 + bt)e^{-bt})$ $a(t) = a$ $b(t) = \dfrac{b^2 t}{1 + bt}$	$\hat{a} = 483.039$ $\hat{b} = 0.06866$	71247
Inflection S-shaped	$m(t) = \dfrac{a(1 - e^{-bt})}{1 + \beta e^{-bt}}$ $a(t) = a$ $b(t) = \dfrac{b}{1 + \beta e^{-bt}}$	$\hat{a} = 482.017$ $\hat{b} = 0.07025$ $\hat{\beta} = 4.15218$	60031
Yamada exponential	$m(t) = a(1 - e^{-r\alpha(1 - e^{(-\beta t)})})$ $a(t) = a$ $b(t) = r\alpha\beta e^{-\beta t}$	$\hat{a} = 67958.8$ $\hat{\alpha} = 0.00732$ $\hat{\beta} = 0.03072$	220702
Yamada Rayleigh	$m(t) = a(1 - e^{-r\alpha(1 - e^{(-\beta t^2/2)})})$ $a(t) = a$ $b(t) = r\alpha\beta t e^{-\beta t^2/2}$	$\hat{a} = 500.146$ $\hat{\alpha} = 3.31944$ $\hat{\beta} = 0.00066$	87251
Yamada imperfect debugging model (1)	$m(t) = \dfrac{ab}{\alpha + b}(e^{\alpha t} - e^{-bt})$ $a(t) = ae^{\alpha t}$ $b(t) = b$	$\hat{a} = 654.963$ $\hat{b} = 0.02059$ $\hat{\alpha} = -0.0027$	155011
Yamada imperfect debugging model (2)	$m(t) = a[1 - e^{-bt}][1 - \dfrac{\alpha}{b}] + \alpha a t$ $a(t) = a(1 + \alpha t)$ $b(t) = b$	$\hat{a} = 591.804$ $\hat{b} = 0.02423$ $\hat{\alpha} = -0.0019$	183157
Pham–Nordmann	$m(t) = \dfrac{a[1 - e^{-bt}][1 - \frac{\alpha}{b}] + \alpha a t}{1 + \beta e^{-bt}}$ $a(t) = a(1 + \alpha t)$ $b(t) = \dfrac{b}{1 + \beta e^{-bt}}$	$\hat{a} = 470.759$ $\hat{b} = 0.07497$ $\hat{\alpha} = 0.00024$ $\hat{\beta} = 4.69321$	63189
Pham–Zhang (P–Z)	$m(t) = \dfrac{1}{(1 + \beta e^{-bt})}[(c + a)(1 - e^{-bt})$ $\quad - \dfrac{ab}{b - \alpha}(e^{-\alpha t} - e^{-bt})]$ $a(t) = c + a(1 - e^{-\alpha t})$ $b(t) = \dfrac{b}{1 + \beta e^{-bt}}$	$\hat{a} = 0.46685$ $\hat{b} = 0.07025$ $\hat{\alpha} = 1.4 \times 10^{-5}$ $\hat{\beta} = 4.15213$ $\hat{c} = 482.016$	59549

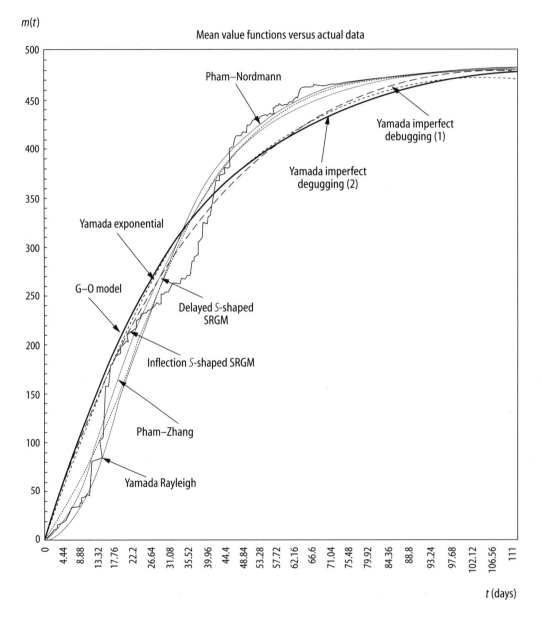

$m(t)$

Mean value functions versus actual data

Pham–Nordmann

Yamada imperfect
debugging (1)

Yamada imperfect
degugging (2)

Yamada exponential

G–O model

Delayed S-shaped
SRGM

Inflection S-shaped SRGM

Pham–Zhang

Yamada Rayleigh

t (days)

FIG. 5.4. The mean value functions of NHPP models.

TABLE 5.9. Prediction comparison of software failure data (CPU execution time).

	Release 1				Release 2			
Testing time (weeks)	CPU hours	Defects found	Predicted total defects by G–O model	Predicted total defects by P–Z model	CPU hours	Defects found	Predicted total defects by G–O model	Predicted total defects by P–Z model
1	519	16	—	—	384	13	—	—
2	968	24	—	—	1,186	18	—	—
3	1,430	27	—	—	1,471	26	—	—
4	1,893	33	—	—	2,236	34	—	—
5	2,490	41	—	—	2,772	40	—	—
6	3,058	49	—	—	2,967	48	—	—
7	3,625	54	—	—	3,812	61	—	—
8	4,422	58	—	—	4,880	75	—	—
9	5,218	69	—	—	6,104	84	—	—
10	5,823	75	98	74.6	6,634	89	203	88.7
11	6,539	81	107	80.7	7,229	95	192	93.4
12	7,083	86	116	85.3	8,072	100	179	99.9
13	7,487	90	123	88.5	8,484	104	178	102.9
14	7,846	93	129	91.4	8,847	110	184	105.2
15	8,205	96	129	94.1	9,253	112	184	107.7
16	8,564	98	134	97.0	9,712	114	183	110.4
17	8,923	99	139	99.4	10,083	117	182	112.6
18	9,282	100	138	101.8	10,174	118	183	113.1
19	9,641	100	135	104.3	10,272	120	184	113.6
20	10,000	100	133	106.7	—	—	—	—

	Release 3				Release 4			
1	162	6	—	—	254	1	—	—
2	499	9	—	—	788	3	—	—
3	715	13	—	—	1,054	8	—	—
4	1,137	20	—	—	1,393	9	—	—
5	1,799	28	—	—	2,216	11	—	—
6	2,438	40	—	—	2,880	16	—	—
7	2,818	48	—	—	3,593	19	—	—
8	3,574	54	163	60.9	4,281	25	—	—
9	4,234	57	107	61.7	5,180	27	—	—
10	4,680	59	93	68.7	6,003	29	84	30.1
11	4,955	60	87	75.7	7,621	32	53	34.9
12	5,053	61	84	82.3	8,783	32	44	38.4
13	—	—	—	—	9,604	36	45	39.7
14	—	—	—	—	10,064	38	46	40.7
15	—	—	—	—	10,560	39	48	41.6
16	—	—	—	—	11,008	39	48	42.6
17	—	—	—	—	11,237	41	50	42.7
18	—	—	—	—	11,243	42	51	42.7
19	—	—	—	—	11,305	42	52	42.7
20	—	—	—	—	—	—	—	—

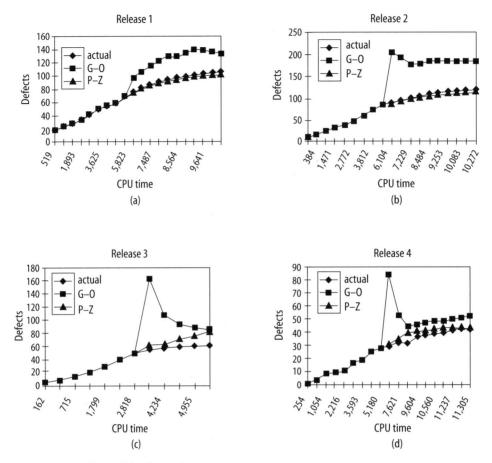

FIG. 5.5. Actual data of the four releases.

TABLE 5.10. Comparing predictions using execution time versus calendar time for Release 4.

Test week	Execution hours	Defects found	Predicted total defects, execution time (G–O)	Predicted total defects, calendar time (G–O)	Predicted total defects, execution time (P–Z)	Predicted total defects, calendar time (P–Z)
10	6,003	29	84	—	30.1	30.1
11	7,621	32	53	—	34.9	33.0
12	8,783	32	44	—	38.4	36.0
13	9,604	36	45	—	39.7	38.7
14	10,064	38	46	No prediction	40.7	41.9
15	10,560	39	48	457	41.6	44.7
16	11,008	39	48	178	42.6	47.6
17	11,237	41	50	125	42.7	50.7
18	11,243	42	51	101	42.7	53.8
19	11,305	42	52	85	42.7	56.7

TABLE 5.11. Model comparisons for Release 1.

Model	Total defects predicted several weeks after the start of system test				
	10 weeks	12 weeks	14 weeks	17 weeks	20 weeks
Goel–Okumoto (G–O)	98	116	129	139	133
G–O S-Shaped	71	82	91	99	102
Hossain–Dahiya/G–O	98	116	129	139	133
Gompertz	96	110	107	114	112
Pareto (Pham, 1999)	757	833	735	631	462
Weibull	98	116	129	139	133
Yamada exponential	152	181	204	220	213
Yamada Raleigh	77	89	98	107	111
Pham–Zhang (P–Z)	74.6	85.3	91.4	99.4	106.7
Actual data	75	86	93	99	100

5.5 Imperfect Debugging Versus Perfect Debugging

It is worthwhile to note that a study by Ohba and Chou (1989) shows that, for an arbitrary $b(t)$, a model with $a(t) = a$ and a model with $a(t) = a + \alpha\, m(t)$ are isomorphic. In other words, both models yield, after maximum likelihood estimation, the same mean value functions. This explains why in many cases, imperfect debugging models do not significantly improve perfect debugging models. Consider the following two models.

Model P: The Perfect Debugging Model
Assume that

$$a(t) = a$$
$$b(t) = \text{arbitrary.}$$

Then, from Eq. (5.15), we obtain

$$m(t) = a(1 - e^{-B(t)}),$$

where

$$B(t) = \int_0^t b(s)\,ds.$$

Model IP: The Imperfect Debugging Model
Assume that

$$a(t) = \alpha m(t) + a$$
$$b(t) = \text{arbitrary} \qquad\qquad (5.41)$$

and from Eq. (5.34), we obtain the result as follows:

$$m(t) = \frac{a}{1-\alpha}(1 - e^{-B(t)(1-\alpha)}). \qquad\qquad (5.42)$$

Let the error detection rates $b_i(t)$ be functions of type

$$b_i(t) = b_i f(\beta_1^{(1)}, \beta_2^{(2)}, \ldots, \beta_1^{(n)}, t),$$

where b_i are positive constants and $f(.,t)$ an arbitrary positive function. Furthermore, let

$$v_1 := (a_1, b_1, \beta_1^{(1)}, \beta_1^{(2)}, \ldots, \beta_1^{(n)})$$

be a vector of valid values for the free parameters in Model P. Then Model P is uniquely defined through the verctor v_1. Analogously, Model IP is uniquely defined through another vector:

$$v_2 := (a_2, b_2, \beta_2^{(1)}, \beta_2^{(2)}, \ldots, \beta_2^{(n)}).$$

By letting $m_1(v_1, t)$ denote the mean value function for Model P, and $m_2(v_2,t)$ denote the mean value function for Model IP, we find:

Lemma 5.1   Let x be a number such that $0 < x < 1$. Then the function

$$\phi_x(a_1, b, \beta_1^{(1)}, \ldots, \beta_1^{n}) := \left(a_1(1-x), \frac{b}{(1-x)}, x, \beta_1^{(1)}, \ldots, \beta_1^{(n)} \right)$$

defines a one-to-one mapping from Model P parameters to Model IP parameters with $\alpha_2 = x$ such that

$$m_1(v_1, t) = m_2(\phi_x(v_1), t).$$

The proof of this lemma is straightforward.

This result shows that the mean value function of the imperfect debugging model equals that of (the) perfect debugging model with an error detection rate of the same type. This observation immediately implies the suspicion that, after substitution of the MLE for the parameters in each of the models, we will have the same mean value function, regardless of which model we begin with. The following results prove this relationship to be true under general conditions in Eq. (5.41).

Lemma 5.2

(1) If

$$v_1^* := (a_1^*, b_1^*, \beta_1^{(1)*}, \beta_1^{(2)*}, \ldots, \beta_1^{(n)*})$$

are MLEs for parameters in Model P, then for every $\alpha_2^* (0 < \alpha_2^* < 1)$,

$$\phi_{\alpha_2^*}(v_1^*)$$

are MLEs for parameters in Model IP.

(2) If

$$v_2^* := (a_2^*, b_2^*, \beta_2^{(1)*}, \beta_2^{(2)*}, \ldots, \beta_2^{(n)*})$$

are MLEs for parameters in Model IP, then

$$\phi_{\alpha_2^*}^{-1}(v_2^*)$$

are MLEs for parameters in Model P.

Proof   Let $\mathrm{MLF}_1(v_1)$ denote Model P's maximum likelihood function (MLF) as a function of the free parameters, $\mathrm{MLF}_2(v_2)$ the MLF for Model IP. Let v_1^* and v_2^* denote the arbitrary MLE of parameters for Model P and Model IP, respectively. We obtain

$$\mathrm{MLF}_1(v_1^*) = \mathrm{MLF}_2(\phi_{\alpha_2^*}(v_1^*)) \leq \mathrm{MLF}_2(v_2^*) = \mathrm{MLF}_1(\phi_{\alpha_2^*}^{-1}(v_2^*))$$
$$\leq \mathrm{MLF}_1(v_1^*).$$

The two "$\leq$" relations hold due to Lemma 5.1. The two "$=$" signs hold because v_1^* is an MLE for Model P and v_2^* is an MLE for Model IP. Since the left term in the above series equals the right, we conclude that all "$\leq$" must be "$=$" signs, and consequently,

$$\mathrm{MLF}_2(\phi_{\alpha_2^*}(v_1^*)) = \mathrm{MLF}_2(v_2^*)$$

and

$$\mathrm{MLF}_1(\phi_{\alpha_2^*}^{-1}(v_2^*)) = \mathrm{MLF}_1(v_1^*).$$

Theorem 5.1   For every $x, 0 < x < 1$, the function $\theta_x(v_1)$ is a one-to-one mapping of the MLE of Model P parameters to Model IP parameters with $\alpha_2 = x$.

Proof   From Lemma 5.2, the result follows.

5.6   A Generalized NHPP Software Reliability Model

The derivation of software reliability models is usually divided into three processes. The counting process $\{N(t), t \geq 0\}$ that represents the cumulative number of software errors detected by time t is a stochastic process. Thus, in the first step, this counting process must be described by statistical means. Basic assumptions about this process lead to the commonly accepted conclusion that, for any fixed $t \geq 0$, $N(t)$ is Poisson-distributed with a time-dependent Poisson-parameter $m(t)$, the so-called mean value function (MVF). The MVF represents the expected number of software errors that have accumulated up to time t. In a second step, this MVF must be defined analytically. This is usually done by expressing the MVF as a function of two other functions: the error content function $a(t)$ and the error detection rate $b(t)$. By making assumptions about the analytical behavior of these two functions, $a(t)$ and $b(t)$ are then defined as functions of time with one or more free parameters. Some of these parameters might be determined through mathematical or physical inferences. In most cases, however, these parameters need to be inferred statistically. Therefore, in a third step, actual test data are analyzed, using the statistical model defined in the first step with the class of MVFs defined in the second step.

The derivation of the generalized mean value function is discussed next. Most of the existing models (Yamada, 1991, 1992; Goel, 1979; Pham, 1996; Ohba, 1984, 1989) for an MVF build upon the assumption that the error detection rate is proportional to the residual error content. Pham and Nordmann (1997) formulate a generalized NHPP software reliability model and provide an analytical expression for the MVF. The generalized form of the MVF can be obtained by solving the following equations:

$$\frac{\partial m(t)}{\partial t} = b(t)[a(t) - m(t)] \tag{5.43}$$

with the initial condition

$$m(t_0) = m_0.$$

In the simplest model, the function $a(t)$ and $b(t)$ are both constants. A constant $a(t)$ stands for the assumption that no new errors are introduced during the debugging process (perfect debugging). A constant $b(t)$ implies that the proportional factor relating the error detection rate $\lambda(t)$ to the total number of remaining errors is constant. This model is known as the Goel–Okumoto NHPP model. Many existing models describe perfect debugging, i.e., $a(t) = a$, with a time-dependent error detection rate $b(t)$ (see Subsection 5.3.2). Other studies deal with an imperfect debugging process and a constant error detection rate $b(t) = b$ (Subsection 5.3.3.)

In a general model, the functions $a(t)$ and $b(t)$ are both functions of time, and for practical purposes, both are increasing with time (Subsection 5.3.4). An increasing $a(t)$ shows that the total number of errors (including those already detected) increases with time because new errors are introduced during the debugging process. An increasing proportional factor $b(t)$ indicates that the error detection rate usually increases as debuggers establish greater familiarity with the software.

Although the relationship above does not yield immediate conclusions about $m(t)$, it relates $m(t)$ to two other functions that, by their definition, possess actual physical meanings. The function $a(t)$ represents the total error content at time t, and $b(t)$ represents the error detection rate. In this way, by introducing functional assumptions about $a(t)$ and $b(t)$, which are more tangible, an analytical expression for $m(t)$ can be derived.

A general solution for the mean value function $m(t)$ of Eq. (5.43) can be obtained using the techniques of differential equations as follows (Pham, 1997):

$$m(t) = e^{-B(t)} \left[m_0 + \int_{t_0}^{t} a(\tau)b(\tau)e^{B(\tau)}d\tau \right], \tag{5.44}$$

where

$$B(t) = \int_{t_0}^{t} b(\tau)d\tau$$

and t_0 is the time to begin the debugging process and $m(t_0) = m_0$.

To obtain this result, let us rewrite Eq. (5.43) as follows:

$$\frac{\partial}{\partial t} m(t) + b(t)m(t) = S(t), \qquad (5.44a)$$

where $S(t) = a(t)b(t)$. Note that

$$\frac{\partial}{\partial t}\left[m(t)e^{\int_{t_0}^{t} b(\tau)d\tau} \right] = \left[\frac{\partial}{\partial t} m(t) + b(t)m(t) \right] e^{\int_{t_0}^{t} b(\tau)d\tau}.$$

Multiplying both sides of Eq. (5.44a) by the integrating factor

$$e^{\int_{t_0}^{t} b(\tau)d\tau},$$

we have

$$\frac{\partial}{\partial t}\left[m(t)e^{\int_{t_0}^{t} b(\tau)d\tau} \right] = S(t)e^{\int_{t_0}^{t} b(\tau)d\tau}.$$

Integrating between t_0 and t, we obtain

$$m(t) = m(t_0)e^{-\int_{t_0}^{t} b(\tau)d\tau} + \int_{t_0}^{t} S(\tau)e^{-\int_{\tau}^{t} b(y)dy}\, d\tau.$$

Note that

$$e^{-\int_{\tau}^{t} b(y)dy} = e^{-\left[\int_{t_0}^{t} b(y)dy - \int_{t_0}^{\tau} b(y)dy\right]}.$$

Therefore,

$$m(t) = m(t_0)e^{-\int_{t_0}^{t} b(\tau)d\tau} + \int_{t_0}^{t} S(\tau)e^{-\left[\int_{t_0}^{t} b(y)dy - \int_{t_0}^{\tau} b(y)dy\right]}\, d\tau.$$

Substituting

$$B(t) = \int_{t_0}^{t} b(\tau)d\tau$$

$$S(t) = a(t)b(t) \text{ and } m(t_0) = m_0$$

into the above equation and after simplifications, we obtain

$$m(t) = e^{-B(t)} \left[m_0 + \int_{t_0}^{t} a(\tau)b(\tau)e^{B(\tau)}d\tau \right].$$

This yields the same result as in Eq. (5.44).

For both given specific functions $a(t)$ and $b(t)$, from Eq. (5.44), the mean value function $m(t)$ can be easily obtained.

Example 5.4   Assume

$$a(t) = a(1 + \alpha t)$$

and

$$b(t) = \frac{b}{1 + \beta e^{-bt}}$$

with the initial condition $m(0) = 0$ which means that no errors are yet detected at time $t = 0$, the starting point of the debugging process. From Eq. (5.44), it can be shown that the mean value function is given as

$$m(t) = \frac{a}{(1 + \beta e^{-bt})} \left[(1 - e^{-bt})\left(1 - \frac{\alpha}{b}\right) + \alpha t \right].$$

This result yields the same as in Eq. (5.36).

Example 5.5   Assume

$$a(t) = c + a(1 - e^{-\alpha t})$$

$$b(t) = \frac{b}{1 + \beta e^{-bt}}$$

with $m(t_0) = m_0$. From Eq. (5.44), we obtain the mean value function as

$$m(t) = \frac{e^{bt_0} + \beta}{e^{bt} + \beta} m_0$$
$$+ \frac{1}{(1 + \beta e^{-bt})} \left((c + a)(1 - e^{b(t_0 - t)}) - \frac{ab}{b - \alpha}[e^{-\alpha t} - e^{[(b - \alpha)t_0 - bt]}] \right).$$

Assume we use the data given in Table 5.6 and also assume that $t_0 = 149$, $m(149) = 22$. Then we obtain the following estimates:

$$a = 0.2231085$$
$$b = 0.004418$$
$$c = 46.6092$$

$$\alpha = 0.0067655$$
$$\beta = 0.0060707$$

which was calculated by the MLE. From this result, we can obtain a simple approach to determine when the next failure will occur. In other words, after substituting all the estimate parameters into the above mean value function $m(t)$, given $m(149) = 22$, we can then determine the time to the next failure, t_{23}, by solving the following equation:

$$\hat{m}(t) = 23$$

therefore, we obtain

$$t_{23} = 158.356.$$

We next discuss an exact method of determining the mean time to failure for NHPP.

5.7   Mean Time Between Failures for NHPP

Let $N(t)$ be an NHPP with the mean value function $m(t)$ where $N(t)$ denotes the random variable of the total number of the events during $[0, t]$. Let T_k denote the random variable of the occurring time for the kth event, and let X_k be the time interval between the $(k-1)$st and kth event. Then

$$X_k = T_k - T_{k-1}$$

where $k = 1, 2, \ldots$ and $T_0 = 0$.

The probability density function of T_k (Nakagawa, 1983) is given by

$$f_{T_k}(t) = \frac{\lambda(t)e^{-m(t)}[m(t)]^{k-1}}{(k-1)!}, \tag{5.45}$$

where

$$\lambda(t) = \frac{\partial}{\partial t}m(t)$$

is the intensity function for the NHPP. In software reliability, we know that the mean value function is bounded, which means $m(t)$ is always finite as t approaches infinity. In this section, we assume that

1. $m(t)$ is a strictly increasing function and uniformly continuous on any closed interval; and
2. $m(0) = 0$, $m(t)$ is a positive, finite, and differentiable function.

Let

$$E^*[T_k] = E[T_k | T_k < \infty]$$

be the conditional expectation. Then (Koshimae *et al.*, 1994).

$$E^*[T_k] = \frac{\int_0^a m^{-1}(z) z^{k-1} e^{-z} dz}{\int_0^a z^{k-1} e^{-z} dz}, \qquad (5.46)$$

where

$$m(\infty) = a.$$

If the expectation of T_k is given, the mean time between failures (MTBFs) are given by

$$E^*[X_k] = E^*[T_k] - E^*[T_{k-1}], \qquad (5.47)$$

where $E^*[X_k]$ is given in Eq. (5.46).

Example 5.6   Consider the mean value function

$$m(t) = a[1 - (1 + bt)e^{-bt}], \qquad (5.48)$$

where parameters a and b denote the expected total number of initial errors and the detection rate per error, respectively. The intensity function is easily obtained as follows:

$$\lambda(t) = ab^2 t e^{-bt}.$$

It is difficult to derive the inverse value function analytically in most mean value functions, for example, in Eq. (5.48). Let us denote

$$u = m^{-1}(z),$$

then we can rewrite Eq. (5.46) as follows:

$$E^*[T_k] = \frac{\int_0^\infty u\lambda(u)[m(u)]^{k-1} e^{-m(u)} du}{\int_0^a z^{k-1} e^{-z} dz}. \qquad (5.49)$$

It should be noted that one can solve the inverse function of $m(t)$ numerically using the Newton method. Given the NTDS software failure data in Table 5.6 and the total number of observed errors as $k = 26$, we obtain the following estimates:

$$a = 27.50$$
$$b = 0.0186$$

which was calculated by the MLE method. Table 5.12 shows numerical examples of the MTBFs using Eqs. (5.47) and (5.49).

TABLE 5.12. Mean time between failures (MTBFs) for $m(t)$ as given in Eq. (5.48).

Failure no. n	Time between failures x_k (days)	MTBF $E^*[X_k]$
Production (checkout) phase		
1	9	14.4
2	12	8.2
3	11	6.7
4	4	6.1
5	7	5.8
6	2	5.6
7	5	5.5
8	8	5.5
9	5	5.6
10	7	5.7
11	1	5.9
12	6	6.2
13	1	6.5
14	9	6.8
15	4	7.2
16	1	7.6
17	3	8.0
18	3	8.4
19	6	8.7
20	1	9.0
21	11	9.1
22	33	9.1
23	7	8.9
24	91	8.7
25	2	8.5
26	1	8.1

Problems

5.1 Assume the total error content function and error detection rate function are

$$a(t) = \alpha_2(1 + \gamma t)$$

$$b(t) = \frac{b^2 t}{bt + 1},$$

respectively. From Eq. (5.44), show that the mean value function is given as follows:

$$m(t) = \alpha_2(1 + \gamma t) - \frac{bt + 1}{e^{bt}}$$
$$- \frac{(1 + bt)\alpha_2 \gamma}{be^{bt+1}}$$
$$\left[\ln(bt + 1) + \sum_{i=0}^{\infty} \frac{(bt + 1)^{i+1} - 1}{(i + 1)!(i + 1)} \right].$$

5.2 Assume the total error content function and error detection rate function are

$$a(t) = \alpha_2(1 + \gamma t)^2$$

$$b(t) = \frac{\gamma^2 t}{\gamma t + 1},$$

respectively. Using Eq. (5.44), show that the mean value function is given as follows:

$$m(t) = \alpha_2 \left(\frac{1 + \gamma t}{\gamma t} \right)[\gamma t e^{\gamma t} + 1 - e^{\gamma t}].$$

5.3 Using the Real-Time Control Software System data as given in Table P.5.1.
   (a) Calculate the maximum likelihood estimates for the parameters a and b of the Goel–Okumoto (G–O) NHPP model based on all available data.
   (b) Obtain the mean value function $m(t)$ and the reliability function.
   (c) What is the probability that a software failure does not occur in the time (hours) interval [10, 12]?
   (d) Choose another NHPP software reliability model and repeat items (a)–(c).

Is this model better than the G–O model? Explain why and justify your results.

5.4 The data set which is given in Table P.5.2 was reported in 1970 by Musa (1987). It shows the successive inter-failure times for a real-time command and control system. The table reads from left to right in rows, and the recorded times are execution times, in seconds, between successive failures. Musa reports that fixes were introduced whenever a failure occurred and execution did not begin again until the identified failure source had been removed. Therefore, we can assume that there are no repeat occurrences of individual faults, although it is possible that an attempt to fix one fault may introduce new ones. Note that there are several zeros in the table, which are apparently accounted for by rounding up the raw times.
   (a) Calculate the maximum likelihood estimates for the unknown parameters of the Musa exponential NHPP model based on all available data.
   (b) Obtain the mean value function $m(t)$ and the reliability function.
   (c) Choose another NHPP S-shaped model in Subsection 5.3.2 and repeat items (a) and (b). Is this model better than the Musa exponential model? Explain why and justify your results.

5.5 Based on the first 25 errors in Table 5.6:
   (a) Calculate the MLE for unknown parameters of five models discussed in Section 5.4
   (b) Obtain the mean value function and reliability function of all five models.
   (c) Analyze and compare the results of all the models.

TABLE P.5.1. Failure in one-hour (execution time) intervals and cumulative failures.

Hour	Number of failures	Cumulative failures
1	27	27
2	16	43
3	11	54
4	10	64
5	11	75
6	7	82
7	2	84
8	5	89
9	3	92
10	1	93

TABLE P.5.2. Execution times in seconds between successive failures (data set #3).

3	30	113	81	115
9	2	91	112	15
138	50	77	24	108
88	670	120	26	114
325	55	242	68	422
180	10	1,146	600	15
36	4	0	8	227
65	176	58	457	300
97	263	452	255	197
193	6	79	816	1,351
148	21	233	134	357
193	236	31	369	748
0	232	330	365	1,222
543	10	16	529	379
44	129	810	290	300
529	281	160	828	1,011
445	296	1,755	1,064	1,783
860	983	707	33	868
724	2,323	2,930	1,461	843
12	261	1,800	865	1,435
30	143	108	0	3,110
1,247	943	700	875	245
729	1,897	447	386	446
122	990	948	1,082	22
75	482	5,509	100	10
1,071	371	790	6,150	3,321
1,045	648	5,485	1,160	1,864
4,116				

References

Goel, A.L. and K. Okumoto, "Time-dependent error-detection rate model for software and other performance measures," *IEEE Trans. Reliability*, Vol. R-28(3), 1979.

Hossain, S.A. and R.C. Dahiya, "Estimating the parameters of a non-homogeneous Poisson process model for software reliability," *IEEE Trans. Reliability*, Vol. 42(4), December 1993, 604–612.

Koshimae, H., H. Tanaka, and S. Osaki, "Some remarks on MTBF's for non-homogeneous Poisson process," *IEICE Trans. Fundamentals*, Vol. E77-A(1), January 1994.

Misra, P.N. "Software reliability analysis," *IBM Systems J.*, Vol. 22(3), 1983.

Musa, J.D., A. Iannino, and K. Okumoto, *Software Reliability: Measurement, Prediction, Application*, McGraw-Hill, New York, 1987.

Musa, J.D. and K. Okumoto, "Application of basic and logarithmic Poisson execution model in software reliability measure," in *The Challenge of Advanced Computing Technology to System Design Method*, NATO Advanced Study Institute, 1985.

Nakagawa, Y., "A connective exponential software reliability growth model based on analysis of software reliability growth curves," *IEICE Trans.*, Vol. J77-D-I(6), June 1994, 433–442.

Ohba, M., "Software reliability analysis models," *IBM. J. Research Development*, Vol. 21(4), 1984.

Ohba, M. and S. Yamada, "S-shaped software reliability growth models," in *Proc. 4th International Conference on Reliability and Maintainability*, Perros Guirec, France, 1984, OR Society, Paris, 1984.

Ohba, M. and X.M., Chou, "Does imperfect debugging affect software reliability growth?," in *Proc. 11th International Conference on Software Engineering*, IEEE Computer Society Press, Los Angeles, 1989.

Pham, H. and M. Pham, "Software reliability models for critical applications", Idaho National Engineering Laboratory, EG&G–2663, 1991.

Pham, H., "Software reliability assessment: Imperfect debugging and multiple failure types in software development," Idaho National Engineering Laboratory, EG&G–RAAM–10737, 1993.

Pham, H. (ed.), *Software Reliability and Testing*, IEEE Computer Society Press, Los Angeles, 1995.

Pham, H., "A software cost model with imperfect debugging, random life cycle and penalty cost," *Int. J. Systems Science*, Vol. 27(5), 1996.

Pham, H. and L. Nordmann, "A generalized NHPP software reliability model," in *Proc. 3rd International Conference on Reliability and Quality in Design*, Anaheim, March 1997, ISSAT Press, Anaheim, 1997.

Pham, H. and X. Zhang, "An NHPP software reliability model and its comparison," *Int. J. Reliability, Quality and Safety Engineering*, Vol. 4(3), 1997, 269–282.

Pham, H., *Software Reliability*, Wiley Encyclopedia Electrical and Electronics, John Wiley & Sons, New York, 1999.

Tohma, Y., H. Yamano, M. Ohba, and R. Jacoby, "The estimation of parameters of the hypergeometric distribution and its application to the software reliability growth model," *IEEE Trans. Software Engineering*, Vol. SE-17(5), 1991.

Yamada, S., "*S*-shaped reliability growth modeling for software error-detection," *IEEE Trans. Reliability*, Vol. R-32(5), 1983.

Yamada, S., M. Ohba, and S. Osaki, "*S*-shaped software reliability growth models and their applications," *IEEE Trans. Reliability*, Vol. R-33, 1984.

Yamada, S. and S. Osaki, "Reliability growth model for hardware and software systems based on nonhomogeneous Poisson process: A survey," *Microelectronics and Reliability*, Vol. 23(1), 1983.

Yamada, S. and S. Osaki, "Software reliability growth modeling: Models and applications," *IEEE Trans. Software Engineering*, Vol. SE-11(12), 1985.

Wood, A., "Predicting software reliability," *IEEE Computer*, November 1996, 69–77.

Zhao, M. and M. Xie, "Recent advances in NHPP software reliability models," in *Proc. IASTED International Conference on Reliability, Quality Control and Safety Assessment*, Washington, DC, 1992, ACTA Press, Anaheim, 1992.

Zhao, M., "Nonhomogeneous Poisson process and their applications in software reliability," Ph.D. Dissertation, Division of Quality Technology, Linkoping University, Linkoping, Sweden, 1994.

Zhao, M. and M. Xie, "On maximum likelihood estimation for a general nonhomogeneous Poisson process," *Scandinavian J. Statistics,* Vol. 23, 1996.

6

Software Cost Models

6.1 Introduction

In recent years, the costs of developing software and software failures have entailed great expenses in a system development. Therefore, it is important to determine when to stop testing, or when to release the software to the users so that the total system cost is minimized, subject to a desired reliability level and other constraints.

The quality of the software system usually depends upon the length of testing and what testing methodologies are used. Generally speaking, the longer the testing takes, the more reliable the software is expected to be. However, the total cost of developing the software is expected to increase. On the other hand, if the testing time is too short, the cost of the software could be reduced, but the customers risk buying unreliable software. This will also increase the cost during the operational phase, as the cost of debugging during the operational phase is much higher than that of the testing phase.

Testing is an efficient way to remove faults in software products, but testing of all possible executable paths in a general program is impractical. Moreover, after reaching a certain level of software refinement, any effort to increase reliability will require an exponential increase in cost. Thus, it is important to determine when to stop testing, or when to release the software to customers, to keep the expected total software cost subject to warranty and risk costs minimal. It is common for software companies to provide a warranty period during which they are still responsible for debugging errors.

In defining important software cost factors, a cost model should help software developers and managers answer the following questions:

(1) How should resources be scheduled to ensure the on-time and efficient delivery of a software product?
(2) Is the software product sufficiently reliable for release (e.g., have we done enough testing?
(3) What information does a manager or software developer need to determine the release of software from current software testing activities?

This chapter aims to discuss software cost models that will help answer these questions. In addition to the costs of traditional cost models, the cost model proposed also considers the testing cost, debugging cost during testing phase, debugging cost during the warranty period, and risk cost due to software failure (see Fig. 6.1). This model can be used to estimate the realistic total software cost for applications such as telecommunications, customer service, etc., and to determine the optimal testing release policies of the software systems.

In this chapter, we present several software cost models based on the NHPP software reliability functions. In addition to the costs of traditional cost models, the cost models will include the cost to perform testing, the cost of removing detected errors, and the risk cost due to software failures. These models can be used to formulate realistic total software cost projects in many applications and to determine the optimal release policies of the software system.

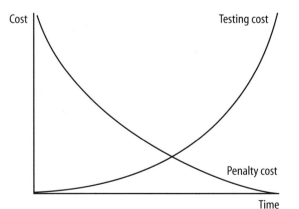

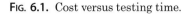
FIG. 6.1. Cost versus testing time.

The following notations and basic assumptions are applied throughout this chapter.

Notation

$m(T)$	expected number of errors to be detected by time T
a	total number of software errors to be eventually detected
b	exponential index
$\lambda(T)$	fault detection rate per unit time or intensity function
x	mission time
$R(x/T)$	reliability function of software by time T for a mission time x
T	software release time
C_1	software test cost per unit time
C_2	cost of removing each error per unit time during testing
$E(T)$	expected total cost of a software system by time T
Y	time to remove an error during testing phase
μ_y	expected time to remove an error during testing phase which is $E(Y)$

General Assumptions
(1) The cost to perform testing is proportional to the testing time.
(2) The cost to remove errors during the testing phase is proportional to the total time of removing all errors detected by the end of the testing phase.
(3) There is a risk cost related to the reliability at each release time point.
(4) The time to remove each error during testing follows a truncated exponential distribution.
(5) Without loss of generality, the Goel–Okumoto NHPP model will be used as a reliability funtion.

Let Y be a random variable of time to remove an error. Based on assumption (4), the probability density distribution of Y is given by

$$s(y) = \frac{\lambda e^{-\lambda y}}{\int\limits_0^{T_0} \lambda e^{-\lambda z} dz} \quad \text{for } 0 \le y \le T_0,$$

where T_0 is the maximum time to remove an error. The expected time to remove each error is

$$\mu_y = E(Y) = \int_0^{T_0} ys(y)dy$$

$$= \int_0^{T_0} \frac{y\lambda e^{-\lambda y}}{\int_0^{T_0} \lambda e^{-\lambda z}dz} dy.$$

After simplifications, we obtain

$$\mu_y = \frac{1 - (\lambda T_0 + 1)e^{-\lambda T_0}}{\lambda(1 - e^{-\lambda T_0})} \qquad (6.1)$$

6.2   A Software Cost Model with Risk Factor

In this section, we discuss a cost model addressing the risk level and the time to remove errors and the optimal release policies that minimize the expected total software cost. The expected software system cost, $E(T)$, is defined as: (1) the cost to perform testing; (2) the cost incurred in removing errors during the testing phase; and (3) a risk cost due to software failure.

(a) The cost to perform testing is given by

$$E_1(T) = C_1 T.$$

(b) The expected total time to remove all $N(T)$ errors is

$$E[\sum_{i=1}^{N(T)} Y_i] = E[N(T)]E[Y_i] = m(T)\mu_y.$$

Hence, the expected cost to remove all errors detected by time T can be expressed as

$$E_2(T) = C_2 E[\sum_{i=1}^{N(T)} Y_i] = C_2\, m(T)\mu_y.$$

(c) The risk cost due to software failure after releasing the software is

$$E_3(T) = C_3[1 - R(x|T)],$$

where C_3 is the cost due to software failure.

Therefore, the expected total software cost can be expressed (Zhang, 1998) as

$$E(T) = C_1 T + C_2 m(T)\mu_y + C_3[1 - R(x|T)].$$

From assumption 5 (see Chapter 5), the mean value function $m(T)$ is

$$m(T) = a(1 - e^{-bT}). \tag{6.2}$$

The error detection rate function is

$$\lambda(T) = abe^{-bT} \tag{6.3}$$

and the reliability of the software is

$$\begin{aligned} R(x|T) &= e^{-[m(T+x)-m(T)]} \\ &= e^{-a[e^{-bT}-e^{-b(T+x)}]}. \end{aligned} \tag{6.4}$$

Let us define

$$\begin{aligned} f(T) &= \lambda(T)[C_3(1 - e^{-bx})R(x|T) - C_2\mu_y] \\ g(T) &= C_3(1 - e^{-bx})R(x|T)[1 - ae^{-bT}(1 - e^{-bx})]. \end{aligned} \tag{6.5}$$

It should be noted that $g(T)$ is a strictly increasing function of T. The optimal software release time T^*, which minimizes the expected total system cost, is given as:

Theorem 6.1 Given C_1, C_2, C_3, x, and μ_y, the optimal value of T, say T^*, which minimizes the expected total cost of the software can be determined as follows:

(1) If $g(0) > C_2 \mu_y$, then
   (a) If $f(0) \le C_1$, then $T^* = 0$.
   (b) If $f(\infty) > C_1$, then $T^* = \infty$.
   (c) If $f(0) > C_1, f(T) \ge C_1$ for any $T \in (0, T')$ and $f(T) < C_1$ for any $T \in (T', \infty)$, then $T^* = T'$ where
$$T' = \inf\{T: f(T) < C_1\}.$$
(2) If $g(\infty) < C_2 \mu_y$, then
   (a) If $f(0) \ge C_1$, then $T^* = \infty$.
   (b) If $f(\infty) < C_1$, then $T^* = 0$.
   (c) If $f(0) < C_1, f(T) \le C_1$ for any $T \in (0, T')$ and $f(T) > C_1$ for any $T \in (T'', \infty)$, then
$$T^* = 0 \text{ if } E(0) < E(\infty)$$
$$T^* = \infty \text{ if } E(0) \ge E(\infty)$$
   where $T'' = \inf\{T: f(T) > C_1\}$.

(3) If $g(0) < C_2\,\mu_y$, $g(T) \le C_2\,\mu_y$ for $T \in (0, T_0)$, and $g(T) > C_2\,\mu_y$ for $T \in (T_0,\infty)$, then

(a) If $f(0) < C_1$, then

$$T^* = 0 \text{ if } E(0) < E(T_b)$$
$$T^* = T_b \text{ if } E(0) \ge E(T_b)$$

where $T_b = \inf\{T: f(T) < C_1, T > T_a\}$.

(b) If $f(0) > C_1$, then $T^* = T_b{}'$ where $T_b{}' = \inf\{T: f(T) < C_1\}$.

Proof   See (Zhang, 1998).

Example 6.1   Assuming a software failure data is given in Table 5.1, the parameters of the Goel–Okumoto model using MLE is given by

$$\hat{a} = 142.32, \ \hat{b} = 0.1246.$$

The mean value function becomes

$$m(T) = a(1 - e^{-bt})$$
$$= 142.32(1 - e^{-0.1246T}).$$

Given $C_1 = \$25$, $C_2 = \$200$, $C_3 = \$7,000$, $\mu_y = 0.1$, and $x = 0.05$. From Theorem 6.1, the results are shown in Table 6.1.

The optimal release time in this case is $T^* = 21.5$ hours and the corresponding expected total cost is \$3,600.49. If we increase the value of C_3 from \$7,000 to \$10,000, we expect to have a longer testing time. In this case ($C_1 = \$25$, $C_2 = \$200$, $C_3 = \$10,000$, $\mu_y = 0.1$, and $x = 0.05$), the optimal release time is $T^* = 27$ hours and the corresponding expected total cost is \$3,723.95.

TABLE **6.1.** Optimal release time for $C_1 = \$25$, $C_2 = \$200$, $C_3 = \$7,000$, $\mu_y = 0.1$, $x = 0.05$

Release time T^*(hours)	Expected total cost $E(T)$(\$)
19.5	3,607.34
20.0	3,604.47
20.5	3,602.39
21.0	3,601.07
21.5*	3,600.49
22.0	3,600.60
22.5	3,601.39
23.0	3,602.81
23.5	3,604.83

6.3   A Generalized Software Cost Model

In this section, in addition to the cost factors presented in Section 6.2, we present a generalized cost model considering the cost of removing errors detected during the warranty period and the risk cost due to software failure. Throughout this section, we define the following notations and assumptions.

Notation

C_0 set-up cost for software testing

C_3 cost of removing an error per unit time during the operational phase

C_4 loss due to software failure

W variable of time to remove an error during the warranty period in the operation phase

μ_w expected time to remove an error during the warranty period in the operation phase, which is $E(W)$

T_w period of warranty time

α the discount rate of the testing cost

Additional Assumptions

(6) There is a set-up cost at the beginning of the software development process.

(7) The cost of testing is a power function of the testing time. This means that at the beginning of the testing, the cost increases with a higher gradient, slowing down later.

(8) The time to remove each error during the warranty period follows a truncated exponential distribution.

(9) The cost to remove errors during the warranty period is proportional to the total time of removing all errors detected between the interval of (T, T_w).

Similarly, from assumption 8, the truncated exponential density function of error removal time during the warranty period is

$$q(w) = \frac{\lambda_w e^{-\lambda_w w}}{\int_0^{T_0} \lambda_w e^{-\lambda_w x} dx} \quad \text{for } 0 \leq w \leq T_0'. \tag{6.6}$$

Therefore, the expected time to remove an error during the warranty period is

$$\mu_w = \frac{1 - (\lambda_w T_0' + 1)e^{-\lambda_w T_0'}}{\lambda_w (1 - e^{-\lambda_w T_0'})}. \tag{6.7}$$

The expected software system cost comprises of the set-up cost, the cost to do testing, the cost incurred in removing errors during the testing phase and during the warranty period, and the risk cost in releasing the software system by time T. Hence, the expected total software system cost $E(T)$ can be expressed as follows (Pham, 1999):

$$E(T) = C_0 + C_1 T^\alpha + C_2 m(T)\mu_y + C_3 \mu_w [m(T + T_w) - m(T)]$$
$$+ C_4 [1 - R(x|T)], \tag{6.8}$$

where $0 \le \alpha \le 1$. Define

$$y(T) = \alpha C_1 T^{(\alpha - 1)} - \mu_w C_3 abe^{-bT}(1 - e^{-bT_w})$$
$$- abe^{-bT}[C_4(1 - e^{-bx})R(x|T) - C_2\mu_y]$$
$$u(T) = ab^2 C_4(1 - e^{-bx})R(x|T)[1 - ae^{-bT}(1 - e^{-bx})]$$
$$+ \alpha(\alpha - 1)C_1 T^{(\alpha - 2)}e^{bT}$$
$$C = C_2\mu_y ab^2 - \mu_w C_3 ab^2(1 - e^{-bT_w}). \tag{6.9}$$

It can be shown that the function $u(T)$ is an increasing function of T (see Problem 6.2). The optimal software release time, T^*, which minimizes the expected total system cost is given below.

Theorem 6.2   Given C_0, C_1, C_2, C_3, C_4, x, μ_y, μ_w, T_w, the optimal value of T, say T^*, which minimizes the expected total cost of the software is as follows:

(1) If $u(0) \ge C$ and
   (a) if $y(0) \ge 0$, then $T^* = 0$;
   (b) if $y(\infty) < 0$, then $T^* = \infty$;
   (c) if $y(0) < 0$, $y(T) < 0$ for $T \in (0, T']$ and $y(T) > 0$ for $T \in (T', \infty)$, then
$$T^* = T'$$
   where $T' = y^{-1}(0)$.
(2) If $u(\infty) < C$ and
   (a) if $y(0) \le 0$, then $T^* = \infty$;
   (b) if $y(\infty) > 0$, then $T^* = 0$;
   (c) if $y(0) > 0$, $y(T) > 0$ for $T \in (0, T'']$ and $y(T) < 0$ for $T \in (T'', \infty)$, then
$$T^* = 0 \text{ if } E(0) \le E(\infty)$$
$$T^* = \infty \text{ if } E(0) > E(\infty)$$
   where $T'' = y^{-1}(0)$.

(3) If $u(0) < C$, $u(T) \leq C$ for $T \in (0, T_0)$, and $u(T) > C$ for $T \in (T_0, \infty)$
where $T_0 = u^{-1}(C)$, then
(a) if $y(0) \geq 0$, then
$$T^* = 0 \text{ if } E(0) \leq E(T_b)$$
$$T^* = T_b \text{ if } E(0) > E(T_b),$$
where $T_b = \inf\{T > T_a: y(T) > 0\};$
(b) if $y(0) < 0$, then $T^* = T_b{}'$ where $T_b{}' = y^{-1}(0)$.

Proof Taking the first derivative of $E(T)$, we obtain

$$\frac{\partial E(T)}{\partial T} = \alpha C_1 T^{(\alpha-1)} - \mu_w C_3 abe^{-bT}(1 - e^{-bT_w})$$
$$- abe^{-bT}[C_4(1 - e^{-bx})R(x|T) - C_2\mu_y]$$
$$= y(T).$$

The second derivative of $E(T)$,

$$\frac{\partial^2 E(T)}{\partial T^2} = e^{-bT}[u(T) - C].$$

Case 1 If $u(0) \geq C$, then $u(T) > C$ for any T. In this case, $y(T)$ is a strictly increasing function of T and

$$\frac{\partial^2 E(T)}{\partial T^2} > 0.$$

There are three subcases:

(a) If $y(0) \geq 0$, then $y(T) \geq 0$ for all T and $E(T)$ is strictly increasing in T. Hence, $T^* = 0$ minimizes $E(T)$.
(b) If $y(\infty) < 0$, then $y(T) < 0$ for all T and $E(T)$ is decreasing in T. Hence, $T^* = \infty$ minimizes $E(T)$.
(c) If $y(0) < 0$, $y(T) < 0$ for any $T \in (0, T']$ and $y(T) > 0$ for any $T \in (T', \infty)$, then $T^* = T'$ where $T' = y^{-1}(0)$.

Case 2 If $u(\infty) < C$, then $u(T) < C$ for any T. In this case, $y(T)$ is a strictly decreasing function of T and

$$\frac{\partial^2 E(T)}{\partial T^2} < 0.$$

There are three subcases:

(a) If $y(0) \leq 0$, then $y(T) \leq 0$ for all T and $E(T)$ is strictly decreasing in T. Hence, $T^* = \infty$ minimizes $E(T)$.

TABLE **6.2.** Optimal release time for $C_0 = \$100$.

T^* (hours)	$E(T)$ ($)
22.5	1,843.31
23.0	1,843.31
23.5	1,839.52
24.0	1,837.17
24.5*	1,836.15
25.0	1,836.39
25.5	1,837.82
26.0	1,840.35
26.5	1,843.93

(b) If $y(\infty) > 0$, then $y(T) > 0$ for all T and $E(T)$ is increasing in T. Hence, $T^* = 0$ minimizes $E(T)$.

(c) If $y(0) > 0$, $y(T) > 0$ for any $T \in (0, T'']$ and $y(T) < 0$ for any $T \in (T'', \infty)$, then $T^* = 0$ if $E(0) < E(\infty)$ and $T^* = \infty$ if $E(0) \geq E(\infty)$, where $T'' = y^{-1}(0)$.

Case 3   See Problem 6.3.

Example 6.2   Considering a set of testing data given in Table 5.1 and Example 6.1, the mean value function is

$$m(T) = 142.32(1 - e^{-0.1246T}).$$

Given $C_1 = \$50$, $C_2 = \$25$, $C_3 = \$100$, $C_4 = \$1,000$, $\mu_y = 0.1$, $\mu_w = 0.5$, $x = 0.05$, $\alpha = 0.05$, and $T_w = 20$. Based on Theorem 6.2, we obtain the following results in Table 6.2.

The optimal release time in this case is $T^* = 24.5$ and the corresponding expected total cost is \$1,836.15.

Example 6.3   Given $C_1 = \$50$, $C_2 = \$25$, $C_3 = \$100$, $C_4 = \$10,000$, $\mu_y = 0.1$, $\mu_w = 0.5$, $x = 0.05$, $\alpha = 0.05$, and $T_w = 20$. Using Theorem 6.2, we obtain the results in Table 6.3.

The optimal release time for this case is $T^* = 30.5$ and the corresponding expected total cost is \$3,017.13.

TABLE **6.3.** Optimal release time for $C_0 = \$1,000$.

T^* (hours)	$E(T)(\$)$
28.5	3,029.72
29.0	3,024.41
29.5	3,020.60
30.0	3,018.20
30.5*	3,017.13
31.0	3,017.30
31.5	3,018.64
32.0	3,021.09
32.5	3,024.57

6.4 A Cost Model with Multiple Failure Errors

In this section, a software cost model is presented under the following assumptions:

(1) The cost of debugging an error during the development phase is lower than in the operational phase.
(2) The cost of removing a particular type of error is constant during the debugging phase.
(3) The cost of removing a particular type of error is constant during the operational phase.
(4) The cost of removing critical errors is more expensive than major errors, and the cost of removing major errors is more expensive than minor errors.
(5) There is a continuous cost incurred during the entire time of the debugging period.

Notation

T software release time
C_{i1} cost of fixing a type i error during the test phase $i = 1, 2, 3$
C_{i2} cost of fixing a type i error during the operation phase ($C_{i2} \geq C_{i1}$, $i = 1, 2, 3$)
C_3 cost of testing per unit time
$E(T)$ expected cost of software
R_0 pre-specified software reliability
T_r debugging time required to attain minimum cost subject to a reliability constraint

T_e debugging time required to attain minimum cost subject to the number of remaining errors constraint

T_{rel} debugging time required to attain maximum reliability subject to a cost constraint

Assume that the duration of the software lifecycle is random. Let t be the random variable of the duration of the software lifecycle and $g(t)$ the probability density function of t. Assume that the cost of testing per unit time and the cost of fixing any type i error during the test phase and the operation phase are given.

The expected software cost is defined as the cost incurred in removing and fixing errors in the software during the software lifecycle measured from the time the testing starts. Hence, the expected software cost $E(T)$ can be formulated as (Pham, 1996)

$$E(T) = \int_0^T [C_3 t + \sum_{i=1}^3 C_{i1} m_i(t)] g(t) dt$$

$$+ \int_T^\infty [C_3 T + \sum_{i=1}^3 C_{i1} m_i(T)$$

$$+ \sum_{i=1}^3 C_{i2}(m_i(t) - m_i(T))] g(t) dt, \qquad (6.10)$$

where $m_i(t)$ is given in Eq. (5.22).

The function $E(T)$ represents testing costs per unit time and of fixing errors during testing incurred if the determination of the software lifecycle is less than or equal to the software release time. On the other hand, if the determination of the software lifecycle is greater than the software release time, then an additional cost factor should be involved, i.e., the cost of fixing errors during the operation phase.

Next, we present the optimal testing time that minimizes the expected software cost. In other words, we wish to find the value of T such that $E(T)$ is minimized. Define

$$h(T) = \sum_{i=1}^3 (c_{i2} - c_{i1}) \lambda_i(T). \qquad (6.11)$$

Theorem 6.3   Given C_3, C_{i1}, and C_{i2} for $i = 1, 2, 3$. There exists an optimal testing time, T^*, for T that minimizes $E(T)$:

$$\text{IF } h(0) \leq C_3, \text{THEN } T^* = 0$$
$$\text{ELSE } T^* = h^{-1}(C_3); \text{ENDIF.}$$

Proof   See Problem 6.4.

Theorem 6.3 shows that if $h(0) \geq C_3$, then the testing time required to attain the minimum cost has already been achieved. Thus, the marginal cost for further testing is an increasing function, and as each additional test and debug increases the cost of the software, no further testing should be done, and the software package should be released for sale. However, if the testing time required to attain the minimum cost has not been achieved, and the marginal cost for further debugging is a decreasing function for an interval of times, testing should be continued until time T^*, where T^* satisfies $h(T^*) = C_3$.

Although the optimal policies in Theorem 6.3 are sound in theory, it seems reasonable in practice that simply minimizing cost should not be the only goal for some applications. In the following subsection, we discuss the optimum release policies that minimize the expected software system cost subject to various constraints.

6.4.1 Cost Subject to Reliability Constraint

Consider the expected software cost $E(T)$ and the software reliability $R(x|T)$ as the evaluation criteria. We determine the optimum release time that minimizes the expected software cost subject to attaining a desired reliability level, R_0. Then the optimization problem can be formulated as

$$\text{Minimize} \quad E(T)$$
$$\text{Subject to} \quad R(x|T) \geq R_0,$$

where $R(x|T)$ and $E(T)$ are given in Eqs. (5.25) and (6.10), respectively.

It can be proven that the software reliability $R(x|T)$ increases as T increases. Let T_1 be the solution of $R(x|T_1) = R_0$.

Lemma 6.1   Given C_3, R_0, C_{i1}, and C_{i2} for $i = 1, 2, 3$. The optimal value of T, say T_r, which minimizes $E(T)$ subject to software reliability not less than a specified value, R_0, is determined from

$$\text{IF } h(0) \leq C_3 \text{ THEN}$$
$$\quad \text{IF } R(x|0) \geq R_0 \text{ THEN } T_r = 0$$
$$\quad \text{ELSE } T_r = T_1$$
$$\text{ELSE IF } R(x|T^*) \geq R_0 \text{ THEN } T_r = T^*$$
$$\quad \text{ELSE } T_r = T_1.$$

The physical interpretation of the result of Lemma 6.1 is as follows. If $h(0) \leq C_3$, then the current amount of debugging has already minimized the expected software system cost. Furthermore, if the current amount of debugging has met the reliability constraint, then no further debugging should be done, and the software should be released. Otherwise, the current amount of debugging does not meet the reliability constraint, and the debugging should be continued until time T_1, where T_1 satisfies $R(x|T_1) = R_0$. The interpretation of the case $h(0) > C_3$ is similar to the above.

6.4.2 Cost Subject to the Number of Remaining Errors Constraint

We now present a method that will allow for the constraining of a particular type of error. The importance of this is that though a program may be able to tolerate a large number of minor errors, it cannot tolerate critical errors. In this case, a constraint can be put on the expected number of remaining critical errors in the system before its release. It is also possible to set up constraints for each of the different types of errors independent of the other types.

Consider both the expected total software system cost, $E(T)$, and the expected number of failure type i errors remaining in the system, $\bar{m}_i(T)$, as the evaluation criteria. The optimal release problem can be formulated as

$$\text{Minimize} \quad E(T)$$
$$\text{Subject to} \quad \bar{m}_i(T) \leq d_i \quad i = 1, 2, 3,$$

where

$$\bar{m}_i(T) = m_i(\infty) - m_i(T)$$
$$= \frac{ap_i}{1 - \beta_i} e^{-(1-\beta_i)b_i T} \tag{6.12}$$

and d_i is the accepted number of remaining type i errors. Define

$$T_{m_i} = \frac{\ln\left[\frac{ap_i}{d_i(1-\beta_i)}\right]}{(1 - \beta_i)b_i}. \tag{6.13}$$

The function $\bar{m}_i(T)$ is, of course, decreasing in T for all T. Then $\bar{m}_i(T) \leq d_i$ if and only if $T \geq T_{mi}$.

Lemma 6.2 Given C_3, d_i, C_{i1}, and C_{i2} for $i = 1, 2, 3$, then the optimal value of T, say T_e, that minimizes $E(T)$ subject to the number of remaining errors constraint is determined from

IF $h(0) \leq C_3$ THEN $T_e = \max 1 \leq i \leq 3 \{0, T_{mi}\}$
ELSE $T_e = \max_{1 \leq i \leq 3} \{T^*, T_{mi}\}$; ENDIF.

Similarly, the physical interpretation of the results of Lemma 6.2 is that if $h(0) \leq C_3$, then the current debugging has already minimized cost, but not all of the expected error constraints have been met. In this situation, the software program should be debugged until all of the expected error constraints have been met. However, if $h(0) > C_3$, then the current amount of debugging has not achieved minimum cost. In this situation, debugging should continue until all constraints have been met and minimum cost has been achieved.

6.4.3 Software Reliability Subject to Cost Constraint

Consider both the software reliability $R(x|T)$ and the expected software cost $E(T)$ as the evaluation criteria. The optimal policies problem can be formulated as

$$\begin{cases} \text{Maximize} & R(x|T) \\ \text{Subject to} & E(T) \leq C_R, \end{cases}$$

where C_R is the maximum amount allowable.

Lemma 6.3   Given C_3, C_R, C_{i1}, and C_{i2} for $i = 1, 2, 3$. The optimal value of T, say T_{rel}, that maximizes $R(x|T)$ subject to the cost constraint is determined from

$$\text{IF } E(T^*) > C_R \text{ THEN there is No solution}$$
$$\text{ELSE } T_{rel} = \{T \geq T^*: T = E^{-1}(C_R)\}; \text{ ENDIF.}$$

These results show that if $E(T^*) > C_R$, the minimum software system cost required to develop and debug the program exceeds the maximum amount allowable. Therefore, it is impossible to produce the software under these conditions. Similarly, if $E(T^*) \leq C_R$, and as the reliability of the software continually improves with testing and debugging time, then the program should be debugged until the cost constraint is binding, implying that additional debugging will violate the constraint.

6.5   Applications

Using a set of data given in Table 5.4, and given the following reliability and error introduction rate parameters values,

$$p_1 = 0.0173 \quad p_2 = 0.3420 \quad p_3 = 0.6407$$
$$\beta_1 = 0.5 \quad \beta_2 = 0.2 \quad \beta_3 = 0.05.$$

Using the maximum likelihood estimate, we obtain

$$a = 428$$
$$b_1 = 0.00024275$$

$$b_2 = 0.00029322$$
$$b_3 = 0.00030495.$$

Given the following cost coefficients,

$$C_{1,1} = 200 \qquad C_{2,1} = 80 \qquad C_{3,1} = 30$$
$$C_{1,2} = 1{,}000 \qquad C_{2,2} = 350 \qquad C_{3,2} = 150$$
$$C_3 = 10.$$

Assume that the mean rate of the software lifecycle length is constant. Given that the mean rate $\mu = 0.00005$, then

$$g(T) = \mu e^{-\mu T} \quad \text{for } T > 0.$$

From Eq. (6.10), the expected software system cost is given by

$$E(T) = \int_0^T [10t + 200m_1(t) + 80m_2(t) + 30m_3(t)]g(t)dt$$

$$+ \int_T^\infty [10T + 200m_1(T) + 80m_2(T) + 30m_3(T)$$

$$+ 1{,}000(m_1(t) - m_1(T)) + 350(m_2(t) - m_2(T))$$
$$+ 150(m_3(t) - m_3(T))]g(t)dt.$$

Substituting $m_i(t)$ in Eq. (5.22) for $i = 1, 2, 3$ into the above equation, we obtain

$$E(T) = \frac{1}{\mu}(1 - e^{-\mu T})C_3$$

$$- \sum_{i=1}^{3}(C_{i2} - C_{i1})A_i[1 - e^{-(1-\beta_i)b_i T}]e^{-\mu T}$$

$$+ \sum_{i=1}^{3} C_{i1}\{A_i[(1 - e^{-\mu T}) - \frac{\mu}{B_i}(1 - e^{B_i T})]\}$$

$$+ \sum_{i=1}^{3} C_{i2}\{A_i[e^{-\mu T} - \frac{\mu}{B_i}e^{-\beta_i T}]\},$$

where

$$A_i = \frac{ap_i}{1 - \beta_i} \quad \text{and} \quad B_i = (1 - \beta_i)b_i + \mu.$$

Substituting and simplifying the cost coefficient values to the above equation, we obtain

$$E(T) = 200{,}000(1 - e^{-0.00005T}) - 11{,}848(1 - e^{-0.0001214T})e^{-0.00005T}$$
$$- 49{,}401.9(1 - e^{-0.0002346T})e^{-0.00005T}$$
$$- 34{,}638(1 - e^{-0.0002897T})e^{-0.00005T}$$
$$+ 2{,}962[1 - e^{-0.00005T} - 0.2917153(1 - e^{-0.0001714T})]$$
$$+ 14{,}637.6[1 - e^{-0.00005T} - 0.1756852(1 - e^{-0.0002846T})]$$
$$+ 8{,}659.5[1 - e^{-0.00005T} - 0.1471887(1 - e^{-0.0003397T})]$$
$$+ 14{,}810[e^{-0.00005T} - 0.2917153e^{-0.0001714T}]$$
$$+ 64{,}039.5[e^{-0.00005T} - 0.1756852e^{-0.0002846T}]$$
$$+ 43{,}297.5[e^{-0.00005T} - 0.1471887e^{-0.0003397T}].$$

It is easy to obtain the optimum total testing time, T^*, that minimizes the expected total software system cost using Theorem 6.3. The results are given as below:

$$T^* = 3{,}366.8 \text{ hours and } E(3{,}366.8) = \$82{,}283.2.$$

If a desired level of reliability is 0.99 for a mission of 10 hours, then by using Lemma 6.1, the optimal software release time, T_r, that minimizes the expected total software system cost is easily obtained as $T_r = 19{,}045$ hours.

If we assume that the remaining error constraints are

type 1 errors: $d_1 \leq 5$
type 2 errors: $d_2 \leq 7$
type 3 errors: $d_3 \leq 10$.

Then using Lemma 6.2, the optimal release time in this situation is given as follows:

$$T_{m1} = 8{,}946 \text{ hours}$$
$$T_{m2} = 13{,}912 \text{ hours, and}$$
$$T_{m3} = 11{,}607 \text{ hours}.$$

Since T_{m2} is the maximum of the four values, T_{m1}, T_{m2}, T_{m3}, and T^*, $T_e = 13{,}912$ hours.

Problems

6.1   Show that the function $g(T)$ in Eq. (6.5) is increasing in T.

6.2   Show that the function $u(T)$ in Eq. (6.9) is increasing in T.

6.3   Complete the proof of Case 3 in Theorem 6.2.

6.4   Prove Theorem 6.3.

6.5   Assume that the risk cost due to software failure after releasing the software beyond the warranty period, $E_4(T)$, is given by

$$E_4(T) = C_4[1 - R(x|(T + T_w))].$$

From Eq. (6.8), the expected total software cost $E(T)$ can be modified as follows:

$$\begin{aligned} E(T) = {}& C_0 + C_1 T^\alpha + C_2 m(T)\mu_y \\ & + C_3\mu_w[m(T + T_w) - m(T)] \\ & + C_4[1 - R(x|(T + T_w))], \end{aligned}$$

where

$$R(x|(T + T_w)) = e^{-ae^{-b(T+T_w)}[1-e^{-bx}]}.$$

Given C_0, C_1, C_2, C_3, C_4, x, μ_y, μ_w, T. Show that the optimal value of T, say T^*, which minimizes the expected total cost of the software, is given as below:

(1) If $v(0) \geq C$, and
  (a) if $y(0) \geq 0$, then $T^* = 0$;
  (b) if $y(\infty) < 0$, then $T^* = \infty$;
  (c) if $y(0) < 0$, $y(T) < 0$ for any $T \in (0, T']$ and $y(T) > 0$ for any $T \in (T', \infty)$, then $T^* = T'$, where $T' = y^{-1}(0)$.

(2) If $v(\infty) < C$, and
  (a) if $y(0) \leq 0$, then $T^* = \infty$;
  (b) if $y(\infty) > 0$, then $T^* = 0$;
  (c) if $y(0) > 0$, $y(T) > 0$ for any $T \in (0, T'']$ and $y(T) < 0$ for any $T \in (T'', \infty)$, then
    $T^* = 0$, if $E(0) \leq E(\infty)$
    $T^* = \infty$, if $E(0) > E(\infty)$,
    where $T'' = \inf\{T: y(T) < 0\}$.

(3) If $v(0) < C$, $v(T) \leq C$ for $T \in (0, T_0]$, and $v(T) > C$ for $T \in (T_0, \infty)$, where $T^0 = \{T: T = v^{-1}(C)\}$, then
  (a) if $y(0) \geq 0$, then
    $T^* = 0$ if $E(0) \leq E(T_b)$

$T^* = T_b$ if $E(0) > E(T_b)$,
where $T_b = \{T > T_a: T = y^{-1}(0)\}$.
  b) if $y(0) < 0$, then $T^* = T_b'$ minimizes $E(T)$, where $T_b' = \{T: T = y^{-1}(0)\}$
where

$$\begin{aligned} y(T) = {}& \alpha C_1 T^{(\alpha-1)} - \mu_w C_3 abe^{-bT}(1 - e^{-bT}) \\ & - abe^{-bT}[C_4(1 - e^{-bx})e^{-b(T+T_w)} \\ & \quad R(x|(T + T_w)) - C_2\mu_y] \end{aligned}$$

$$\begin{aligned} v(T) = {}& ab^2 C_4(1 - e^{-bx})e^{-bT_w} R(x|(T + T_w)) \\ & [1 - ae^{-bT}(1 - e^{-bx})] \\ & + \alpha(\alpha - 1)C_1 T^{(\alpha-2)}e^{bT} \end{aligned}$$

$$C = C_2\mu_y ab^2 - \mu_w C_3 ab^2(1 - e^{-bT_w}).$$

6.6   Given $C_0 = \$100$, $C_1 = \$50$, $C_2 = \$25$, $C_3 = \$100$, $C_4 = \$1,000$, $\mu_y = 0.1$, $\mu_w = 0.5$, $x = 0.05$, and $T_w = 20$, $\alpha = 0.95$. Using the results in Problem 6.5, show that the release times and the corresponding expected total software cost are given in Table P.6.1.

TABLE P.6.1. Optimal release time.

Release time T^* (hours)	Expected total cost $E(T)$ ($)
21.5	1,806.69
22.0	1,801.17
22.5	1,797.18
23.0	1,794.65
23.5*	1,793.47
24.0	1,793.56
24.5	1,794.85
25.0	1,797.27
25.5	1,800.74

6.7 $C_0 = \$100$, $C_1 = \$10$, $C_2 = \$5$, $C_3 = \$100$, $C_4 = \$1,000$, $\mu_y = 0.1$, $\mu_w = 0.5$, $x = 0.05$, and $T_w = 20$, $\alpha = 0.95$. Using the results in Problem 6.5, show that the optimal release time and the corresponding expected total software cost are 37 and \$545.2, respectively.

References

Kapur, P.K. and V.K. Bhalla, "Optimal release policies for a flexible software reliability growth model," *Reliability Engineering and System Safety*, Vol. 35, 1992, 45–54.

Leung, Y.-W., "Optimum software release time with a given cost budget," *J. Systems and Software*, Vol. 17, 1992, 233–242.

Misra, P.N., "Software reliability analysis," *IBM Systems J.*, Vol. 22(5), 1983, 262–270.

Pham, H., "Software reliability assessment: Imperfect debugging and multiple failure types in software development," EG&GRAAM-10737, April 1993.

Pham, H., "A software cost model with imperfect debugging, random life cycle and penalty cost," *Int. J. Systems Science*, Vol. 27(5), 1996.

Pham, H. and X. Zhang, "A software cost model with warranty and risk costs," *IEEE Trans. Computers*, Vol. 48(1), 1999.

Yamada, S. and S. Osaki, "Optimum software release policies for a non-homogeneous software error detection rate model," *Microelectronics and Reliability*, Vol. 26, 1986, 691–702.

Zhang, X. and H. Pham, "A software cost model with error removal times and risk costs," *Int. J. Systems Science*, Vol. 29(4), 1998.

7

Fault-Tolerant Software

"I think and think for months and years. Ninety-nine times,
the conclusion is false. The hundreth time I am right."
Albert Einstein (1879–1955)

7.1   Introduction

Computer systems are now applied to many important areas such as defense, transportation, and air traffic control. Therefore, fault-tolerance has become one of the major concerns of computer designers. Fault-tolerant computer systems are defined as systems capable of recovery from hardware or software failure to provide uninterrupted real-time service.

It is important to provide very high reliability to critical applications such as aircraft controller and nuclear reactor controller software systems. No matter how thorough the testing, debugging, modularization, and verification of software are, design bugs still plague the software. After reaching a certain level of software refinement, any effort to increase the reliability, even by a small margin, will increase exponential cost. Consider, for example, a fairly reliable software subjected to continuous testing and debugging, and guaranteed to have no more than 10 faults throughout the lifecycle. In order to improve the reliability such that, for example, only seven faults may be tolerated, the effort and cost to guarantee this may be enormous. A way of handling unpredictable software failure is through fault-tolerance. Over the last three decades, there has been considerable research in the area of fault-tolerant software.

Fault-tolerant software has been considered for use in a number of critical areas of application. For example, in traffic control systems, the Computer Aided Traffic Control System (COMTRAC) (Lala, 1985) is a fault-tolerant computer system designed to control the Japanese railways. It consists of three symmetrically interconnected computers. Two computers are synchronized at the program task level while the third acts as an active-standby. Each computer can be in one of the following states: on-line control, standby, or off-line. The COMTRAC software has a symmetric configuration. The configuration system contains the configuration control program and the dual monitor system contains the state control program. When one of the computers under dual operation has a fault, the state control program switches the system to single operation and reports the completion of the system switching to the configuration control program. The configuration control program commands the state control program to switchover to dual operation with the standby computer. The latter program then executes the system switchover, transferring control to the configuration control program, which judges the accuracy of the report and indicates its own state to the other computers. The COMTRAC shows that it failed seven times during a three-year period — once due to hardware failure, five times due to software failure, and once for unknown causes.

Another example is the NASA space shuttle. The shuttle carries a configuration of four identical flight computers, each loaded with the same software, and a fifth computer developed by a different manufacturer and running dissimilar (but functionally equivalent) software. This software is executed only if the ones in the other four processors cannot reach consensus during critical phases of the flight (Spector, 1984).

Software fault-tolerance is achieved through special programming techniques that enable the software to detect and recover from failure incidents. The method requires redundant software elements that provide alternative means of fulfilling the same specifications. The different versions must be such that they will not all fail in response to the same circumstances. Many researchers have investigated and suggested that diverse software versions developed using different specifications, designs, programming teams, programming languages, etc., might fail in a statistically independent manner. Empirical evidence questions that hypothesis (Leveson, 1990). On the other hand, almost all software fault-tolerance experiments have reported some degree of reliability improvement (Avizienis, 1988). Software fault-tolerance is the reliance on design redundancy to mask residual design faults present in software programs (Pham, 1992). Fault-tolerance, however, incurs costs due to the redundancy in hardware and software resources required to provide backup for system components. We must weigh the cost of fault-tolerance against the cost of software failure. With the current growth of software system complexity, we cannot afford to postpone the implementation of fault-tolerance in critical areas of software application.

In this chapter, we provide a basic concept for fault-tolerant software techniques and also discuss some advanced techniques including self-checking systems. We then give the reliability analysis of fault-tolerant software schemes such as recovery block (RB), N-version programming (NVP), and hybrid fault-tolerant systems. Basically, in the last system, an RB can be embedded within an NVP by applying the RB approach to each version of the NVP. Similarly, an NVP can be nested within an RB. The above structures of embedding can recursively be applied to form various multilevel hybrid fault-tolerant systems.

7.2   Basic Fault-Tolerant Software Techniques

The study of software fault-tolerance is relatively new as compared with the study of fault-tolerant hardware. In general, fault-tolerant approaches can be classified into fault-removal and fault-masking approaches. Fault-removal techniques can be either forward error recovery or backward error recovery. Forward error recovery aims to identify the error and, based on this knowledge, correct the system state containing the error. Exception handling in high-level languages, such as Ada and PL/1, provides a system structure that supports forward recovery. Backward error recovery corrects the system state by restoring the system to a state which occurred prior to the manifestation of the fault. The recovery block (RB) scheme provides such a system structure. Another fault-tolerant software technique commonly used is error masking. The N-version programming (NVP) scheme uses several independently developed versions of an algorithm. A final voting system is applied to the results of these N-versions and a correct result is generated.

Fault-tolerant software assures system reliability by using protective redundancy at the software level. There are two basic techniques for obtaining fault-tolerant software:

- RB scheme
- NVP.

Both schemes are based on software redundancy assuming that the events of coincidental software failures are rare.

7.2.1 Recovery Block Scheme

The recovery block scheme, proposed by Randell (1975), consists of three elements: primary module, acceptance tests, and alternate modules for a given task. The simplest scheme of the recovery block is as follows:

$$\text{Ensure } T$$
$$\text{By } P$$
$$\text{Else by } Q_1$$
$$\text{Else by } Q_2$$
$$\cdot$$
$$\cdot$$
$$\cdot$$
$$\text{Else by } Q_{n-1}$$
$$\text{Else Error}$$

where T is an acceptance test condition that is expected to be met by successful execution of either the primary module P or the alternate modules $Q_1, Q_2, \ldots, Q_{n-1}$. The process begins when the output of the primary module is tested for acceptability. If the acceptance test determines that the output of the primary module is not acceptable, it recovers or rolls back the state of the system before the primary module is executed. It allows the second module Q_1 to execute. The acceptance test is repeated to check the successful execution of module Q_1. If it fails, then module Q_2 is executed, etc. The alternate modules are identified by the keywords "else by". When all alternate modules are exhausted, the recovery block itself is considered to have failed and the final keywords "else error" declares the fact. In other words, when all modules execute and none produce acceptable outputs, then the system fails.

A reliability optimization model has been studied by Pham (1989) to determine the optimal number of modules in a recovery block scheme that minimizes the total system cost given the reliability of the individual modules. Details of the model can be obtained in (Pham, 1989).

7.2.2 N-Version Programming

The NVP was proposed by Chen and Avizienis (1978) for providing fault-tolerance in software. In concept, the NVP scheme is similar to the N-modular redundancy scheme used to provide tolerance against hardware faults (Lala, 1985).

The NVP is defined as the independent generation of $N \geq 2$ functionally equivalent programs, called *versions*, from the same initial specification. *Independent generation of programs* means that the programming efforts are carried out by N individuals or groups that do not interact with respect to the programming process. Whenever possible, different algorithms, techniques, programming languages, environments, and tools are used in each effort. In this technique, N program versions are executed in parallel on identical input and the results are obtained by voting on the outputs from the individual programs. The advantage of NVP is that when a version failure occurs, no additional time is required for reconfiguring the system and redoing the computation.

The main difference between the recovery block scheme and the N-version programming is that the modules are executed sequentially in the former. The recovery block generally is not applicable to critical systems where real-time response is of great concern. The N-version programming and the recovery block techniques have been discussed in detail by Anderson and Lee (1980).

7.2.3 Other Advanced Techniques

N-version programming has been researched thoroughly during the past decade. Correlated errors form a main source of failure of the N-version programs (Nicola, 1990) and can be minimized by design diversity. A design paradigm has been developed to assure design diversity in N-version software. Several experiments (see Lyu, 1991; Leveson, 1990) have been conducted to validate the assumption of error independence in multiple versions, to analyze the types of faults, to investigate the use of self-checks and voting in error detection, and to establish the need for a complete and unambiguous specification. There has been some effort on modeling the reliability of such fault-tolerant software approaches (Arlat, 1990; Belli, 1990; Tso, 1986; Vouk, 1990). Pham (1995) has given a cost model to obtain the optimal number of program versions that minimizes the expected cost of the NVP scheme.

However, in critical systems with real-time deadlines, voting at the end of the program, as in the basic N-version programming, may not be acceptable. Therefore, voting at intermediate points is called for. Such a scheme, where the comparison of results is done at intermediate points, is called the community error recovery (CER) scheme (Nicola, 1990) and is shown to offer a higher degree of fault-tolerance compared to the basic N-version programming. This approach, however, requires the synchronization of the various versions of the software at the comparison points.

Another scheme which adopts intermediate voting is the N-program, self-checking scheme (Yau, 1975) where each version is subject to an acceptance test or checking by comparison. When $N = 2$, it is a two-version, self-checking scheme or self-checking duplex scheme (see Section 7.3). Whenever a particular version raises an exception, the correct result is obtained from the remaining versions and execution is continued. This method is similar to the CER approach, with the only difference being the on-line detection in the former by an acceptance test rather than a comparison.

7.3   Self-Checking Duplex Scheme

In this section, we discuss an approach, called a self-checking duplex scheme, for the enhancement of software reliability. This scheme incorporates redundancy at two levels and can increase the reliability of software in critical systems significantly.

If individual versions are made highly reliable, an ultra-high reliability can be achieved merely by having two versions. These two versions should be made self-checking and work simultaneously as a duplex system as shown in Fig. 7.1. For example, if module i raises an exception, correct results can be obtained from the other version. We call this new approach a self-checking duplex system. Figure 7.1 shows a simple architecture of the system scheme where both versions are represented by a sequence of N_s modules.

Although N self-checking versions can be used, Pham (1991) reported a preliminary study that two versions are sufficient to raise the reliability to acceptable levels using our new approach. It is also more practical to have as few self-checking versions as possible because of the high cost of developing N different self-checking versions.

Self-checking software can be developed in a various ways. Self-checking provides an on-line detection of errors and prevents the contamination of the software by not letting the errors manifest. Software integrity can be assured by testing for illegal branching,

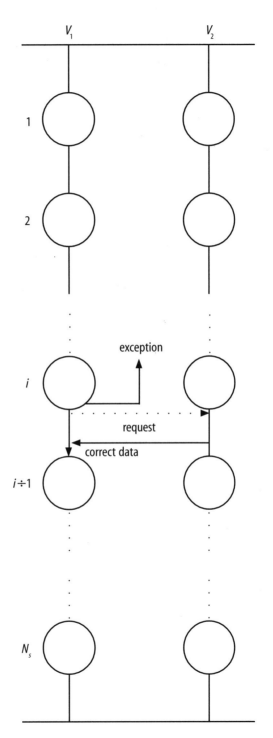

FIG. 7.1. Duplex system.

infinite loops, wrong branching, etc., and testing for functionality and validity of the results. It is easier to incorporate self-checking assertions into the software during the design stage since the team that develops the software is expected to have the best understanding of the problem. A good understanding of the application and the algorithms is deemed important for creating and placing meaningful assertions in the code. Both local and global self-checking assertions need to be incorporated to guarantee a high reliability. Hua and Abraham (1986) provide a systematic method for developing the self-checking assertions. In the following paragraphs, we show how self-checking assertions can provide ultra-high reliability.

Let the executable assertions be inserted in a module both locally and globally. By inserting local and global assertions, it is possible to check not only the internal states of the modules, but also the input/output specifications. As the inputs to an intermediate module such as $i + 1$ (see Fig. 7.1) are reset to the correct value by the corresponding module of the other version if and only if an error is detected, any undetected error at module i will propagate to the next module $i + 1$. Let p represent the probability of detecting an error in module i of a self-checking version. Now, given that an error goes undetected at the ith, $(i + 1)$th, ... , $(i + k - 1)$th module, the probability of this error being detected at the $i + 1$th module of the version is

$$p_k = p \sum_{j=i}^{i+k} (1 - p)^{j-i}.$$ (7.1)

Example 7.1 Suppose that the probability of detecting a design error at a particular module by self-checking is 0.9. If an error is not detected at this module, the probability of this error being detected at the following module, using Eq. (7.1), is 0.99 and the probability of detecting it at the next module is 0.999 and so on. This establishes that self-checking assertions could be a very powerful tool in increasing software reliability.

7.4 Reliability Modeling

A fundamental way of improving the reliability of software systems depends on the principle of design diversity where different versions of the functions are implemented. In order to prevent software failure caused by unpredicted conditions, different programs (alternative programs) are developed separately, preferably based on different programming logic, algorithm, computer language, etc. This diversity is normally applied under the form of recovery blocks or *N*-version programming.

7.4.1 Recovery Block

In a recovery block, a programming function is realized by n alternative programs, P_1, P_2, ... ,P_n. The computational result generated by each alternative program is checked by an acceptance test, T. If the result is rejected, another alternative program is then executed. The program will be repeated until an acceptable result is generated by one of the n alternatives or until all the alternative programs fail.

The probability of failure of the RB scheme, P_{rb}, is as follows:

$$P_{rb} = \prod_{i=1}^{n}(e_i + t_{2i}) + \sum_{i=1}^{n} t_{1i}e_i \left(\prod_{j=1}^{i-1}(e_j + t_{2j}) \right), \qquad (7.2)$$

where

e_i = probability of failure for version P_i

t_{1i} = probability that acceptance test i judges an incorrect result as correct

t_{2i} = probability that acceptance test i judges a correct result as incorrect.

The first term of Eq. (7.2) corresponds to the case when all versions fail the acceptance test. The second term corresponds to the probability that acceptance test i judges an incorrect result as correct at the ith trial of the n versions.

7.4.2 *N*-Version Programming

The NVP method consists of n programs and a voting mechanism, V. As opposed to the RB approach, all n alternative programs are usually executed simultaneously and their results are sent to a decision mechanism which selects the final result. The decision mechanism is normally a voter when there are more than two versions (or, more than k versions, in general), and it is a comparator when there are only two versions (k versions). The syntactic structure of NVP is as follows:

$$seq$$
$$par$$
$$P_1 \text{ (version 1)}$$
$$P_2 \text{ (version 2)}$$
$$\cdot$$
$$\cdot$$
$$\cdot$$
$$P_n \text{ (version } n)$$
$$\text{decision } V$$

Assume that a correct result is expected where there are at least two correct results. The probability of failure of the NVP scheme, P_{nv}, can be expressed as

$$P_{nv} = \prod_{i=1}^{n} e_i + \sum_{i=1}^{n}(1 - e_i)e_i^{-1}\prod_{j=1}^{n} e_j + d. \tag{7.3}$$

The first term of Eq. (7.3) is the probability that all versions fail. The second term is the probability that only one version is correct. The third term, d, is the probability that there are at least two correct results but the decision algorithm fails to deliver the correct result.

Eckhardt and Lee (1985) also developed a statistical model of NVP. In their model, "independently developed versions" are modeled as programs randomly selected from the input space of possible program versions that support problem-solving. Assume that the aggregate fails whenever at least m versions fail. Let $q(x)$ be the proportion of versions failing when executing on input state x. Let $Q(A)$ be the usage distribution, the probability that the subset of input state A is selected. Then the reliability of the NVP aggregate is given as

$$R = 1 - \int \sum_{i=m}^{N}\binom{N}{i}[q(x)]^i[1 - q(x)]^{N-i}dQ(A), \tag{7.4}$$

where $m = \lfloor 1(N+1)/2 \rfloor$, a majority of the N versions. Eckhardt and Lee also noted that independently developed versions do not necessarily fail independently. It is worthwhile to note that the goal of the NVP approach is to ensure that multiple versions will unlikely fail on the same inputs. With each version independently developed by a different programming team, design approach, etc., the goal is that the versions will be different enough in order that they will not fail too often on the same inputs. However, multiversion programming is still a controversial topic.

7.4.3 Hybrid Fault-Tolerant Scheme

A hybrid fault-tolerant system is defined as a software system which combines the RB and NVP schemes in order to improve the reliability of software systems. For simplicity, we only discuss two-level hybrid systems. We use the notation (NVP_n, RB_m) to represent an n-version NVP with each version being an m-version RB. Similarly, (RB_n, NVP_m) represents an n-version RB with each version being an m-version NVP. For example, two of the possible combinations of recovery block and NVP using four versions P_1, P_2, P_3, and P_4, are shown in Fig. 7.2.

Without loss of generality, we discuss here the hybrid fault-tolerant schemes with only two levels: RB embedded in NVP or NVP embedded in RB. Figure 7.3 shows the basic structure of a two-level hybrid fault-tolerant scheme. The first level consists of P_i basic program versions which form the second level composite program modules M_j where $1 \leq i \leq n$ and $1 \leq j \leq m$. If RB (or NVP) is used at the first level, NVP (or RB) is used at the second level. The composite version failure rates of the program version are ei, acceptance test error probabilities are t_1 and t_2, and the decision error probability is d. The hybrid fault-tolerant scheme's reliability can be obtained by calculating the reliability of the lower level program versions or composite versions, and then using the lower level reliabilities as inputs to the higher level composite versions. This process is repeated until the total system reliability is obtained. Mathematically, the probability of failure of the hybrid system, P_h, is calculated by using Eqs. (7.2) and (7.3) where the program version's failure rates e_i are substituted by $P_{rb}(i)$ or $P_{nv}(i)$. We now obtain

$$P_h(rb) = \prod_{i=1}^{n}(P_{nv}(i) + t_{2i}) + \sum_{i=1}^{n} t_{1i}P_{nv}(i)\left(\prod_{j=1}^{i-1}(P_{nv}(j) + t_{2j})\right) \quad (7.5)$$

for hybrid systems with a recovery block where each version is an NVP scheme and

$$P_h(nv) = \prod_{i=1}^{n} P_{rb}(i) + \sum_{i=1}^{n}(1 - P_{rb}(i)P_{rb}(i)^{-1}\prod_{j=1}^{n} P_{rb}(j) + d \quad (7.6)$$

for hybrid systems with an NVP where each version is a recovery block.

```
/* NVP */
seq
    par
        /* RB */
        ensure T
        by P1 (version 1)
        else by P2 (version 2)
        else error;
        /* RB */
        ensure T
        by P3 (version 3)
        else by P4 (version 4)
        else error;
```

FIG. 7.2(a) (NVP_2, RB_2) configuration.

```
/* RB */
ensure T
by /* NVP */
    seq
        par
            P1 (version 1);
            P2 (version 2);
    decision V
else by /* NVP */
    seq
        par
            P3 (version 3);
            P4 (version 4);
    decision V
```

FIG. 7.2(b) (RB_2, NVP_2) configuration.

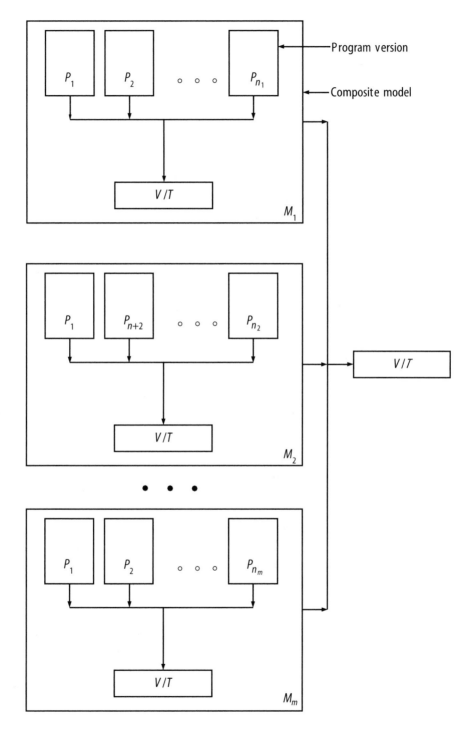

FIG. 7.3. A two-level hybrid fault-tolerant scheme.

7.5 Reduction of Common-Cause Failures

Before the *N*-version programming schemes can be applied to enhance the reliability of critical software, such as the nuclear reactor controller system and the fly-by-wire aircraft, their feasibility should be determined. Both the nuclear reactor controller and fly-by-wire software require ultra-high reliability. The existing *N*-version schemes may be unable to offer the required reliability because of their vulnerability to failures due to identical causes. If the majority of the versions fail because of common design errors, then a wrong result may be given by voting on incorrect outputs. The likelihood of common-cause failures in nuclear controller software cannot be ruled out because of its complexity.

Methods to alleviate the common-cause failures include the development of diverse versions by independent teams so as to minimize the commonalities between the various versions. According to Knight and Leveson (1986), experiments have shown that the use of different languages and design philosophy has little effect on the reliability in *N*-version programming because people tend to make similar logical mistakes in a difficult-to-program part of the software. Thus, in the presence of a common-cause failure, all the different variations of the *N*-version programming prove to be equally useless. It seems beneficial to have a single version in order to minimize cost.

According to the latest research on software reliability, fault-tolerance is a highly recommended application for critical systems. However, a new approach that could alleviate the weakness of the existing fault-tolerant software reliability models is even more desirable. An empirical study by Leveson *et al.* (1990) suggests not to utilize the *N*-version if it is known that the probability of making common mistakes during programming is unavoidable. Furthermore, Pham *et al.* (1991) recommend (1) developing fewer versions, (2) minimizing the errors in the individual versions, and (3) minimizing or eliminating the incidence of common-cause failures in these versions.

Clearly, a complex software is developed in a modular fashion and not all the modules are equally complex and difficult to design. Therefore, it is accurate to conclude that the common-cause failures are confined to the "difficult to logically understand and design" part of the problem. The common-cause failures can be reduced if such critical parts are identified and certain design guidelines are followed.

We suggest the following design guidelines to reduce or eliminate the common-cause failures:

- Techniques to identify critical parts in a program. Generally, the control flow complexity of an algorithm indicates the level of difficulty. We can therefore use the McCabe measure to identify the critical parts of a program.
- The manager of the project should identify the critical sections of the problem, meet the development teams, and steer them to different techniques for solving the critical parts. Suppose that the critical part involves sorting a file. Then one team should be asked to use Quicksort, the other teams should be asked not to use Quicksort but to use some other naïve scheme. In this way, the probability of committing identical logical mistakes can be reduced or eliminated.

Further research on the development of additional design guidelines to minimize the common-cause failures should also be studied.

7.6 Summary

The self-checking duplex scheme discussed in Section 7.3 incorporates fault-tolerance in two layers. The first layer of protection is provided by self-checking assertions. The second layer is duplication. In the self-checking system, if one of the versions detects an error at the end of the current module, results are obtained from the other version. After exchanging the correct results, both versions will continue execution in a lock-step fashion. Finally, the output of the duplicated versions are compared for consistency before accepting the result as correct. By keeping the size of the modules sufficiently small, a larger number of errors can be masked by this approach. However, too small a size for the module will increase the overhead of implanting the self-checking assertions. The analysis of the reliability and the optimal module size selection requires further research.

Common-cause failure is still a problem in the self-checking duplex system. There is no known technique to address this in N-version programming. Therefore, it can only be attempted to reduce the common-cause failures by design diversity. A guideline

TABLE **7.1.** Summary of references.

Group models	References
General fault tolerant systems	Abbott (1990), Anderson (1980, 1985), Arlat (1990), Avizienis (1984), Geist (1990), Hecht (1979), Iyer (1985), Kim (1989), Kljaich (1989), Laprie (1990), Leveson (1983), Pham (1989, 1992), Siewiorek (1990)
N-version programming	Anderson (1980, 1985), Avizienis (1982, 1984, 1988), Chen (1978), Gersting (1991), Kelly (1988), Knight (1986), Lyu (1991), Pham (1995), Shimeall (1991), Tso (1987), Vouk (1990)
Recovery block	Kim (1988), Laprie (1990), Pham (1989), Randell (1975)
Other fault-tolerant techniques	Eckhardt (1985), Hua (1986), Kim (1989), Nicola (1990), Taylor (1980), Upadhyaya (1986), Vouk (1990)

to reduce the probability of common-cause failures is discussed in Section 7.5. A summary of references on fault-tolerant systems is given in Table 7.1.

Problems

7.1 The cost of the fault-tolerant software for a new product (XZY) includes development and design, testing, implementation, and operation costs. In general, the reliability of the system can be increased by adding more redundant programs or modules. However, the extra cost and complexity may not justify the small gains in reliability. Let us consider the following problem.

Suppose $P_h(s)$ is the probability of failure for the hybrid scheme based on configuration s, then $[1 - P_h(s)]$ is the reliability of the fault-tolerant software. Let

C = the total amount of resources available

C_{ei} = the amount of resources needed for program version i

C_{vj} = the resources needed for voting version j

C_{tk} = the amount of resources needed for testing version k.

(a) Show that the reliability optimization model, given all the information above, can be formulated as follows:

Objective $\text{Max}_{s \in S}[1 - P_h(s)]$
Subject to

$$\sum_i C_{ei} + \sum_j C_{vj} + \sum_k C_{tk} \leq C$$

$$C_{ei} \geq 0, C_{vj} \geq 0, C_{tk} \geq 0$$

where S is a set of all the possible configurations of the hybrid scheme.

(b) Develop a heuristic algorithm to determine the optimal solution of the above optimization problem.

7.2 Continuing with Problem 7.1, assume that all programs and their testing versions have the same reliability and costs. Given that

Cost of a program version $C_e = \$15{,}000$
Cost of test $C_t = 85\%$ of $C_e = \$12{,}750$
Cost of voter $C_v = 10\%$ of $C_e = \$1{,}500$
Total amount or resources available $C = \$120{,}000$

Probability of program version failure $e = 0.05$
Probability of test failure $t = 0.02$
Probability of voter failure $d = 0.002$,

calculate the system reliability and cost of each of the following possible configurations:

(a) 7 RB
(b) 7 NVP
(c) 3 RB 2 NVP
(d) 2 NVP 3 RB
(e) 2 RB 3 NVP

References

Abbott, R.J., "Resourceful systems for fault tolerance, reliability, and safety," *ACM Computing Survey,* Vol. 22(1), March 1990.

Anderson, T., P.A. Barrett, D.N. Halliwell, and M.R. Moulding, "Software fault tolerance: An evaluation," *IEEE Trans. Software Engineering,* Vol. SE-11(12), 1985.

Anderson, T. and P.A. Lee, *Fault Tolerance: Principles and Practice*, Prentice-Hall, Englewood Cliffs, 1980.

Arlat, J., K. Kanoun, and J.C. Laprie, "Dependability modeling and evaluation of software fault tolerant systems," *IEEE Trans. Computers,* Vol. 39(4), April 1990.

Avizienis, A., M. Lyu, and W. Schutz, "In search of effective diversity: A six-language study of fault tolerant flight control software," in *Digest of 18th International Symposium on Fault Tolerant Computing,* Tokyo, Japan, 1988, IEEE Computer Society Press, Los Angeles, 1988.

Avizienis, A. and J.P.J. Kelly, "Fault tolerance by design diversity: Concepts and experiments," *IEEE Computers,* Vol. 17, August 1984.

Avizienis, A., "Design diversity — The challenge of the eighties," in *Proc. 12th Annual International Symposium Fault-Tolerant Computing,* Santa Monica, CA, June 1982.

Belli, F. and P. Jedrzejowics, "Fault-tolerant programs and their reliability," *IEEE Trans. Reliability,* Vol. R-39(2), 1990.

Chen, L. and A. Avizienis, "N-version programming: A fault tolerance approach to the reliability of software," in *Proc. 8th International Symposium on Fault-Tolerant Computing,* Toulouse, France, June 1978, IEEE Computer Society Press, Los Angeles, 1978.

Eckhardt, D.E., Jr., and L.D. Lee, "A theoretical basis for the analysis of multiversion software subject to coincident errors," *IEEE Trans. Software Engineering,* Vol. SE-11(12), 1985.

Geist, R. and K. Trivedi, "Reliability estimation of fault-tolerant systems: Tools and techniques," *IEEE Computers,* Vol. 23(7), July 1990.

Gersting, J.L., R.L. Nist, D.B. Roberts, and R.V. Valkenburg, "A comparison of voting Algorithms for N-version programming," in *Proc. 24th Annual Hawaii International Conference on System Sciences*, Vol. II, January 1991.

Hecht, H., "Fault-tolerant software," *IEEE Trans. Reliability*, Vol. R-28(3), 1979.

Hua, K.A. and J.A. Abraham, "Design of systems with concurrent error detection using software redundancy," in *Joint Fault Tolerant Computer Conference*, 1986, IEEE Computer Society Press, Los Angeles, 1986.

Iyer, R.K. and P. Velardi, "Hardware-related software errors: Measurement and analysis," *IEEE Trans. Software Engineering*, Vol. SE-11(2), 1985.

Kelly, J., D. Eckhardt, M. Vouk, D. McAllister, and A. Caglayan, "A large scale second generation experiment in multi-version software: Description and early results," in *Digest Papers, FTCS-18*, Tokyo, Japan, 1988, IEEE Computer Society Press, Los Angeles, 1988.

Kim, K.H., "An approach to experimental evaluation of real time fault tolerant distributed computing schemes," *IEEE Trans. Software Engineering*, Vol. SE-15(6), June 1989.

Kim, K.H. and H.O. Welch, "Distributed execution of recovery blocks: An approach for uniform treatment of hardware and software faults in real time applications," *IEEE Trans. Computers*, Vol. C-38(5), May 1989.

Kim, K. and J. Yoon, "Approaches to implementation of a reparable distributed recovery block scheme," in *Digest Papers, FTCS-18*, Tokyo, Japan, 1988, IEEE Computer Society Press, Los Angeles, 1988.

Kljaich, J., B.T. Smith, and A.S. Wojcik, "Formal verification of fault tolerance using theorem-proving techniques," *IEEE Trans. Computers*, Vol. C-38(3), March 1989.

Knight, J.C. and N.G. Leveson, "An experimental evaluation of the assumption of independence in multiversion programming," *IEEE Trans. Software Engineering*, Vol. SE-12, January 1986.

Lala, P.K., *Fault Tolerant & Fault Testable Hardware Design*, Prentice-Hall, London, 1985.

Laprie, J.C., J. Arlat, C. Beounes, and K. Kanoun, "Definition and analysis of hardware- and software-fault tolerant architectures," *IEEE Computers*, Vol. 23(7), July 1990.

Leveson, N.G., S.S. Cha, J.C. Knight, and T.J. Shimeall, "The use of self checks and voting in software error detection: An empirical study," *IEEE Trans. Software Engineering*, Vol. 16(4), April 1990.

Leveson, N.G. and P.R. Harvey, "Analyzing software safety," *IEEE Trans. Software Engineering*, Vol. SE-9(5), 1983.

Lyu, M.R. and A. Avizienis, "Assuring design diversity in *N*-version software: a design paradigm for *N*-version programming," in *Proc. Second International Working Conference on Dependable Computing for Critical Applications*, 1991, IEEE Computer Society Press, Los Angeles, 1991.

Nicola, V.F. and A. Goyal, "Modeling of correlated failures and community error recovery in multiversion software," *IEEE Trans. Software Engineering*, Vol. 16(3), 1990.

Pham, H., *Fault-Tolerant Software Systems: Techniques and Applications*, IEEE Computer Society Press, Los Angeles, 1992.

Pham, H., "On the optimal design of fault tolerant N-version programming software systems subject to constraints," *Systems and Software*, Vol. 27(1), 1994.

Pham, H. and M. Pham, "Software reliability models for critical applications," Idaho National Engineering Laboratory, EG&G2663, December 1991.

Pham, H. and S.J. Upadhyaya, "Reliability analysis of a class of fault tolerant systems," *IEEE Trans. Reliability*, Vol. 38(3), August 1989.

Randell, B., "System structure for software fault tolerance," *IEEE Trans. Software Engineering*, Vol. SE-1(2), 1975.

Shimeall, T.J. and N.G. Leveson, "An empirical comparison of software fault tolerance and fault elimination," *IEEE Trans. Software Engineering*, Vol. 17(2), February 1991.

Siewiorek, D.P., "Fault tolerance in commercial computers," *IEEE Computers*, Vol. 23(7), July 1990.

Spector, A. and D. Gifford, "The space shuttle primary computer system," *Commun. ACM*, Vol. 27(8), 1984, 874–900.

Taylor, D.J., "Redundancy in data structure: Improving software fault-tolerance," *IEEE Trans. Software Engineering*, Vol. 6(6), 1980.

Tso, K.S. and A. Avizienis, "Community error recovery in N-version software: A design study with experimentation," in *Proc. Fault Tolerant Computing Symposium, FTCS-17*, 1987, IEEE Computer Society Press, Los Angeles, 1987.

Tso, K.S., A. Avizienis, and J.P.J. Kelly, "Error recovery in multi-version software development," in *Proc. SAFECOMP '86*, Sarlat, France, October 1986, IEEE Computer Society Press, Los Angeles, 1986.

Upadhyaya, S.J. and K.K. Saluja, "A watchdog processor based general rollback technique with multiple retries," *IEEE Trans. Software Engineering*, Vol. SE-12, January 1986.

Vouk, M.A., A.M. Paradkar, and D.F. McAllister, "Modeling execution time of multi-stage N-version fault tolerant software," in *IEEE Computer Software and Applications Conference*, November 1990, IEEE Computer Society Press, Los Angeles, 1990.

Yau, S.S. and R.C. Cheung, "Design of self-checking software," in *1975 Reliable Software*, April 1975 , IEEE Press, New York, 1975.

8

Software Reliability Models with Environmental Factors

> *"The whole of science is nothing more*
> *than a refinement of everyday thinking."*
> *Albert Einstein (1879–1955)*

8.1 Introduction

The software reliability models which use testing time as the only influencing factor may not be appropriate for predicting the software reliability assessment. It is necessary to develop a software reliability model which incorporates the environmental factors during development. Several researchers (Takahashi and Kamayachi, 1985; Putnum, 1978) have indicated in their studies that many environmental factors, such as programmer skill, programming language, techniques, reuse of existing code, etc., have some influence on error characteristics.

Some of the suggested error environmental factors are briefly discussed in this chapter. They need to be considered and incorporated into the software reliability model to predict an accurate reliability measurement. We will then discuss several new software reliability models that incorporate environmental factors.

8.2 Definition of Environmental Factors

The environmental factors can be defined and measured as follows:

Difficulty of Programming (PDIF)
Difficulty of programming is defined as follows:

$$\text{PDIF} = \frac{k}{t^2} (\text{man years/year})^2,$$

where
k = amount of programming effort
t = amount of programming time.

Program Categories

The program categories that indicate the system complexity, for example, operating system, communication control program, data base system, etc.

Program Complexity

McCabe's $V(G)$, Halstead's E, and program size are well-known complexity measures. There is a significant relationship between these measures. However, it has been proven that none of these are significantly better than program size. For this reason, program size (kiloline of code (KLOC)) is used as a measure of program complexity. If the level of this factor is "high", it means that the program size is greater than 50 KLOC, otherwise, the level is "low".

Programmer's Skill

Programmer's Skill (PS) can be defined as the average number of years of programming experience. PS can be calculated as follows:

$$PS = \frac{\sum_{i=1}^{n} Y_i}{n},$$

where
Y_i = number of years of experience per programmer i
n = total number of programmers involved in the project.

Frequency of Program Specification and Requirements Change

The frequency of program specification change can be calculated by the total number of pages of problem reports generated to change the program design specifications and/or requirements during the software development.

Percentage of Reused Code (PORC)

When new software products are developed or when the old version of software products is updated, users usually keep some of the module of the code which can be reused, and add in some new ones. This is why the measure is important to include.

$$\mathrm{PORC} = \frac{S_0}{S_0 + S_N},$$

where

$S_0 =$ KLOC for the existing modules

$S_N =$ KLOC for new modules.

Programming Languages

Different programming languages have different complexity and structure. Thus, the possibility for different languages to introduce errors are different.

Design Methodologies

Different design methodologies for the same software may have a different impact on the quality of the final software products. There are two types of design methodologies: structural design and functional design.

Requirements Analysis

Requirements are provided by customers. Based on the requirements, software developers generate specifications. Usually, customers and developers meet to verify the requirements and achieve an understanding of the problem definitions. Requirement analysis is fundamental to the ensuing design and coding work.

Relationship of Detailed Design to Requirements

At the end of the design phase, detailed design is compared with requirements. Inspections are performed to verify whether the functions designed meet the requirements. Modifications can be made to eliminate the differences between the customers and developers.

Work Standard

Work standard is the norm that the developing team needs to follow. This could be company standard or group standard. Work standard indicates products to be made at each phase, design document format, design document description level and content, and the items to be checked in verifying the design documents.

Development Management

Development management includes all the organization and decision-making activities. From the specification phase to design, code, testing, and even operational phase, development managers schedule the meeting time, keep all participants in touch and track the development progress and work standards, give instructions, and make decisions.

Programmer Organization (ICON)
ICON is defined as the percentage of high-quality programmers and is computed as follows:

$$\text{ICON} = \frac{n_h}{n},$$

where
n_h = number of programmers whose programming experience is more than six years
n = number of programmers.

Development Team Size
Team size has an impact on the quality of software products. Some believe that a large team will improve the quality of the software since there are more people involved in the development process. However, others claim that a smaller, but experienced, development team is better.

Program Work Contents (Stress)
During software development, stress factors in terms of "work contents", such as schedule pressure and too much work, are the major factors. This includes the developer's mental stress and physical stress. Stress can be classified into several degree groups.

Domain Knowledge
Domain knowledge refers to the programmer's knowledge of the input space and output target. Insufficient knowledge may cause problems in coding and testing procedures.

Testing Environment
In order to find more errors during the testing phase, the testing environment should mimic the operational environment. This can be defined as the degree of the compatibility of testing and operational environments.

Testing Efforts
Testing effort can be defined as the number of testing cases generated, testing expenditure, or the number of human years that the testing takes.

Testing Methodologies
Different testing methodologies have different impacts on the quality of software products. Good testing methodologies may test more paths and need less time.

Testing Resource Allocation
Testing resource allocation refers to different schemes to allocate the testing resources in terms of testers, facilities, and schedules of the testing activities.

Testing Coverage (TCVG)
Test coverage is defined as the percentage of the source codes which are covered by the test cases. It can be expressed as

$$\text{TCVG} = \frac{S_c}{S_t},$$

where
S_c = KLOC which is covered by testing
S_t = total KLOCs.

Testing Tools
Many different testing tools exist. These include the software package testers utilize to carry out the testing tasks. Different tools also provide different quality and testing measures.

Programming Difficulty
Putnum (1978) proposed this concept and suggested that the level of difficulty of programming (PD) can be determined as follows:

$$\text{PD} = \frac{E}{t^2} (\text{man years/years})^2,$$

where
E = amount of programming effort
t = amount of programming time.

Programming Effort
The programming effort may be regarded as effective for reducing the number of errors.

Volume of Program Design Documents
Program design documents which lack sufficient content for the program produce errors. The volume of program design documents

can be calculated based on the number of pages of new and modified program design documents.

Levels of Programming Technologies (LPT)
The levels of programming technologies can be computed by using the rating scores as follows:

$$\text{LPT} = \sum_{i=1}^{n} R_i,$$

where
$n =$ number of categories
$R_i =$ rating score of the category i.

Given a sample data, four categories can be classified: design techniques, documentation techniques, programming techniques, and development computer access environment. Each of the categories may have a rating scale (low, medium, high, and extremely high) and the corresponding rating scores $R = 1, 5, 10,$ and 20, respectively.

Percentage of Reused and Modified Modules (PRMM)
The percentage of reused and modified modules (PRMM) can be computed as follows:

$$\text{PRMM} = \frac{\text{LOC}m}{\text{LOC}n}$$

where
$\text{LOC}m =$ line of code for modified modules
$\text{LOC}n =$ line of code for new modules.

Program Complexity (PC)
McCabe's measure, Halstead's measure, and program size are commonly used to determine the program complexity.

Documentation
Documentation includes all paperwork from specification to design, coding, and testing. This can serve as a resource for developers to allocate changes and problems, for programmers to review the codes, and for testers to examine the codes and detect bugs.

Mental Stress and Human Nature
Mental stress from deadlines or short development time causes imperfect survey, investigation, documentation, etc. Human nature

causes developers to skip some part of the requirement procedures because of their experience. A study showing the ratios of stress factor effect to human nature factor effect for each error category are as follows:

(1) imperfect investigation   6:1 (stress factors are dominant)
(2) imperfect documentation   4:2 (stress factors are relatively effective)
(3) imperfect survey   4:6 (stress factors are less effective)

8.3   Environmental Factors Analysis

The ranking and evaluation of environmental factors in the survey form in Appendix C are discussed in this section. The survey form consists of two parts. Part I is a survey on environmental factors. The purpose is to obtain the perceptions of software developers with regard to the accuracy of software reliability assessment in light of incorporating these environmental factors.

Part I data, collected using a formal survey questionnaire, was provided by the software developers or managers. Table 8.1 shows the results by the relative weight method. The ten most important environmental factors are classified as factors in the analysis phase (3 factors), coding (1 factor), testing (4 factors), and general (2 factors). The column "Normalized priorities" gives the contribution of each environmental factor. For example, program complexity factor contributes approximately 3.7%. A higher priority value indicates a higher ranking. The application of this finding in Table 8.1 is not to discard the environmental factors belonging to lower ranking classes. On the contrary, it will help software developers or managers prioritize their tasks. Based on the survey information, as seen from Fig. 8.1, the analysis, design, coding, and testing phases take, respectively, about 25%, 18%, 36%, and 21% of the development efforts. Analysis and design testing phases together take about 64% of the total development time. Figure 8.2 shows that the percentage of reusable code is about 43.6% and 41.2%, respectively. In general, the percentage of reusable code is about 40.8%.

Correlation analysis is also studied based on the survey information. The purpose is to find out the correlation of environmental factors and determine if they are independent or not. (If not, then which factors are related to each other?) Table 8.2 shows the correlation of the environmental factors (Pham, 1999).

The results of this study also indicate that it is desirable to incorporate environmental factors into the software reliability modeling. More research is needed for the development of software reliability engineering (Pham, 1999).

TABLE 8.1. Results ranking based on the relative weight method.

Rank	Rank factors	Factor name	Normalized priorities
1	f1	Program complexity	0.03767585
2	f15	Programmer skills	0.03693248
3	f25	Testing coverage	0.0367487
4	f22	Testing effort	0.03650041
5	f21	Testing environment	0.03532612
6	f8	Frequency of specification change	0.03483333
7	f24	Testing methodologies	0.03433004
8	f11	Requirements analysis	0.03416625
9	f6	Percentage of reused code	0.03369112
10	f12	Relationship of detailed design and requirement	0.03329922
11	f5	Level of programming technologies	0.03314973
12	f27	Documentation	0.03280882
13	f18	Program workload	0.03274672
14	f26	Testing tools	0.03226921
15	f16	Programmer organization	0.03210433
16	f19	Domain knowledge	0.03179907
17	f3	Difficulty of programming	0.03171322
18	f10	Design methodologies	0.0317048
19	f20	Human nature (mistake and omission)	0.03168718
20	f14	Development management	0.03165692
21	f23	Testing resource allocation	0.03096135
22	f4	Amount of programming effort	0.03072114
23	f2	Program categories	0.030576
24	f13	Work standards	0.02984902
25	f32	System software	0.02839442
26	f9	Volume of program design documents	0.02750213
27	f17	Development team size	0.02738358
28	f7	Programming language	0.02711005
29	f28	Processor	0.02414395
30	f31	Telecommunication device	0.02404204
31	f30	Input/Output device	0.02290799
32	f29	Storage device	0.02126481

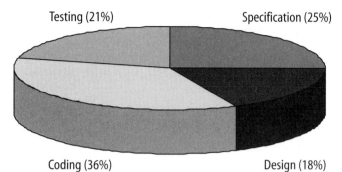

Testing (21%)   Specification (25%)

Coding (36%)   Design (18%)

FIG. 8.1. Percentage time in each development process phase.

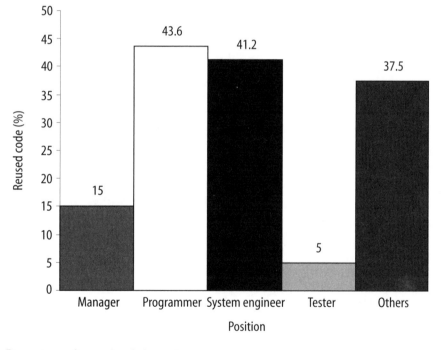

FIG. 8.2. Percentage of reused code by each group.

TABLE **8.2.** Correlation of the environmental factors.

Factor number	Name of factors	Correlated factors
f1	Program complexity	f17 Development team size
f15	Programmer skills	f3 Difficulty of programming
		f18 Program workload
f25	Testing coverage	f2 Program categories
		f24 Testing methodologies
f22	Testing effort	f5 Level of programming technologies
		f13 Work standards
		f14 Development management
		f21 Testing environment
f21	Testing environment	f5 Level of programming technologies
		f12 Relationship of detailed design and requirement
		f13 Work standards
		f22 Testing effort
f8	Frequency of specification change	
f24	Testing methodologies	f25 Testing coverage
		f26 Testing tools
f11	Requirements analysis	f12 Relationship of detailed design and requirement
		f30 Input/Output device
		f32 System software
f6	Percentage of reused code	f7 Programming language
		f9 Volume of program design documents
		f14 Development management
		f18 Program workload
		f23 Testing resource allocation
		f24 Testing methodologies
		f26 Testing tools
		f27 Documentation
f12	Relationship of detailed Design and requirement	f11 Requirements analysis
		f21 Testing environment
f5	Level of programming technologies	f21 Testing environment
		f22 Testing effort
f27	Documentation	f6 Percentage of reused code
		f26 Testing tools
		f31 Telecommunication device
f18	Program workload	f3 Difficulty of programming
		f4 Amount of programming effort
		f7 Programming language
		f13 Work standards
		f15 Programmer skills
		f23 Testing resource allocation
		f26 Testing tools
		f30 Input/Output device

TABLE **8.2.** *continued*

Factor number	Name of factors	Correlated factors
f 26	Testing tools	f 6 Percentage of reused code f 18 Program workload f 27 Documentation
f 16	Programmer organization	
f 19	Domain knowledge	
f 3	Difficulty of programming	f 6 Percentage of reused code f 15 Programmer skills f 18 Program workload
f 10	Design methodologies	f 2 Program categories
f 20	Human nature (mistake and omission)	
f 14	Development management	f 22 Testing effort
f 23	Testing resource allocation	f 6 Percentage of reused code f 24 Testing methodologies
f 4	Amount of programming effort	f 6 Percentage of reused code f 7 Programming language f 18 Program workload
f 2	Program categories	f 10 Design methodologies f 25 Testing coverage
f 13	Work standards	f 14 Development management f 18 Program workload f 21 Testing environment f 22 Testing effort
f 32	System software	f 11 Requirements analysis f 28 Processor f 30 Input/Output device f 31 Telecommunication device
f 9	Volume of program design documents	f 6 Percentage of reused code
f 17	Development team size	f 1 Program complexity f 18 Program workload
f 7	Programming language	f 4 Amount of programming effort f 6 Percentage of reused code f 18 Program workload
f 28	Processor	f 29 Storage device f 31 Telecommunication device
f 31	Telecommunication device	f 27 Documentation f 28 Processor f 32 System software
f 30	Input/Output device	f 28 Processor
f 29	Storage device	f 30 Input/Output device

8.4 A Generalized Model with Environmental Factors

In this section, we discuss several newly developed software reliability models that consider environmental factors by combining the proportional hazard model (Cox, 1975) and existing software reliability models (Pham, 1999). Such factors are, e.g., the complexity metrics of the software, the development and environmental conditions, the effect of mental stress and human nature, the level of the test-team members, and the facility level during testing. The proportional hazard model has been widely used in medical applications to estimate the survival rate of patients.

Based on the proportional hazard (PH) model, let us consider the failure intensity function of a software system as the product of an unspecified baseline failure intensity $\lambda_0(t)$, a function that only depends on time, and an exponential function term incorporating the effects of a number of environmental factors. The basic assumption is that the ratio of the failure intensity functions of any two errors observed at any time t associated with any environmental factor sets z_{1i} and z_{2i} is a constant with respect to time and they are proportional to each other. In other words, $(t_i; z_{1i})$ is directly proportional to $(t_i; z_{2i})$.

A generalized failure intensity function of the software reliability model that considers environmental factors can be written as

$$\lambda(t_i; z_i) = \lambda_0(t_i)e^{\left(\sum_{j=1}^{m} \beta_j z_{ji}\right)}, \tag{8.1}$$

where
z_{ji} = environmental factor j of the ith error
β_j = regression coefficient of the jth factor
t_i = failure time between the $(i-1)$th error and ith error, $i = 1, 2, \ldots, n$
z_i = environmental factor of the ith error
m = number of environmental factors.

It is easy to see that $\lambda_0(t)$ is a baseline failure intensity function that represents the failure intensity when all environmental factors variables are set to zero.

Let Z be a column vector consisting of the environmental factors and B represents a row vector consisting of the corresponding regression parameters. Then the above failure intensity model can be rewritten as

$$\lambda(t; Z) = \lambda_0(t)e^{(BZ)}. \tag{8.2}$$

Therefore, the reliability of the software systems can be written, in a general form, as follows:

$$R(t; Z) = e^{-\int_0^t \lambda_0(s)e^{BZ}ds}$$

$$= \left[e^{-\int_0^t \lambda_0(s)ds} \right]^{e^{(BZ)}}$$ (8.3)

$$= [R_0(t)]^{e^{(BZ)}},$$

where $R_0(t)$ is the time-dependent software reliability.

The pdf of the software system is given by

$$f(t; Z) = \lambda(t; Z) \cdot R(t; Z)$$

$$= \lambda_0(t)e^{BZ}[R_0(t)]^{e^{BZ}}.$$ (8.4)

The regression coefficient B can be estimated, using either the maximum likelihood estimation (MLE) method (see Appendix A) or the maximum partial likelihood approach, which is discussed later, without assuming any specific distributions about the failure data and estimating the baseline failure intensity function. A direct generalization of the above model in Eq. (8.1) is that one may want to consider the environmental factor variables Z_{ji} as a function of time. In this case, a mathematical generalized form of the failure intensity function is given by

$$\lambda(t_i; z_i) = \lambda_0(t_i)e^{\left[\sum_{j=1}^{m} \beta_j z_{ji}(t) \right]}.$$

8.4.1 Environmental Factors Estimation Using Maximum Likelihood Estimation

Assume that there are p unknown parameters in the baseline failure intensity function $\lambda_0(t)$, say $\alpha_1, \alpha_2, \ldots, \alpha_p$, and there are m environmental factors $\beta_1, \beta_2, \ldots, \beta_m$. Let $A = (\alpha_1, \alpha_2, \ldots, \alpha_p)$ be a set of unknown parameters $\alpha_2, \ldots, \alpha_p$, and let B be a set of $\beta_1, \beta_2, \ldots, \beta_m$. Then the likelihood function is given by

$$L(A, B) = \prod_{i=1}^{n} f(t_i; z_i)$$

$$= \prod_{i=1}^{n} \left(\lambda_0(t_i)e^{\left(\sum_{j=1}^{m} \beta_j z_{ji} \right)} [R_0(t_i)]^{e^{\left(\sum_{j=1}^{m} \beta_j z_{ji} \right)}} \right).$$ (8.5)

The log likelihood function is given by

$$\ln L(A, B) = \sum_{i=1}^{n} \ln[\lambda_0(t_i)] + \sum_{i=1}^{n}\sum_{j=1}^{m} \beta_j z_{ji} + \sum_{i=1}^{n} e^{\left(\sum\limits_{j=1}^{m} \beta_j z_{ji}\right)} \ln[R_0(t_i)].$$

Taking the first partial derivatives of the log likelihood function with respect to $(m + p)$ parameters, we obtain

$$\frac{\partial}{\partial \alpha_k} \ln L(A, B)] = \sum_{i=1}^{n} \frac{\frac{\partial}{\partial \alpha_k}[\lambda_0(t_i)]}{\lambda_0(t_i)} + \sum_{i=1}^{n} e^{\left(\sum\limits_{j=1}^{m} \beta_j z_{ji}\right)} \frac{\frac{\partial}{\partial \alpha_k}[R_0(t_i)]}{R_0(t_i)}$$

$$\frac{\partial}{\partial \beta_s} \ln L(A, B) = \sum_{i=1}^{n} z_{si} + \sum_{i=1}^{n} z_{si} e^{\left(\sum\limits_{j=1}^{m} \beta_j z_{ji}\right)} \ln[R_0(t_i)],$$

where $k = 1, 2, \ldots, p$ and $s = 1, 2, \ldots, m$.

Setting the previous equations equal to zero, we can obtain all the $(m + p)$ parameters by solving the following system of $(m + p)$ equations simultaneously:

$$\sum_{i=1}^{n} \left[\frac{\frac{\partial}{\partial \alpha_k}[\lambda_0(t_i)]}{\lambda_0(t_i)} + e^{\left(\sum\limits_{j=1}^{m} \beta_j z_{ji}\right)} \frac{\frac{\partial}{\partial \alpha_k}[R_0(t_i)]}{R_0(t_i)} \right] = 0 \quad \text{for } k = 1, 2, \ldots, p$$

$$\sum_{i=1}^{n} z_{si} \left[1 + e^{\left(\sum\limits_{j=1}^{m} \beta_j z_{ji}\right)} \ln[R_0(t_i)] \right] = 0 \quad \text{for } s = 1, 2, \ldots, m.$$

8.4.2 Environmental Factors Estimation Using Maximum Partial Likelihood Approach

According to the idea of Cox's proportional hazard model, we can use the maximum partial likelihood method to estimate environmental factors without assuming any specific distributions about the failure data and estimating the baseline failure intensity function. The only basic assumption of this model is that the ratio of the failure intensity functions of any two errors observed at any time t associated with any environmental factor sets z_{1i} and z_{2i} is constant with respect to time and they are proportional to each other.

First we estimate the environmental factor parameters based on the partial likelihood function. The partial likelihood function of this model is given by

$$L(B) = \prod_{i=1}^{n} \frac{e^{(\beta_1 z_{1i} + \beta_2 z_{2i} + \cdots + \beta_m z_{mi})}}{\sum_{k \in R_i} e^{(\beta_1 z_{1k} + \beta_2 z_{2k} + \cdots + \beta_l z_{lk})}}, \tag{8.6}$$

where R_i is the risk set at t_i. Take the derivatives of the log partial likelihood function with respect to $\beta_1, \beta_2, \ldots, \beta_m$ and let them equal to zero. Therefore, we can obtain all of the estimated βs by solving these equations simultaneously using numerical methods. After estimating the factor parameters $\beta_1, \beta_2, \ldots, \beta_m$, the remaining task is to estimate the unknown parameters of the baseline failure intensity function $\lambda_0(t)$.

8.5   Enhanced Proportional Hazard Jelinski–Moranda (EPJM) Model

Recall that the Jelinski–Moranda (J–M) model is one of the earliest models developed for predicting software reliability (see Chapter 4). The failure intensity of the software at the ith failure interval of this model is given by

$$\lambda(t_i) = \phi[N - (i - 1)] \quad i = 1, 2, \ldots, N$$

and the probability density function is given by

$$f(t_i) = \phi[N - (i - 1)]e^{-\phi[N-(i-1)]t_i}.$$

The enhanced proportional hazard J–M model (Pham, 1999), called the EPJM model, which is based on the proportional hazard and J–M model, is expressed as

$$\lambda(t_i; z_i) = \phi[N - (i - 1)]e^{\left(\sum_{j=1}^{m} \beta_t z_{ji}\right)}$$

and the pdf corresponding to (t_i, z_i) is given by

$$f(t_i; z_i) = \phi[N - (i - 1)]e^{\left(\sum_{j=1}^{m} \beta_t z_{ji}\right)} e^{\left[-\phi[N-(i-1)]t_i e^{\left(\sum_{j=1}^{m} \beta_t z_{ji}\right)}\right]}. \tag{8.7}$$

Now we wish to estimate the parameters of the EPJM model using the two methods discussed in Section 4, the maximum likelihood method and the maximum partial likelihood method. There are $(m + 2)$ unknown parameters in this model.

The Maximum Likelihood Method

From Eq. (8.7), the likelihood function of the model is given by

$$L(B, N, \phi) = \prod_{i=1}^{n} f(t_i; z_i)$$

$$= \prod_{i=1}^{n} (\phi[N - (i-1)]e^{\left(\sum_{j=1}^{m} \beta_t z_{ji}\right)} e^{\{-\phi[N-(i-1)]t_i e^{\left(\sum_{j=1}^{m} \beta_t z_{ji}\right)}\}}).$$

The log likelihood function is given by

$$\ln L(B, N, \phi) = n \ln \phi + \sum_{i=1}^{n} \ln[N - (i-1)] + \sum_{i=1}^{n} \left(\sum_{j=1}^{m} \beta_j z_{ji}\right)$$

$$- \sum_{i=1}^{n} \phi[N - (i-1)]t_i e^{\sum_{j=1}^{m}(\beta_j z_{ji})}.$$

Taking the first partial derivatives of the log likelihood function with respect to $(m + 2)$ parameter: $\beta_1, \beta_2, \ldots, \beta_m, N$, and Φ, we obtain the following:

$$\frac{\partial \log L}{\partial \phi} = \frac{n}{\phi} - \sum_{i=1}^{n}[N - (i-1)]t_i e^{\sum_{j=1}^{m}(\beta_j z_{ji})}$$

$$\frac{\partial \log L}{\partial N} = \sum_{i=1}^{n} \frac{1}{[N - (i-1)]} - \phi \sum_{i=1}^{n} t_i e^{\sum_{j=1}^{m}(\beta_j z_{ji})}$$

and

$$\frac{\partial \ln L}{\partial \beta_j} = \sum_{i=1}^{n} z_{ji} - \sum_{i=1}^{n} \phi[N - (i-1)]t_i z_{ji} e^{\sum_{j=1}^{m}(\beta_j z_{ji})}.$$

Setting all of these equations equal to zero, we can obtain the estimated $(m + 2)$ parameters by solving the following system equations simultaneously using a numerical method:

$$\sum_{i=1}^{n}[N - (i-1)]t_i e^{\sum_{j=1}^{m}(\beta_j z_{ji})} = \frac{n}{\phi}$$

$$\sum_{i=1}^{n} \frac{1}{[N - (i-1)]} = \phi \sum_{i=1}^{n} t_i e^{\sum_{j=1}^{m}(\beta_j z_{ji})}$$

(8.8)

$$\sum_{i=1}^{n} \phi[N - (i-1)]t_i z_{ji} e^{\sum_{j=1}^{m}(\beta_j z_{ji})} = \sum_{i=1}^{n} z_{ji} \quad \text{for } j = 1, 2, \ldots, m.$$

The Maximum Partial Likelihood Method

Assume that the baseline failure intensity has the form of the J–M model. That means that the basic assumption of this model (see Section 4) is satisfied and that the ratio of the failure intensity functions of any two errors observed at any time t, associated with any environmental factor sets z_{1i} and z_{2i}, is a constant with respect to time and they are proportional to each other.

Having estimated the factor parameters $\beta_1, \beta_2, \ldots, \beta_m$, the remaining task is to estimate the unknown parameters of the baseline failure intensity function. Note that the failure intensity function model has the form

$$\lambda(t_i; z_i) = \phi[N - (i-1)]e^{\hat{\beta}_1 z_{1i} + \hat{\beta}_2 z_{2i} + \cdots + \hat{\beta}_m z_{mi})}$$
$$= \phi[N - (i-1)]E_i,$$

where

$$E_i = e^{\hat{\beta}_1 z_{1i} + \hat{\beta}_2 z_{2i} + \cdots + \hat{\beta}_m z_{mi})}.$$

The pdf is given by

$$f(t_i; z_i) = \phi E_i[N - (i-1)]e^{-(\phi E_i[N-(i-1)]t_i)}.$$

The likelihood function is given by

$$L(N, \phi) = \prod_{i=1}^{n}(\phi E_i[N - (i-1)]e^{-(\phi E_i[N-(i-1)]t_i)}).$$

By taking the log of the likelihood function and its derivatives with respect to N and ϕ, and setting them equal to zero, we obtain the following equations:

$$\frac{\partial \ln L}{\partial N} = \sum_{i=1}^{n} \frac{1}{N - (i-1)} - \sum_{i=1}^{n} \phi E_i t_i = 0$$

and

$$\frac{\partial \ln L}{\partial \phi} = \frac{n}{\phi} - \sum_{i=1}^{n} E_i[N - (i-1)]t_i = 0.$$

The estimated N and ϕ can be obtained as follows. First, the parameter N can be obtained by solving the following equation:

$$\left(\sum_{i=1}^{n} E_i[N-(i-1)]t_i\right)\left(\sum_{i=1}^{n}\frac{1}{[N-(i-1)]}\right)=n\sum_{i=1}^{n}E_it_i. \qquad (8.9)$$

After finding N, the parameter can be easily obtained and is given by

$$\phi=\frac{\displaystyle\sum_{i=1}^{n}\frac{1}{[N-(i-1)]}}{\displaystyle\sum_{i=1}^{n}E_it_i}. \qquad (8.10)$$

8.6   An Application with Environmental Factors

Almost all software reliability engineering models need one of two basic types of input data: time-domain data and interval-domain data. One can possibly transform between the two types of data domains. The time-domain approach is characterized by recording the individual times at which the failure occurred. The interval-domain approach is characterized by counting the number of failures that occurred over a given period.

There is, however, no records of corresponding environmental factor measures in most, if not all, existing available data. To illustrate the EPJM model, we use the software failure data reported by Musa (1975) and also referred to data set #3 in Chapter 5. The data is related to a real-time command and control system. To demonstrate the use of this model, we generate a failure-cluster factor and give its value which is logically realistic based on the failure data and consultation with several local software firms by the author.

One of the assumptions of the J–M model is that the time between failures is independent. As in many real testing environments, the failure times indeed occur in a cluster, i.e., the failure time within a cluster is relatively shorter than that between the clusters. Data set #3 shows that it is reasonable in that particular application. This may indicate that the assumption of independent failure time is not correct. We can enhance the J–M model considering the failure-cluster factor by generating this factor based on the failure data.

We assume that if the present failure time, compared to the previous failure time, is relatively short, then some correlation may exist between them. Let us define a failure-cluster factor, such as

$$z_i = \begin{cases} 1 & \text{when} \quad \frac{t_{i-1}}{t_i} \geq 7 \quad \text{or} \quad \frac{t_{i-2}}{t_i} \geq 5 \\ 0 & \text{otherwise.} \end{cases}$$

The data used in this model include both the failure time data and the explanatory environmental factor data (see Table 8.3). The

TABLE 8.3. Musa's failure time data with a generated covariate.

Num	time	z	Num	time	z	Num	time	z	Num	time	z	Num	time	z
1	3	0	31	36	1	61	0	1	91	724	0	121	75	1
2	30	0	32	4	1	62	232	0	92	2,323	0	122	482	0
3	113	0	33	0	1	63	330	0	93	2,930	0	123	5,509	0
4	81	0	34	8	0	64	365	0	94	1,461	0	124	100	1
5	115	0	35	227	0	65	1,222	0	95	843	0	125	10	1
6	9	1	36	65	0	66	543	0	96	12	1	126	1,071	0
7	2	1	37	176	0	67	10	1	97	261	0	127	371	0
8	91	0	38	58	0	68	16	1	98	1,800	0	128	790	0
9	112	0	39	457	0	69	529	0	99	865	0	129	6,150	0
10	15	1	40	300	0	70	379	0	100	1,435	0	130	3,321	0
11	138	0	41	97	0	71	44	1	101	30	1	131	1,045	1
12	50	0	42	263	0	72	129	0	102	143	1	132	648	1
13	77	0	43	452	0	73	810	0	103	108	0	133	5,485	0
14	24	0	44	255	0	74	290	0	104	0	1	134	1,160	0
15	108	0	45	197	0	75	300	0	105	3,110	0	135	1,864	0
16	88	0	46	193	0	76	529	0	106	1,247	0	136	4,116	0
17	670	0	47	6	1	77	281	0	107	943	0			
18	120	0	48	79	0	78	160	0	108	700	0			
19	26	1	49	816	0	79	828	0	109	875	0			
20	114	0	50	1,351	0	80	1,011	0	110	245	0			
21	325	0	51	148	1	81	445	0	111	729	0			
22	55	0	52	21	1	82	296	0	112	1,897	0			
23	242	0	53	233	0	83	1,755	0	113	447	0			
24	68	0	54	134	0	84	1,064	0	114	386	0			
25	422	0	55	357	0	85	1,783	0	115	446	0			
26	180	0	56	193	0	86	860	0	116	122	0			
27	10	1	57	236	0	87	983	0	117	990	0			
28	1,146	0	58	31	1	88	707	0	118	948	0			
29	600	0	59	369	0	89	33	1	119	1,082	0			
30	15	1	60	748	0	90	868	0	120	22	1			

explanatory variable data is dynamic, that is, it changes depending on the failure time. For example, in Table 8.3, the time between the fourth and fifth errors is 115 seconds; the time between the fifth and sixth errors is nine seconds. Therefore, z_5 is assigned to 0 and z_6 is equal to 1.

For the J–M model, using the MLE, we obtain the estimate of the two parameters, N and ϕ, as follows:

$$\hat{N} = 142$$
$$\hat{\phi} = (3.48893)10^{-5}.$$

Therefore, the current reliability of the software system is given by

$$R(t_{137}) = e^{-\hat{\phi}[\hat{N}-(137-1)]t_{137}}.$$

Now, we want to predict the future failure behavior using only data collected in the past after 136 errors have been found. For example, the reliability of the software for the next 100 seconds after 136 errors are detected is given by

$$R(t_{137} = 100) = e^{-\hat{\phi}[\hat{N}-(137-1)]t_{137}}$$
$$= e^{-(0.0000348893)[142-136](100)}$$
$$= 0.979284.$$

Similarly, the reliability of the software for the next 1,000 seconds is given by

$$R(t_{137} = 1000) = e^{-(0.0000348893)[142-136](1,000)}$$
$$= 0.811123.$$

Assume that we use the partial likelihood approach to estimate the environmental factor parameter for the EPJM model. As there is only one factor in this example, we can easily obtain the estimated parameter using the statistical software package SAS:

$$\hat{\beta}_1 = 1.767109$$

with a significance level of 0.0001. Then the estimates of N and ϕ are given as follows:

$$\hat{N} = 141$$
$$\hat{\phi} = (3.28246)10^{-5}.$$

Therefore,

$$E_i = e^{\beta_1 z_{1i}} = \begin{cases} 5.853905235 & \text{for } z = 1 \\ 1 & \text{for } z = 0. \end{cases}$$

The current reliability of the software system is given by

$$R(t_{137}) = e^{-\hat{\phi}E_{137}[\hat{N}-(137-1)]t_{137}}$$

$$= \begin{cases} e^{-9.6076048.10^{-4}t_{137}} & \text{for } z = 1 \\ e^{-1.64123.10^{-4}t_{137}} & \text{for } z = 0. \end{cases}$$

Assuming that

$$P(Z = 1) = \frac{28}{136} = 0.20588$$

$$P(Z = 0) = \frac{108}{136} = 0.79412.$$

The reliability of the software for the next 100 seconds is given by

$$R(t_{137} = 100) = \begin{matrix} 0.908394931 & \text{for } z = 1 \text{ with probability} = 0.20588 \\ 0.983721648 & \text{for } z = 0 \text{ with probability} = 0.79412 \end{matrix}$$

or, equivalently, that

$$R(t_{137} = 100) = 0.95375.$$

Similarly, the reliability of the software for the next 1,000 seconds is given by

$$R(t_{137} = 1,000) = \begin{cases} 0.382601814 & \text{for } z = 1 \text{ with probability} = 0.20588 \\ 0.848637633 & \text{for } z = 0 \text{ with probability} = 0.79412 \end{cases}$$

or

$$R(t_{137} = 1,000) = 0.74021.$$

Problems

8.1 Using the real-time control system as in Table 5.7, calculate the MLE for unknown parameters of the EPJM model discussed in Section 8.5.

8.2 Based on the first 60 days in Table 5.7, calculate the MLE for unknown parameters of the EPJM model.

References

Furuyama, T., Y. Arai, and K. Iio, "Fault generation model and mental stress effect analysis," in *Proc. 2nd International Conference Achieving Quality in Software*, Venice, 18–20 October , 1993, IEEE Computer Society Press, Los Angeles, 1993.

Halstead, M.H., *Elements of Software Science*, Elsevier North-Holland, New York, 1977.

McCabe, T.J., "A complexity measure," *IEEE Trans. Software Engineering*, Vol. SE-2(4), December 1976, 308–320.

Musa, J.D., "A theory of software reliability and its applications," *IEEE Trans. Software Engineering*, Vol. SE-1(3), 1975, 312–327.

Pham, H., *Software Reliability*, J.C. Webster (ed.), Wiley Encyclopedia of Electrical and Electronics Engineering, John Wiley & Sons, New York, 1999.

Putnum, L.H., "A general empirical solution to the macro software sizing and estimation problem," *IEEE Trans. Software Eng.*, Vol. SE-4(4), July 1978, 345–361.

Takahashi, M. and Y. Kamayachi, "An empirical study of a model for program error prediction," in *Proc. 8th International IEEE Conference on Software Engineering*, August 1985, IEEE Computer Society Press, Los Angeles, 1985, pp. 330–336.

Appendix A

Theory of Estimation

A.1 Point Estimation

The problem of point estimation is that of estimating the parameters of a population, e.g., λ or θ from an exponential, μ and σ^2 from a normal, etc. It is assumed that the population distribution by type is known, but the distribution parameters are unknown and they have to be estimated by using collected failure data. This appendix is devoted to the theory of estimation and discusses several common estimation techniques such as maximum likelihood, method of moments, least squared, and Baysian methods. We also discuss the confidence interval estimates and tolerance limit estimates. For example, assume that n independent samples from the exponential density $f(x;\lambda) = \lambda e^{-\lambda x}$ for $x > 0$ and $\lambda > 0$, then the joint probability density function (pdf) or sample density (for short) is given by

$$f(x_1, \lambda) \cdot f(x_2, \lambda) \cdots f(x_n, \lambda) = \lambda^n e^{-\lambda \sum_{i-1}^{n} x_i}.$$

The problem here is to find a "good" point estimate of λ which is denoted by $\hat{\lambda}$. In other words, we shall find a function $h(X_1, X_2, \ldots, X_n)$ such that, if $x_1, x_2, \ldots, x_n$ are the observed experimental values of $X_1, X_2, \ldots, X_n$, then the value $h(x_1, x_2, \ldots, x_n)$ will be a good point estimate of λ. By "good" we mean the following properties shall be implied:

(a) unbiasedness,
(b) consistency,
(c) efficiency (i.e., minimum variance), and
(d) sufficiency.

In other words, if $\hat{\lambda}$ is a good point estimate of λ, then one can select the function $h(X_1, X_2, \ldots, X_n)$ such that $h(X_1, X_2, \ldots, X_n)$ is not only an unbiased estimator of λ but also the variance of $h(X_1, X_2, \ldots, X_n)$ is a minimum. We will now present the following definitions.

Definition A.1   For a given positive integer n, the statistic $Y = h(X_1, X_2, \ldots, X_n)$ is called an unbiased estimator of the parameter θ if the expectation of Y is equal to a parameter θ, that is,

$$E(Y) = \theta.$$

Definition A.2   The statistic Y is called a consistent estimator of the parameter θ if Y converges stochastically to a parameter θ as n approaches infinity. If ϵ is an arbitrarily small positive number when Y is consistent, then

$$\lim_{n \to \infty} P(|Y - \theta| \leq \epsilon) = 1.$$

Definition A.3   The statistic Y will be called the minimum variance unbiased estimator of the parameter θ if Y is unbiased and the variance of Y is less than or equal to the variance of every other unbiased estimator of θ.

In other words, an estimator which has the property of minimum variance in large samples is said to be efficient.

Definition A.4   The statistic Y is said to be sufficient for θ if the conditional distribution of X, given $Y = y$, is independent of θ.

From Definition A.3, it is useful in finding a lower bound on the variance of all unbiased estimators. We now establish a lower bound inequality known as the Cramér–Rao inequality. Let X_1, $X_2, \ldots, X_n$ denote a random sample from a distribution with pdf $f(x;\theta)$ for $\theta_1 < \theta < \theta_2$, where θ_1 and θ_2 are known. Let $Y = h(X_1, X_2, \ldots, X_n)$ be an unbiased estimator of θ. The inequality on the variance of Y, $\text{Var}(Y)$, is given by

$$\text{Var}(Y) \geq \frac{1}{nE\left\{\left[\frac{\partial \ln f(x;\theta)}{\partial \theta}\right]^2\right\}}.$$

This inequality is known as the Cramér–Rao inequality.

An estimator $\hat{\theta}$ is said to be asymptotically efficient if

$$\sqrt{n}\hat{\theta}$$

has a variance that approaches the Cramér–Rao lower bound for large n, that is,

$$\lim_{n \to \infty} \text{Var}(\sqrt{n}\hat{\theta}) = \frac{1}{nE\left\{\left[\frac{\partial \ln f(x;\theta)}{\partial \theta}\right]^2\right\}}.$$

We now discuss some basic methods of parameter estimation.

A.1.1 Maximum Likelihood Estimation Method

The method of maximum likelihood estimation (MLE) is one of the most useful techniques for deriving point estimators. As a lead-in to this method, a simple example will be considered. The assumption that the sample is representative of the population will be exercised both in the example and later discussions.

Example A.1 Consider a sequence of 25 Bernoulli trials (binomial situation) where each trial results in either success or failure. From the 25 trials, 6 failures and 19 successes result. Let p be the probability of success, and $1 - p$ the probability of failure. Find the estimator of p, $\hat{p}$, which maximizes that particular outcome.

The sample density function can be written as

$$g(19) = \binom{25}{19} p^{19}(1-p)^6.$$

The maximum of $g(19)$ occurs when

$$p = \hat{p} = \frac{19}{25}$$

so that

$$g\left(19 \mid p = \frac{19}{25}\right) \geq g\left(19 \mid p \neq \frac{19}{25}\right).$$

Now, $g(19)$ is the probability or "likelihood" of six failures in a sequence of 25 trials. Select $p = \hat{p} = 19/25$ as the probability or likelihood maximum value, and hence, $\hat{p}$ is referred to as the maximum likelihood estimate. The reason for maximizing $g(19)$ is that the sample contained six failures, and hence, if it is representative of the population, it is desired to find an estimate which maximizes this sample result. Just as $g(19)$ was a particular sample estimate, in general, one deals with a sample density

$$f(x_1, x_2, \ldots, x_n) = f(x_1; \theta) f(x_2; \theta) \cdots f(x_n; \theta),$$

where $x_1, x_2, \ldots, x_n$ are random, independent observations from a population with density function $f(x)$.

For the general case, it is desired to find an estimate or estimates, $\hat{\theta}_1, \hat{\theta}_2, \ldots, \hat{\theta}_m$ (if such exist) where

$$f(x_1, x_2, \ldots, x_n; \theta_1, \ldots, \theta_m) > f(x_1, x_2, \ldots, x_n; \theta'_1, \theta'_2, \ldots, \theta'_m).$$

Notation $\theta'_1, \theta'_2, \ldots, \theta'_m$ refers to any other estimates different than $\hat{\theta}_1, \hat{\theta}_2, \ldots, \hat{\theta}_m$.

Let us now discuss the method of maximum likelihood. Consider a random sample $X_1, X_2, \ldots, X_n$ from a distribution having pdf $f(x; \theta)$. This distribution has a vector $\theta = (\theta_1, \theta_2, \ldots, \theta_m)'$ of unknown parameters associated with it, where m is the number of unknown parameters. Assuming that the random variables are independent, then the likelihood function, $L(X; \theta)$, is the product of the probability density function evaluated at each sample point

$$L(X, \theta) = \prod_{i=1}^{n} f(X_i; \theta),$$

where $\mathbf{X} = (X_1, X_2, \ldots, X_n)$. The maximum likelihood estimator $\hat{\theta}$ is found by maximizing $L(\mathbf{X}; \theta)$ with respect to θ. In practice, it is often easier to maximize $\ln[L(\mathbf{X}; \theta)]$ to find the vector of MLEs, which is valid because the logarithm function is monotonic.

The log likelihood function is given by

$$\ln L(X; \theta) = \sum_{i=1}^{n} \ln f(X_i; \theta)$$

and is asymptotically normally distributed since it consists of the sum of n independent variables and the implication of the central limit theorem. Since $L(\mathbf{X}; \theta)$ is a joint probability density function for $X_1, X_2, \ldots, X_n$, it must integrate equal to 1, that is,

$$\int_0^\infty \int_0^\infty \cdots \int_0^\infty L(X; \theta) dX = 1.$$

Assuming that the likelihood is continuous, the partial derivative of the left-hand side with respect to one of the parameters, θ_i, yields

$$\frac{\partial}{\partial \theta_i} \int_0^\infty \int_0^\infty \cdots \int_0^\infty L(X; \theta)dX = \int_0^\infty \int_0^\infty \cdots \int_0^\infty \frac{\partial}{\partial \theta_i} L(X; \theta)dX$$

$$= \int_0^\infty \int_0^\infty \cdots \int_0^\infty \frac{\partial \log L(X; \theta)}{\partial \theta_i} L(X; \theta)dX$$

$$= E\left[\frac{\partial \log L(X; \theta)}{\partial \theta_i}\right]$$

$$= E[U_i(\theta)] \quad \text{for } i = 1, 2, \ldots, m,$$

where $\mathbf{U}(\theta) = (U_1(\theta), U_2(\theta), \ldots, U_m(\theta))'$ is often called the score vector and the vector $\mathbf{U}(\theta)$ has components

$$U_i(\theta) = \frac{\partial [\log L(X, \theta)]}{\partial \theta_i} \quad \text{for } i = 1, 2, \ldots, m$$

which, when equated to zero and solved, yields the MLE vector θ.

Suppose that we can obtain a non-trivial function of $X_1, X_2, \ldots, X_n$, say $h(X_1, X_2, \ldots, X_n)$, such that, when θ is replaced by $h(X_1, X_2, \ldots, X_n)$, the likelihood function L will achieve a maximum. In other words,

$$L(X, h(X)) \geq L(X, \theta)$$

for every θ. The statistic $h(X_1, X_2, \ldots, X_n)$ is called a maximum likelihood estimator of θ and will be denoted as

$$\hat{\theta} = h(x_1, x_2, \ldots, x_n).$$

The observed value of $\hat{\theta}$ is called the MLE of θ. In general, the mechanics for obtaining the MLE can be obtained as follows:

Step 1. Find the joint density function $L(X, \theta)$.
Step 2. Take the natural log of the density $\ln L$.
Step 3. Take the partial derivatives of $\ln L$ with respect to each parameter.
Step 4. Set partial derivatives to "zero".
Step 5. Solve for parameter(s).

Example A.2 Let $X_1, X_2, \ldots, X_n$ be a random sample from the exponential distribution with pdf

$$f(x; \lambda) = \lambda e^{-\lambda x} \quad x > 0, \lambda > 0.$$

The joint pdf of $X_1, X_2, \ldots, X_n$ is given by

$$L(X, \lambda) = \lambda^n e^{-\lambda \sum_{i=1}^{n} x_i}$$

and

$$\ln L(X; \lambda) = n \ln \lambda - \lambda \sum_{i=1}^{n} x_i.$$

The function $\ln L$ can be maximized by setting the first derivative of $\ln L$, with respect to λ, equal to zero and solving the resulting equation for λ. Therefore,

$$\frac{\partial \ln L}{\partial \lambda} = \frac{n}{\lambda} - \sum_{i=1}^{n} x_i = 0.$$

This implies that

$$\hat{\lambda} = \frac{n}{\sum_{i=1}^{n} x_i}.$$

The observed value of $\hat{\lambda}$ is the maximum likelihood estimate of λ.

Example A.3   In an exponential censored case, the non-conditional joint pdf that r items have failed is given by

$$f(x_1, x_2, \ldots, x_r) = \lambda^r e^{-\lambda \sum_{i=1}^{r} x_i} \qquad (r \text{ failed items})$$

and the probability distribution that $(n - r)$ items will survive is

$$P(X_{r+1} > t_1, X_{r+2} > t_2, \ldots, X_n > t_{n-r}) = e^{-\lambda \sum_{j=1}^{n-r} t_j}.$$

Thus, the joint density function is

$$L(X, \lambda) = f(x_1, x_2, \ldots, x_r) P(X_{r+1} > t_1, \ldots, X_n > t_{n-r})$$

$$= \frac{n!}{(n-r)!} \lambda^r e^{-\lambda \left(\sum_{i=1}^{r} x_i + \sum_{j=1}^{n-r} t_j \right)}.$$

Let

$$T = \sum_{i=1}^{r} x_i + \sum_{j=1}^{n-r} t_j,$$

then

$$\ln L = \ln\left(\frac{n!}{(n-r)!}\right) + r\ln\lambda - \lambda T$$

$$\frac{\partial \ln L}{\partial \lambda} = \frac{r}{\lambda} - T = 0.$$

Hence,

$$\hat{\lambda} = \frac{r}{T}.$$

Note that with the exponential, regardless of the censoring type or lack of censoring, the MLE of λ is the number of failures divided by the total operating time.

Example A.4   Let $X_1, X_2, \ldots, X_n$ represent a random sample from the distribution with pdf

$$f(x; \theta) = e^{-(x-\theta)} \qquad \text{for } \theta \le x \le \infty, \qquad \text{and } -\infty < \theta < \infty.$$

The likelihood function is given by

$$L(\theta; X) = \prod_{i=1}^{n} f(x_i; \theta) \qquad \text{for } \theta \le x_i < \infty \text{ all } i$$

$$= \prod_{i=1}^{n} e^{-(x_i - \theta)} = e^{-\sum_{i=1}^{n} x_i + n\theta}.$$

For fixed values of $x_1, x_2, \ldots, x_n$, we wish to find that value of θ which maximizes $L(\theta; X)$. Here we cannot use the techniques of calculus to maximize $L(\theta; X)$. Note that $L(\theta; X)$ is largest when θ is as large as possible. However, the largest value of θ is equal to the smallest value of X_i in the sample. Thus,

$$\hat{\theta} = \min\{X_i\} \quad 1 \le i \le n.$$

Example A.5   Let $X_1, X_2, \ldots, X_n$ denote a random sample from the normal distribution $N(\mu, \sigma^2)$. Then

$$L(X, \mu, \sigma^2) = \left(\frac{1}{2\pi}\right)^{\frac{n}{2}} \frac{1}{\sigma^n} e^{-\frac{1}{2\sigma^2} \sum_{i=1}^{n}(x_i - \mu)^2}$$

and

$$\ln L = -\frac{n}{2}\log(2\pi) - \frac{n}{2}\log\sigma^2 - \frac{1}{2\sigma^2}\sum_{i=1}^{n}(x_i - \mu)^2.$$

Thus, we have

$$\frac{\partial \ln L}{\partial \mu} = \frac{1}{\sigma^2} \sum_{i=1}^{n} (x_i - \mu) = 0$$

$$\frac{\partial \ln L}{\partial \sigma^2} = -\frac{n}{2\sigma^2} + \frac{1}{2\sigma^4} \sum_{i=1}^{n} (x_i - \mu)^2 = 0.$$

Solving the two equations simultaneously, we obtain

$$\hat{\mu} = \frac{\sum\limits_{i=1}^{n} x_i}{n}$$

$$\hat{\sigma}^2 = \frac{1}{n} \sum_{i=1}^{n} (x_i - \bar{x})^2.$$

Note that the MLEs, if they exist, are both sufficient and efficient estimates. They also have an additional property called invariance, i.e., for an MLE of θ, then $\mu(\theta)$ is the MLE of $\mu(\theta)$. However, they are not necessarily unbiased, i.e., $E(\hat{\theta}) = \theta$. The point in fact is σ^2

$$E(\hat{\sigma}^2) = \left(\frac{n-1}{n}\right)\sigma^2 \neq \sigma^2.$$

Therefore, for small n, σ^2 is usually adjusted for its bias and the best estimate of σ^2 is

$$\hat{\sigma}^2 = \left(\frac{1}{n-1}\right) \sum_{i=1}^{n} (x_i - \bar{x})^2.$$

Sometimes it is difficult, if not impossible, to obtain maximum likelihood estimators in a closed form, and therefore, numerical methods must be used to maximize the likelihood function. For illustration,

Example A.6   Suppose that $X_1, X_2, \ldots, X_n$ is a random sample from the Weibull distribution with pdf

$$f(x, \alpha, \lambda) = \alpha \lambda x^{\alpha-1} e^{-\lambda x^{\alpha}}.$$

The likelihood function is

$$L(X, \alpha, \lambda) = \alpha^n \lambda^n \prod_{i=1}^{n} x_i^{\alpha-1} e^{-\lambda \sum\limits_{i=1}^{n} x_i^{\alpha}}.$$

Then

$$\ln L = n\log\alpha + n\log\lambda + (\alpha - 1)\sum_{i=1}^{n}\log x_i = -\lambda\sum_{i=1}^{n}x_i^{\alpha}$$

$$\frac{\partial \ln L}{\partial \alpha} = \frac{n}{\alpha} + \sum_{i=1}^{n}\log x_i - \lambda\sum_{i=1}^{n}x_i^{\alpha}\log x_i = 0$$

$$\frac{\partial \ln L}{\partial \lambda} = \frac{n}{\lambda} - \sum_{i=1}^{n}x_i^{\alpha} = 0.$$

As noted, solutions of the above two equations for α and λ and are extremely difficult and require either graphical or numerical methods.

Example A.7   Let X_1, X_2, ... , X_n be a random sample from the gamma distribution with pdf

$$f(x, \lambda, \alpha) = \frac{\lambda^{(\alpha+1)}x^{\alpha}e^{-\lambda x}}{(\alpha!)^n},$$

then

$$L(X, \lambda, \alpha) = \frac{\lambda^{n(\alpha+1)}\prod_{i=1}^{n}x_i^{\alpha}e^{-\lambda\sum_{i=1}^{n}x_i}}{(\alpha!)^n}$$

$$\ln L = n(\alpha + 1)\log\lambda + \alpha\sum_{i=1}^{n}\log x_i - \lambda\sum_{i=1}^{n}x_i - n\log(\alpha!).$$

Taking the partial derivatives, we obtain

$$\frac{\partial \ln L}{\partial \alpha} = n\log\lambda + \sum_{i=1}^{n}\log x_i - n\frac{\partial}{\partial \alpha}[\log\alpha!] = 0$$

$$\frac{\partial \ln L}{\partial \lambda} = \frac{n(\alpha + 1)}{\lambda} - \sum_{i-1}^{n}x_i = 0.$$

As can be seen by the form of the maximum likelihood equations, they can only be solved by trial and error. Some methods along this line are given by Bowker (1972). Another method that could be employed for some applications is the method of moments.

A.1.2 Method of Moments

This is an older method for estimating the value of a parameter which was done by "matching" the population moments to the sample moments. For any positive integer r, the rth moment of a random variable X about the origin is defined by

$$\mu_r = E(X^r).$$

Also, the rth moment of X about the mean is defined by

$$\mu'_r = E(X - \mu)^r,$$

where μ is the first moment of X which is the mean.

Consider a random sample of size n $X_1, X_2, \ldots, X_n$ from a given population. The sample rth moment about the origin is defined analogously to population moments as follows:

$$\hat{\mu}_r = \frac{1}{n}\sum_{i=1}^{n} X_i^r.$$

Similarly, the rth sample moment about the mean is defined as

$$\hat{\mu}'_r = \frac{1}{n}\sum_{i=1}^{n}(X_i - \overline{X})^r$$

where $\overline{X}$ is the sample mean. Using this method, a set of unknown parameters is estimated by equating expected moments to their corresponding sample moments. For example, the first moment was defined as the mean or μ, and hence, equating the sample mean about the origin yields

$$\hat{\mu} = \frac{1}{n}\sum_{i=1}^{n} X_i.$$

Obviously, this first sample moment is an unbiased estimator of the mean μ. It can be shown that the second moment of the random variable about the mean is not an unbiased estimator for the variance σ^2.

A.2 Goodness of Fit Techniques

The problem at hand is to compare some observed sample distribution with a theoretical distribution. Two common techniques that will be discussed are the χ^2 goodness of fit test and the Kolmogorov–Smirnov "d" test.

A.2.1
Chi-Squared Test

It can be shown that the statistic

$$\chi^2 = \sum_{i=1}^{k} \left(\frac{x_i - \mu_i}{\sigma_i} \right)^2$$

has a chi-squared(χ^2) distribution with k degrees of freedom. The technique is applied as follows:

1. Divide the sample data into the mutually exclusive cells (normally 8–12) such that the range of the random variable is covered. (See Rule 3 for additional criteria.)
2. Determine the frequency, f_i, of sample observations in each cell.
3. Determine the theoretical frequency, F_i, for each cell (area under density function between cell boundaries X_n — total sample size). Note that the theoretical frequency for each cell should be greater than 1. To do this step, it normally requires estimates of the population parameters which can be obtained from the sample data.
4. Form the statistic

$$S = \sum_{i=1}^{k} \frac{(f_i - F_i)^2}{F_i}.$$

5. From the χ^2 tables, choose a value of χ^2 with the desired significance level and with degrees of freedom ($= k - 1 - r$), where r is the number of population parameters estimated.
6. Reject the hypothesis that the sample distribution is the same as theoretical distribution if

$$S > \chi^2_{1-\alpha, k-1-r},$$

where α is called the significance level.

Example A.8   Given the following data in Table A.1, can the data be represented by the exponential distribution?

From the above calculation, $\hat{\lambda} = 0.00263$, $R_i = e^{-\lambda t i}$ and $Q_i = 1 - R_i$. Given that a value of significance level α is 0.1, we obtain

$$S = \sum_{i=1}^{11} \frac{(f_i - F_i)^2}{F_i} = 6.165.$$

From Table D.3 in Appendix D, the value of λ^2 with 9 degrees of freedom is 14.68, that is,

$$\chi^2_{9df}(.90) = 14.68.$$

TABLE A.1. Sample observations in each cell boundary.

Cell boundries	f_i	$Q_i = (1 - R_i)X60$	$F_i = Q_i - Q_{i-1}$
0–100	10	13.86	13.86
100–200	9	24.52	10.66
200–300	8	32.71	8.19
300–400	8	39.01	6.30
400–500	7	43.86	4.85
500–600	6	47.59	3.73
600–700	4	50.45	2.86
700–800	4	52.66	2.21
800–900	2	54.35	1.69
900–1,000	1	55.66	1.31
>1,000	1	58.83	2.17

Since $S = 6.165 < 14.68$, we would not reject the hypothesis of exponential with $\lambda = 0.00263$.

If in the following statistic

$$S = \sum_{i=1}^{k} \left(\frac{f_i - F_i}{\sqrt{F_i}} \right)^2, \qquad \left(\frac{f_i - F_i}{\sqrt{F_i}} \right)$$

is approximately normal for large samples, then S also has a χ^2 distribution. This is the basis for the goodness of fit test.

A.2.2 Kolmogorov–Smirnov d Test

Both the χ^2 and "d" tests are non-parameters. However, the χ^2 assumes large sample normality of the observed frequency about its mean while the "d" only assumes a continuous distribution. Let $X_1 \leq X_2 \leq X_3 \leq \cdots \leq X_n$ denote the ordered sample values. Define the observed distribution function, $F_n(x)$, as follows:

$$F_n(X) = \begin{cases} 0 & \text{for } x \leq x_1 \\ \frac{i}{n} & \text{for } x_i < x \leq x_{i+1} \\ 1 & \text{for } x > x_n . \end{cases}$$

Assume the testing hypothesis

$$H_0 : F(x) = F_0(x),$$

where $F_0(x)$ is a given continuous distribution and $F(x)$ is an unknown distribution. Let

$$d_n = \sup_{-\infty < x < \infty} |F_n(x) - F_0(x)|.$$

Since $F_0(x)$ is a continuous increasing function, we can evaluate $|F_n(x) - F_0(x)|$ for each n. If $d_n \leq d_{n,\alpha}$, then we would not reject the hypothesis H_0; otherwise, we would reject it when $d_n > d_{n,\alpha}$. The value $d_{n,\alpha}$ can be found in Table D.7 in Appendix D, where n is the sample size and α is the level of significance.

A.2.3 Least Squared Estimation

A problem of curve fitting, which is unrelated to normal regression theory and MLE estimates of coefficients but uses identical formulas, is called the method of least squares. This method is based on minimizing the sum of the squared distance from the best fit line and the actual data points. It just so happens that finding the MLEs for the coefficients of the regression line also involves these sums of squared distances.

Normal Linear Regression

Regression considers the distributions of one variable when another is held fixed at each of several levels. In the bivariate normal case, consider the distribution of X as a function of given values of Z as illustrated in Fig. A.1. Take a sample of n observations (x_i, z_i) — we obtain the likelihood and its natural log for the normal distribution

$$f(x_1, x_2, \ldots, x_n) = \frac{1}{2\pi^{\frac{n}{2}}}(\frac{1}{\sigma^2})^{\frac{n}{2}} e^{-\frac{1}{2\sigma^2}\sum_{i=1}^{n}(x_i - \alpha - \beta z_i)^2}$$

$$\ln L = -\frac{n}{2}\log 2\pi - \frac{n}{2}\log \sigma^2 - \frac{1}{2\sigma^2}\sum_{i=1}^{n}[x_i - \alpha - \beta z_i]^2.$$

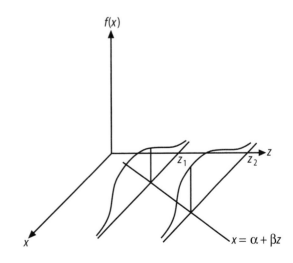

FIG. A.1. Distribution of x as a function of values of Z.

Taking the partial derivatives of $\ln L$ with respect to α and β, we have

$$\frac{\partial \ln L}{\partial \alpha} = \sum_{i=1}^{n}(x_i - \alpha - \beta z_i) = 0$$

$$\frac{\partial \ln L}{\partial \beta} = \sum_{i=1}^{n} z_i(x_i - \alpha - \beta z_i) = 0.$$

The solution of the simultaneous equations is

$$\hat{\alpha} = \overline{X} - \beta \overline{Z}$$

$$\hat{\beta} = \frac{\sum_{i=1}^{n}(X_i - \overline{X})(Z_i - \overline{Z})}{\sum_{i=1}^{n}(Z_i - \overline{Z})^2}.$$

Least Squared Straight Line Fit

Assume there is a linear relationship between X and $E(Y/x)$, that is, $E(Y/x) = a + bx$. Given a set of data, we want to estimate the coefficients a and b that minimizes the sum of the squares. Suppose the desired polynomial, $p(x)$, is written as

$$\sum_{i=0}^{m} a_i x^i,$$

where $a_0, a_1, \ldots, a_m$ are to be determined. The method of least squares chooses as "solutions" those coefficients minimizing the sum of the squares of the vertical distances from the data points to the presumed polynomial. This means that the polynomial termed "best" is the one whose coefficients minimize the function L, where

$$L = \sum_{i=1}^{n}[y_i - p(x_i)]^2.$$

Here, we will treat only the linear case, where $X = \alpha + \beta Z$. The procedure for higher-order polynomials is identical, although the computations become much more tedious. Assume a straight line of the form $X = \alpha + \beta Z$. For each observation $(x_i, z_i): X_i = \alpha + \beta Z_i$, let

$$Q = \sum_{i=1}^{n}(x_i - \alpha - \beta z_i)^2.$$

We wish to find α and β estimates such as to minimize Q. Taking the partial differentials, we obtain

$$\frac{\partial Q}{\partial \alpha} = -2 \sum_{i=1}^{n} (x_i - \alpha - \beta z_i) = 0$$

$$\frac{\partial Q}{\partial \beta} = -2 \sum_{i=1}^{n} z_i(x_i - \alpha - \beta z_i) = 0.$$

Note that the above are the same as the MLE equations for normal linear regression. Therefore, we obtain the following results:

$$\hat{\alpha} = \bar{x} - \beta \bar{z}$$

$$\hat{\beta} = \frac{\sum\limits_{i=1}^{n} (x_i - \bar{x})(z_i - \bar{z})}{\sum\limits_{i=1}^{n} (z_i - \bar{z})^2}.$$

The above gives an example of least squares applied to a linear case. It follows the same pattern for higher-order curves with solutions of 3, 4, and so on, linear systems of equations.

A.3   Interval Estimation

A point estimate is sometimes inadequate in providing an estimate of an unknown parameter since it rarely coincides with the true value of the parameter. An alternative way is to obtain a confidence interval estimation of the form $[\theta_L, \theta_U]$, where θ_L is the lower bound and θ_U is the upper bound. Point estimates can become more useful if some measure of their error can be developed, i.e., some sort of tolerance on their high and low values could be developed. Thus, if an interval estimator is $[\theta_L, \theta_U]$ with a given probability $1 - \alpha$, then θ_L and θ_U will be called $100(1 - \alpha)\%$ confidence limits for the given parameter θ and the interval between them is a $100(1 - \alpha)\%$ confidence interval and $(1 - \alpha)$ is also called the confidence coefficient.

A.3.1 Confidence Intervals for the Normal Parameters

The one-dimensional normal distribution has two parameters: mean μ and variance σ^2. The simultaneous employment of both parameters in a confidence statement concerning percentages of the population will be discussed in the next section on tolerance limits. Hence, individual confidence statements about μ and σ^2 will be discussed here.

Confidence Limits for the Mean μ with Known σ^2

It is easy to show that the statistic

$$Z = \frac{\overline{X} - \mu}{\frac{\sigma}{\sqrt{n}}}$$

is a standard normal distribution where

$$\overline{X} = \frac{1}{n}\sum_{i=1}^{n} X_i.$$

Hence, a $100(1 - \alpha)\%$ confidence interval for the mean μ is given by

$$p\left[\overline{X} - Z_{\frac{\alpha}{2}}\frac{\sigma}{\sqrt{n}} < \mu < \overline{X} + Z_{\frac{\alpha}{2}}\frac{\sigma}{\sqrt{n}}\right] = 1 - \alpha.$$

In other words,

$$\mu_L = \overline{X} - Z_{\frac{\alpha}{2}}\frac{\sigma}{\sqrt{n}} \quad \text{and} \quad \mu_U = \overline{X} + Z_{\frac{\alpha}{2}}\frac{\sigma}{\sqrt{n}}.$$

Example A.9   Draw a sample of size 4 from a normal distribution with known variance $= 9$, say $x_1 = 2$, $x_2 = 3$, $x_3 = 5$, $x_4 = 2$. Determine the location of the true mean (μ). The sample mean can be calculated as

$$\overline{x} = \frac{\sum_{i=1}^{n} x_i}{n}$$
$$= \frac{2 + 3 + 5 + 2}{4} = 3.$$

Assuming that $\alpha = 0.05$ and from the standard normal distribution (Table D.1 in Appendix D), we obtain

$$P\left[3 - 1.96\frac{3}{\sqrt{4}} < \mu < 3 + 1.96\frac{3}{\sqrt{4}}\right] = 0.95$$
$$P[0.06 < \mu < 5.94] = 0.95.$$

The above example shows that there is a 95% probability that the true mean is somewhere between 0.06 and 5.94. Now, μ is a fixed parameter and does not vary, so how do we interpret the probability? If samples of size 4 are repeatedly drawn, a different set of limits would be constructed each time. With this as the case, the interval becomes the random variable and the interpretation is that, for 95% of the time, the interval so constructed will contain the true (fixed) parameter.

Confidence Limits for the Mean μ with Unknown σ^2

Let

$$S = \sqrt{\frac{1}{n-1}\sum_{i=1}^{n}(X_i - \overline{X})^2}.$$

It can be shown that the statistic

$$T = \frac{\overline{X} - \mu}{\frac{S}{\sqrt{n}}}$$

has a t distribution with $(n-1)$ degrees of freedom (see Appendix D). Thus, for a given sample mean and sample standard deviation, we obtain

$$P\left[|T| < t_{\frac{\alpha}{2},n-1}\right] = 1 - \alpha.$$

Hence, a $100(1-\alpha)\%$ confidence interval for the mean μ is given by

$$P\left[\overline{X} - t_{\frac{\alpha}{2},n-1}\frac{S}{\sqrt{n}} < \mu < \overline{X} + t_{\frac{\alpha}{2},n-1}\frac{S}{\sqrt{n}}\right] = 1 - \alpha.$$

Example A.10   A problem on the variability of a new product was encountered. An experiment was run using a sample of size $n = 25$; the sample mean was found to be $\overline{X} = 50$ and the variance $\sigma^2 = 16$. A 95% confidence limit for μ is given by

$$P\left[50 - 1.708\sqrt{\frac{16}{25}} < \mu < 50 + 1.708\sqrt{\frac{16}{25}}\right] = 0.90$$

$$P[48.634 < \mu < 51.366] = 0.90.$$

Note that, for one-sided limits, choose t_α or $t_{1-\alpha}$.

Confidence Limits on σ^2

Note that $n\hat{\sigma}^2/\sigma^2$ has a χ^2 distribution with $(n-1)$ degrees of freedom. Correcting for the bias in $\hat{\sigma}^2$, then $(n-1)\hat{\sigma}^2/\sigma^2$ has this same distribution. Hence,

$$P\left[\chi^2_{\frac{\alpha}{2},n-1} < \frac{(n-1)S^2}{\sigma^2} < \chi^2_{1-\frac{\alpha}{2},n-1}\right] = 1 - \alpha$$

or

$$P\left[\frac{\sum(x_i - \bar{x})^2}{\chi^2_{1-\frac{\alpha}{2},n-1}} < \sigma^2 < \frac{\sum(x_i - \bar{x})^2}{\chi^2_{\frac{\alpha}{2},n-1}}\right] = 1 - \alpha.$$

Similarly, for one-sided limits, choose $\chi^2(\alpha)$ or $\chi^2(1-\alpha)$.

A.3.2 Confidence Intervals for the Exponential Parameters

The pdf and cdf of the exponential distribution are given as

$$f(x) = \lambda e^{-\lambda x} \qquad x > 0, \ \lambda > 0$$

and

$$F(x) = 1 - e^{-\lambda x},$$

respectively. From Subsection A.1.1, it was shown that the distribution of a function of the estimate

$$\hat{\lambda} = \frac{r}{\sum\limits_{i=1}^{n} x_i + (n-r)x_r}$$

derived from a test of n identical components with common exponential failure density (failure rate λ), whose testing was stopped after the rth failure, was chi-squared (χ^2), i.e.,

$$2r\frac{\lambda}{\hat{\lambda}} = 2\lambda T \quad (\chi^2 \text{ distribution with } 2r \text{ degrees of freedom})$$

where T is the total accrued time on all units. Knowing the distribution of $2\lambda T$ allows us to obtain the confidence limits on the parameter as follows:

$$P\left[\chi^2_{1-\frac{\alpha}{2},2r} < 2\lambda T < \chi^2_{\frac{\alpha}{2},2r}\right] = 1 - \alpha$$

or, equivalently, that

$$P\left[\frac{\chi^2_{1-\frac{\alpha}{2},2r}}{2T} < \lambda < \frac{\chi^2_{\frac{\alpha}{2},2r}}{2T}\right] = 1 - \alpha.$$

This means that in $(1-\alpha)\%$ of samples with a given size n, the random interval

$$\left(\frac{\chi^2_{1-\frac{\alpha}{2},2r}}{2T}, \frac{\chi^2_{\frac{\alpha}{2},2r}}{2T}\right)$$

will contain the population of constant failure rate. In terms of $\theta = 1/\lambda$ or the mean time between failures (MTBF), the above confidence limits change to

$$P\left[\frac{2T}{\chi^2_{\frac{\alpha}{2},2r}} < \theta < \frac{2T}{\chi^2_{1-\frac{\alpha}{2},2r}}\right] = 1 - \alpha.$$

If testing is stopped at a fixed time rather than a fixed number of failures, the number of degrees of freedom in the lower limit increases by two. Table A.2 show the confidence limits for θ, the mean of an exponential density.

Example A.11 (Two-sided) From the goodness of fit example, $T = 22{,}850$, testing stopped after $r = 60$ failures. We can obtain $\hat{\lambda} = 0.00263$ and $\hat{\theta} = 380.833$. Assuming that $\alpha = 0.1$, then, from the above formula, we obtain

$$P\left[\frac{2T}{\chi^2_{.05,120}} < \theta < \frac{2T}{\chi^2_{.95,120}}\right] = 0.90$$

$$P\left[\frac{45,700}{146.568} < \theta < \frac{45,700}{95.703}\right] = 0.90$$

$$P[311.80 < \theta < 477.52] = 0.90.$$

Example A.12 (One-sided Lower) Assuming that testing stopped after 1,000 hours with four failures, then

$$P\left[\frac{2T}{\chi^2_{.10,10}} < \theta\right] = 0.90$$

$$P\left[\frac{2,000}{15.987} < \theta\right] = 0.90$$

$$P[125.1 < \theta] = 0.90.$$

TABLE A.2. Confidence limits for θ.

	Fixed no. of failures	Fixed time
One-Sided$_{(Lower)}$	$\dfrac{2T}{\chi^2_{\alpha,2r}}$	$\dfrac{2T}{\chi^2_{\alpha,2r+2}}$
One-Sided$_{(Upper)}$	$\dfrac{2T}{\chi^2_{1-\alpha,2r}}$	$\dfrac{2T}{\chi^2_{1-\alpha,2r}}$
Two-Sided $_{(Lower\ Upper)}$	$\dfrac{2T}{\chi^2_{\alpha/2,2r}}$, $\dfrac{2T}{\chi^2_{1-\alpha/2,2r}}$	$\dfrac{2T}{\chi^2_{\alpha/2,2r+2}}$, $\dfrac{2T}{\chi^2_{1-\alpha/2,2r}}$

A.3.3
Confidence
Intervals for the
Binomial
Parameters

Consider a sequence of n Bernoulli trials with k successes and $(n - k)$ failures. We now determine one-sided upper and lower and two-sided limits on the parameter p, the probability of success. For the lower limit, the binomial sum is set up such that the chance probability of k or more successes with a true p as low as p_L is only $\alpha/2$. This means the probability of k or more successes with a true p higher than p_L is $1 - \alpha/2$.

$$\sum_{i=k}^{n} \binom{n}{i} p_L^i (1 - p_L)^{n-i} = \frac{\alpha}{2}.$$

Similarly, the binomial sum for the upper limit is

$$\sum_{i=k}^{n} \binom{n}{i} p_U^i (1 - p_U)^{n-i} = 1 - \frac{\alpha}{2}$$

or, equivalently, that

$$\sum_{i=0}^{k-1} \binom{n}{i} p_U^i (1 - p_U)^{n-i} = \frac{\alpha}{2}.$$

Solving for p_L and p_U in the above equations,

$$P[p_L < p < p_U] = 1 - \alpha.$$

For the case of one-sided limits, merely change $\alpha/2$ to α.

Example A.13   Given $n = 100$, 25 successes, and 75 failures, an 80% two-sided confidence limits on p can be obtained as follows:

$$\sum_{i=25}^{100} \binom{100}{i} p_L^i (1 - p_L)^{100-i} = 0.10$$

$$\sum_{i=0}^{25-1} \binom{100}{i} p_U^i (1 - p_U)^{100-i} = 0.10.$$

Solving the above two equations simultaneously, we obtain

$$p_L \approx 0.194 \quad \text{and} \quad p_U \approx 0.313$$
$$P[0.194 < p < 0.313] = 0.80.$$

Example A.14   Continuing with Example A.13, find an 80% one-sided confidence limit on p. We now can set the top equation to 0.20 and solve for p_L. It is easy to obtain $p_L = 0.211$ and $P[p > 0.211] =$

0.80. Let us define $\bar{p} = k/n$, the number of successes divided by the number of trials. For large values of n and if $np > 5$ and $n(1-p) > 5$, and from the central limit theorem, the statistic

$$Z = \frac{(\bar{p} - p)}{\sqrt{\frac{\bar{p}(1-\bar{p})}{n}}}$$

approximates to the standard normal distribution. Hence,

$$P[-z_{\frac{\alpha}{2}} < Z < z_{\frac{\alpha}{2}}] = 1 - \alpha.$$

Then

$$P\left[\bar{p} - z_{\frac{\alpha}{2}}\sqrt{\frac{\bar{p}(1-\bar{p})}{n}} < p < \bar{p} + z_{\frac{\alpha}{2}}\sqrt{\frac{\bar{p}(1-\bar{p})}{n}}\right] = 1 - \alpha.$$

Example A.15   Given $n = 900$, $k = 180$, and $\alpha = 0.05$, then $\bar{p} = 180/900 = 0.2$ and

$$P\left[0.2 - 1.96\sqrt{\frac{0.2(0.8)}{900}} < p < 0.2 + 1.96\sqrt{\frac{0.2(0.8)}{900}}\right] = 0.95$$

$$P[.174 < p < .226] = 0.95.$$

A.3.4 Confidence Intervals for the Poisson Parameters

Limits for the Poisson parameters are completely analogous to the binomial except that the sample space is denumerable instead of finite. The lower and upper limits can be solved simultaneously in the following equations:

$$\sum_{i=k}^{\infty} \frac{\lambda_L^i e^{-\lambda_L}}{i!} = \frac{\alpha}{2}$$

$$\sum_{i=k}^{\infty} \frac{\lambda_U^i e^{-\lambda_U}}{i!} = 1 - \frac{\alpha}{2},$$

or, equivalently, such that

$$\sum_{i=k}^{\infty} \frac{\lambda_L^i e^{-\lambda_L}}{i!} = \frac{\alpha}{2}$$

$$\sum_{i=0}^{k-1} \frac{\lambda_U^i e^{-\lambda_U}}{i!} = \frac{\alpha}{2}.$$

Example A.16   One thousand article lots are inspected resulting in an average of 10 defects per lot. Find 90% limits on the average number of defects per 1,000 article lots. Assume $\alpha = 0.1$,

$$\sum_{i=10}^{\infty} \frac{\lambda_L^i e^{-\lambda_L}}{i!} = 0.05$$

$$\sum_{i=0}^{10-1} \frac{\lambda_U^i e^{-\lambda_U}}{i!} = 0.05.$$

Solving for λ_L and λ_U from the Poisson tables in Appendix C, we obtain

$$P[5.45 < \lambda < 16.95] = 0.90.$$

The one-sided limits are constructed similarly to the case for binomial limits.

A.4   Tolerance Limits

Tolerance limits are confidence statements placed on the location of a percentage of the population rather than on any particular parameter, e.g., $P[L < p\% < U] = 1 - \alpha$.

A.4.1   Tolerance Limits for Normal Populations

Inferences about the location of population percentages for the normal distribution requires simultaneous involvement of both μ and σ^2. Setting limits individually on μ and σ^2 and taking a "worst case" approach is not appropriate. Tolerance limits for the normal population are given by

$$P[\bar{x} - k_{\frac{\alpha}{2}}S < p\% < \bar{x} + k_{\frac{\alpha}{2}}S] = 1 - \alpha,$$

or

$$P[\bar{x} - k_{\alpha}S < p\%] = 1 - \alpha,$$
$$P[p\% < \bar{x} + k_{\alpha}S] = 1 - \alpha$$

for one-sided lower and upper limits, respectively. Tables D.8 and D.9 in Appendix D provide values of k for the parameters $n = 2, \ldots, 50$; $\gamma = 0.75, 0.90, 0.95, 0.99$; and $\alpha = 0.25, 0.10, 0.05, 0.01, 0.001$. Estimates of μ and σ, x and σ, are assumed available from the sample.

Example A.17   A manufacturer of transistors would like to construct tolerant limits which he can be assured, with the probability of 0.90, that 95% of his production will not fail. A sample of 25 transistors is observed and the sample mean and sample variance are found to be 200 and 64, respectively.

From Table D.8 in Appendix D, the value of k corresponding to $n = 25$, $\gamma = 0.90$, and $\alpha = 0.05$ is 2.474. We obtain

$$P[200 - 2.474(8) < 95\% < 200 + 2.474(8)] = 0.90$$
$$P[180.21 < 95\% < 219.79] = 0.90.$$

Therefore, the tolerance limits are given by [180.21, 219.79].

A.4.2   Tolerance Limits for Exponential Populations

Define x_p as $-P[\,X \geq x_p] = p$, where X is an exponential random variable with mean $1/\lambda = \theta$. x_p is a quantile, i.e., for $p = 0.50, 0.75, 0.90$, etc., x_p defines that point where $p\%$ of the population lies beyond x_p. The interest here is usually one-sided, e.g., percentage of remaining life, etc.

$$P[X \geq x_p] = e^{-\frac{x_p}{\theta}} = p$$

or

$$x_p = -\theta \, \ln p = \theta \, \ln\left(\frac{1}{p}\right).$$

Recalling the one-sided confidence limits from Subsection A.3.2,

$$P\left[\frac{2T}{\chi^2_{1-\alpha,*}} < \theta\right] = 1 - \alpha,$$

or, equivalently, that

$$P\left[\frac{2T \ln\left(\frac{1}{p}\right)}{\chi^2_{1-\alpha,*}} < \theta \log\left(\frac{1}{p}\right)\right] = 1 - \alpha.$$

Here, "$*$" takes a value either $2r$ or $2r + 2$ degrees of freedom. Therefore,

$$P\left[\frac{2T \ln\left(\frac{1}{p}\right)}{\chi^2_{1-\alpha,*}} < x_p\right] = 1 - \alpha,$$

where

$$x_p = \theta \ln\left(\frac{1}{p}\right).$$

Example A.18   A manufacturer of vacuum tubes wishes to specify a single lower limit with a probability of 0.50, that 50% of his product will fail. From the experiment, 1,000 hours of testing produce one failure. Find the time at which 50% of the items will survive with 90% confidence.

It is easy to obtain the results as follows. Here, $T = 1,000$, $p = 0.5$, and $\alpha = 0.1$. From the Chi-squared Table D.3 in Appendix D, we have $\chi_4^2(0.90) = 7.78$ and $\log(1/p) = 0.693$. Hence, $x_p = 178.149$.

A.5   Non-Parametric Tolerance Limits

Non-parametric tolerance limits are based on the smallest and largest observation in the sample, designated as X_S and X_L, respectively. Due to their non-parametric nature, these limits are quite insensitive and to gain precision proportional to the parametric methods requires much larger samples. An interesting question here is to determine the sample size required to include at least $100(1 - \alpha)\%$ of the population between X_S and X_L with given probability γ.

For two-sided tolerance limits, if $(1 - \alpha)$ is the minimum proportion of the population contained between the largest observation X_L and smallest observation X_S with confidence $(1 - \gamma)$, then it can be shown that

$$n(1 - \alpha)^{n-1} - (n - 1)(1 - \alpha)^n = \gamma.$$

Therefore, the number of observations required is given by

$$n = \left\lfloor \frac{(2 - \alpha)}{4\alpha} \chi_{1-\gamma,4}^2 + \frac{1}{2} \right\rfloor + 1.$$

Example A.19   Determine tolerance limits which include at least 90% of the population with probability 0.95. Here,

$$\alpha = 0.1, \gamma = 0.95, \qquad \text{and } \chi_{0.05,4}^2 = 9.488,$$

therefore, a sample of size

$$n = \left(\frac{2 - 0.1}{0.1} \right) \left(\frac{0.488}{4} \right) + \frac{1}{2} = 45.6 \equiv 46$$

is required.

For a one-side tolerance limit, the number of observations required is given by

$$n = \left\lfloor \frac{\log(1 - \gamma)}{\log(1 - \alpha)} \right\rfloor + 1.$$

Example A.20   As in Example A.19, we wish to find a lower tolerance limit, that is, the number of observations required so that the probability is 0.95, that at least 90% of the population will exceed X_s, is given by

$$n = \left\lfloor \frac{\log(1 - 0.95)}{\log(1 - 0.1)} \right\rfloor + 1 = 29.4 \equiv 30.$$

Table D.9 in Appendix D gives the sample size required to include a given percentage of the population between X_S and X_L with given confidence, or sample size required to include a given percentage of the population above or below X_S or X_L, respectively.

A.6   Sequential Sampling

A sequential sampling scheme is one in which items are drawn one at a time and the results at any stage determine if sampling or testing should stop. Thus, any sampling procedure for which the number of observations is a random variable can be regarded as sequential sampling. Sequential tests derive their name from the fact that the sample size is not determined in advance, but allowed to "float" with a decision (accept, reject, or continue test) after each trial or data point.

In general, let us consider the hypothesis

$$H_0 : f(x) = f_0(x) \quad \text{versus} \quad H_1 : f(x) = f_1(x).$$

For an observation test, say X_1, if $X_1 \leq A$, then we will accept the testing hypothesis ($H_0: f(x) = f_0(x)$); if $X_1 \geq A$, then we will reject H_0 and accept $H_1: f(x) = f_1(x)$. Otherwise, we will continue to perform at least one more test. The interval $X_1 \leq A$ is called the acceptance region. The interval $X_1 \geq A$ is called the rejection or critical region. See Fig. A.2.

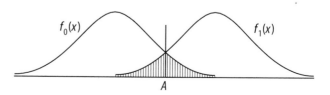

FIG. A.2.   A sequential sampling scheme.

A "good" test is one that makes the α and β errors as small as possible. However, there is not much freedom to do this without increasing the sample size. Common procedure is to fix the β error and then choose a critical region to minimize the error or maximize the "power" (power $= 1 - \beta$) of the test, or to choose the critical region so as to equalize the α and β errors to reasonable levels.

A criterion, similar to the MLE, for constructing tests is called the "probability ratio", which is the ratio of the sample densities under H_1 over H_0. Consider the ratio of probabilities

$$\lambda_m = \frac{f_1(x_1) \cdot f_1(x_2) \cdots f_1(x_n)}{f_0(x_1) \cdot f_0(x_2) \cdots f_0(x_n)} > k.$$

Here, $x_1, x_2, \ldots, x_n$ are n independent random observations and k is chosen to give the desired α error.

Recall from the MLE discussion in Section A.1 that $f_1(x_1), f_1(x_2)$, $\ldots, f_1(x_n)$ are maximized under H_1 when the parameter(s), e.g., $\theta = \theta_1$ and, similarly, $f_0(x_1), f_0(x_2), \ldots, f_0(x_n)$, are maximized when $\theta = \theta_0$. Thus, the ratio will become large if the sample favors H_1 and will become small if the sample favors H_0. Therefore, the test will be called a sequential probability ratio test if we

(i) stop sampling and conclude to reject H_0 as soon as $\lambda_m \geq A$;
(ii) stop sampling and conclude to accept H_0 as soon as $\lambda_m \leq B$;
(iii) continue sampling as long as $B < \lambda_m < A$, where $A > B$.

The choice of A and B with the above test, suggested by Wald (1947), can be determined as follows:

$$B = \frac{\beta}{1 - \alpha} \quad \text{and} \quad A = \frac{1 - \beta}{\alpha}.$$

The basis for α and β are therefore:

$$P[\lambda_m > A | H_0] = \alpha$$
$$P[\lambda_m < B | H_1] = \beta.$$

Exponential Case
Let

$$V(t) = \sum_{i=1}^{r} X_i + \sum_{j=1}^{n-r} t_j,$$

where X_i are the times to failure and t_j are the times to test termination without failure. Thus, $V(t)$ is merely the total operating time accrued on both successful and unsuccessful units where the total number of units is n. The hypothesis to be tested is

$$H_0 : \theta = \theta_0 \quad \text{versus} \quad H_1 : \theta = \theta_1.$$

For the failed items,

$$g(x_1, x_2, \ldots x_r) = \left(\frac{1}{\theta}\right)^r e^{-\sum_{i=1}^{r} x_i}.$$

For the non-failed items,

$$P(X_{r+1} > t_1, X_{r+2} > t_2, \ldots, P(X_n > t_{n-r}) = e^{-\sum_{j=1}^{n-r} t_j}.$$

The joint density for the first r failures among n items is

$$f(x_1, x_2, \ldots, x_r, t_{r+1}, \ldots, t_n) = \left(\frac{1}{\theta}\right)^r e^{-\sum_{i=1}^{r} \frac{x_i}{\theta} - \sum_{j=1}^{n-r} \frac{t_j}{\theta}}$$

$$= \left(\frac{1}{\theta}\right)^r e^{-\frac{V(t)}{t}}$$

and

$$\lambda_m = \frac{\left(\frac{1}{\theta_1}\right)^r e^{-\frac{V(t)}{\theta_1}}}{\left(\frac{1}{\theta_0}\right)^r e^{-\frac{V(t)}{\theta_0}}}$$

$$= \left(\frac{\theta_0}{\theta_1}\right)^r e^{-V(t)\left[\frac{1}{\theta_1} - \frac{1}{\theta_0}\right]}.$$

Now, it has been shown that for sequential tests, the reject and accept limits, A and B, can be equated to simple functions of α and β. Thus, we obtain the following test procedures:

$$\text{continue test:} \quad \frac{\beta}{1-\alpha} \equiv B < \lambda_m < A \equiv \frac{1-\beta}{\alpha}$$

$$\text{reject } H_0 : \quad \lambda_m > A \equiv \frac{1-\beta}{\alpha}$$

$$\text{accept } H_0 : \quad \lambda_m < B \equiv \frac{\beta}{1-\alpha}.$$

Working with the continue test inequality, we now have

$$\frac{\beta}{1-\alpha} < \left(\frac{\theta_0}{\theta_1}\right)^r e^{-V(t)\left[\frac{1}{\theta_1} - \frac{1}{\theta_0}\right]} < \frac{1-\beta}{\alpha}.$$

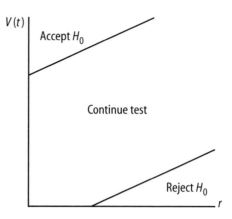

FIG. A.3. Test procedure.

Taking natural ln of the above inequality, we obtain

$$\log\left(\frac{\beta}{1-\alpha}\right) < r\log\left(\frac{\theta_0}{\theta_1}\right) - V(t)\left[\frac{1}{\theta_1} - \frac{1}{\theta_0}\right] < \log\left(\frac{1-\beta}{\alpha}\right).$$

The above inequality is linear in $V(t)$ and r (see Fig. A.3), therefore, the rejection line $V(t)$ can be obtained by setting

$$r\ln\left(\frac{\theta_0}{\theta_1}\right) - V(t)\left[\frac{1}{\theta_1} - \frac{1}{\theta_0}\right] = \ln\left(\frac{1-\beta}{\alpha}\right),$$

or, equivalently, that

$$V(t) = \frac{r\ln\left(\frac{\theta_0}{\theta_1}\right)}{\left[\frac{1}{\theta_1} - \frac{1}{\theta_0}\right]} - \frac{\ln\left(\frac{1-\beta}{\alpha}\right)}{\left[\frac{1}{\theta_1} - \frac{1}{\theta_0}\right]}.$$

Similarly, the acceptance line $V(t)$ (see Fig. A.3) can be obtained by setting

$$r\log\left(\frac{\theta_0}{\theta_1}\right) - V(t)\left[\frac{1}{\theta_1} - \frac{1}{\theta_0}\right] = \log\left(\frac{\beta}{1-\alpha}\right).$$

This implies that

$$V(t) = \frac{r\ln\left(\frac{\theta_0}{\theta_1}\right)}{\left[\frac{1}{\theta_1} - \frac{1}{\theta_0}\right]} - \frac{\ln\left(\frac{\beta}{1-\alpha}\right)}{\left[\frac{1}{\theta_1} - \frac{1}{\theta_0}\right]}.$$

Example A.21 Given that $H_0: \theta = 500$ versus $H_1: \theta = 250$ and $\alpha = \beta = 0.1$. Then the acceptance line is

$$V(t) = 346.6r + 1098.6$$

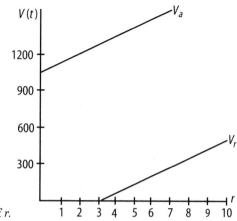

and the rejection line

$$V(t) = 346.6r - 1098.6.$$

Figure A.4 shows these two linear functions in terms of r, the number of first r failures in the test.

For an exponential distribution

θ	$P(A)$
0	0
θ_1	β
$\dfrac{\log\left(\frac{\theta_0}{\theta_1}\right)}{\left[\frac{1}{\theta_1} - \frac{1}{\theta_0}\right]}$	$\dfrac{\log\left(\frac{1-\beta}{\alpha}\right)}{\log\left(\frac{1-\beta}{\alpha}\right) - \log\left(\frac{\beta}{1-\alpha}\right)}$
θ_0	$1 - \alpha$
∞	$1.$

From the information given in the above example (see Fig. A.5), we obtain

θ	$P(A)$
0	0
250	0.10
346.5	0.5
500	0.90
∞	1.00

Since there is no pre-assigned termination to a regular sequential test, it is customary to draw a curve called the "average sample

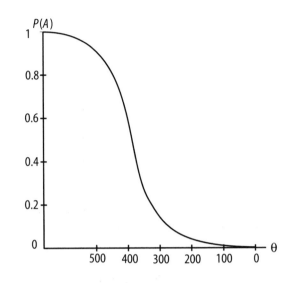

FIG. A.5. The function $P(A)$ versus θ for the exponential distribution.

number" (ASN). This curve shows the expected sample size as a function of the true parameter value. It is known that the test will be terminated with a finite observation. (It should be noted that "on the average", the sequential tests utilizes significantly smaller samples than fixed sample plans.)

$$
\begin{array}{cc}
\theta & E(r) = ASN \\[2mm]
0 & 0 \\[4mm]
\theta_1 & \dfrac{\theta_0\beta\log\left(\dfrac{\beta}{1-\alpha}\right) + (1-\beta)\log\left(\dfrac{1-\beta}{\alpha}\right)}{\left[\log\left(\dfrac{\theta_0}{\theta_1}\right) - \left(\dfrac{\theta_0 - \theta_1}{\theta_1}\right)\right]\theta_1} \\[8mm]
\dfrac{\log\left(\dfrac{\theta_0}{\theta_1}\right)}{\left[\dfrac{1}{\theta_1} - \dfrac{1}{\theta_0}\right]} & \dfrac{\log\left(\dfrac{1-\beta}{\alpha}\right)\log\left(\dfrac{\beta}{1-\alpha}\right)}{\left[\log\left(\dfrac{\theta_0}{\theta_1}\right)\right]^2} \\[8mm]
\theta_0 & \dfrac{\left[(1-\alpha)\log\left(\dfrac{\beta}{1-\alpha}\right) + \alpha\log\left(\dfrac{1-\beta}{\alpha}\right)\right]\theta_1}{\left[\log\left(\dfrac{\theta_0}{\theta_1}\right) - \left(\dfrac{\theta_0 - \theta_1}{\theta_1}\right)\right]\theta_0} \\[8mm]
\infty & 0
\end{array}
$$

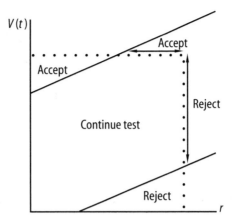

FIG. **A.6.** A truncated test procedure.

An approximate formula for $E(t)$, the expected time to reach a decision, is

$$E(t) \equiv \theta \log\left(\frac{n}{n - E(r)}\right),$$

where n is the total number of units on test (assuming no replacement of failed units). If replacements are made, then

$$E(t) = \frac{\theta}{n} E(r).$$

Occasionally, it is desired to "truncate" a sequential plan such that, if no decision is made before a certain point, testing is stopped and a decision is made on the basis of data acquired up to that point. Graphically, truncation is accomplished as shown in Fig. A.6.

There are a number of rules and theories on optimum truncation. In the reliability community, a $V(t)$ truncation point at $10\theta_0$ is often used to determine the $V(t)$ and r lines for truncation and the corresponding exact α and β errors (these will in general be larger for truncated tests than for the non-truncated). An approximate method draws the $V(t)$ truncation line to the center of the continue test band and constructs the r truncation line perpendicular to that point (see Fig. A.7).

Bernoulli Case

$$H_0 : p = p_0 \quad \text{versus} \quad H_1 : p = p_1$$

and α and β are pre-assigned.

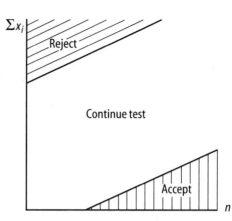

FIG. A.7. Test procedure for Σx_i versus n.

$$\lambda_m = \frac{p_1^{x_1}(1-p_1)^{1-x_1}p_1^{x_2}(1-p_1)^{1-x_2}\cdots p_1^{x_n}(1-p_1)^{1-x_1}}{p_0^{x_1}(1-p_0)^{1-x_1}p_0^{x_2}(1-p_0)^{1-x_2}\cdots p_0^{x_n}(1-p_0)^{1-x_n}}$$

$$= \frac{p_1^{\sum_{i=1}^{n}x_i}(1-p_1)^{n-\sum_{i=1}^{n}x_i}}{p_0^{\sum_{i=1}^{n}x_i}(1-p_0)^{n-\sum_{i=1}^{n}x_i}}$$

$$= \left(\frac{p_1}{p_0}\right)^{\sum_{i=1}^{n}x_i}\left(\frac{1-p_1}{1-p_0}\right)^{n-\sum_{i=1}^{n}x_i}.$$

The following inequality will be determined to continue the test region:

$$\frac{\beta}{1-\alpha} < \left(\frac{p_1}{p_0}\right)^{\sum_{i=1}^{n}x_i}\left(\frac{1-p_1}{1-p_0}\right)^{n-\sum_{i=1}^{n}x_i} < \frac{1-\beta}{\alpha}.$$

Taking logs through the above inequality produces linear relationships in Σx_i (e.g., number of failures or defects) and n, the total number of units or trials, that is,

$$\log\left(\frac{\beta}{1-\alpha}\right)$$
$$< \sum_{i=1}^{n}x_i\log\left(\frac{p_1}{p_0}\right) + \left(n - \sum_{i=1}^{n}x_i\right)\log\left(\frac{1-p_1}{1-p_0}\right) < \log\left(\frac{1-\beta}{\alpha}\right).$$

Similar tests can be constructed for other distribution parameters following the same general scheme.

A.7 Bayesian Methods

The Bayesian approach to statistical inference is based on a theorem first presented by the Reverend Thomas Bayes. To demonstrate the approach, let X have a pdf $f(x)$, which is dependent on θ. In the traditional statistical inference approach, θ is an unknown parameter, and hence, is a constant. We now describe our prior belief in the value of θ by a pdf $h(\theta)$. This amounts to quantitatively assessing subjective judgment and should not be confused with the so-called objective probability assessment derived from the long-term frequency approach. Thus, θ will now essentially be treated as a random variable θ with pdf $h(\theta)$.

Consider a random sample $X_1, X_2, \ldots, X_n$ from $f(x)$ and define a statistic Y as a function of this random sample. Then there exists a conditional pdf $g(y|\theta)$ of Y for a given θ. The joint pdf for y and θ is

$$f(\theta, y) = h(\theta)g(y|\theta).$$

If θ is continuous, then

$$f_1(y) = \int_\theta h(\theta)g(y|\theta)d\theta$$

is the marginal pdf for the statistic y. Given the information y, the conditional pdf for θ is

$$k(\theta|y) = \frac{h(\theta)g(y|\theta)}{f_1(y)} \quad \text{for } f_1(y) > 0$$

$$= \frac{h(\theta)g(y|\theta)}{\int_0 h(\theta)g(y|\theta)d\theta}.$$

If θ is discrete, then

$$f_1(y) = \sum_k P(\theta_k)P(y|\theta_k)$$

and

$$P(\theta_i|y_j) = \frac{P(\theta_i)P(y_i|\theta_i)}{\sum_k P(\theta_k)P(y_j|\theta_k)},$$

where $P(\theta_i)$ is a prior probability of event θ_i and $P(\theta_i|y_j)$ is a posterior probability of event y_j given θ_i. This is simply a form of Bayes' theorem. Here, $h(\theta)$ is the prior pdf that expresses our belief in the value of θ before the data $(Y = y)$ became available. Then $k(\theta|y)$ is the posterior pdf of given the data $(Y = y)$. Note that the change in the shape of the prior pdf $h(\theta)$ to the posterior pdf $k(\theta|y)$ due to the information is a result of the product of $g(y|\theta)$ and $h(\theta)$ because $f_1(y)$ is simply a normalization constant for a fixed y. The idea in reliability is to take "prior" data and combine it with current data to gain a better estimate or confidence interval or test than would be possible with either singularly. As more current data is acquired, the prior data is "washed out".

Case 1: Binomial Confidence Limits — Uniform Prior   Results from ten missile tests are used to form a one-sided binomial confidence interval of the form:

$$P[R \geq R_L] = 1 - \alpha.$$

From Subsection A.3.3, we have

$$\sum_{i=k}^{10} \binom{10}{i} R_L^i (1 - R_L)^{10-i} = \alpha.$$

Choosing $\alpha = 0.1$, lower limits as a function of the number of missile test successes are shown in Table A.3.

Assume from previous experience that it is known that the true reliability of the missile is somewhere between 0.8 and 1.0 and furthermore that the distribution through this range is uniform. The prior density on R is then

$$g(R) = 5 \qquad 0.8 < R < 1.0.$$

TABLE A.3. Lower limits as a function of the number of missile test successes.

k	R_L	Exact level
10	0.79	0.905
9	0.66	0.904
8	0.55	0.900
7	0.45	0.898
6	0.35	0.905

From the current tests, results are k successes out of ten missile tests, so that for the event A that contained k successes is given by

$$P(A|R) = \binom{10}{k} R^k (1-R)^{10-k}.$$

Applying Bayes' theorem, we obtain

$$g(R|A) = \frac{g(R)P(A|R)}{\int\limits_R g(R)P(A|R)dR}$$

$$= \frac{5\binom{10}{k} R^k (1-R)^{10-k}}{\int\limits_{0.8}^{1.0} 5\binom{10}{k} R^k (1-R)^{10-k}dR}.$$

For the case of $k = 10$,

$$g(R|A) = \frac{R^{10}}{\int\limits_{0.8}^{1.0} R^{10}dR}$$

$$= \frac{11 R^{10}}{0.914} = 12.035\ R^{10}.$$

To obtain confidence limits incorporating the "new" or current data,

$$\int\limits_{R_L}^{1.0} g(R|A)dR = 0.9$$

$$\int\limits_{R_L}^{1.0} 12.035\ R^{10}dR = 0.9.$$

After simplifications, we have

$$R_L^{11} = 0.177$$
$$R_L = 0.855.$$

Limits for the 10/10, 9/10, 8/10, 7/10, and 6/10 cases employing the Bayesian method are given in Table A.4 along with a comparison with the previously calculated limits not employing the prior assumption. Note that the lower limit of 0.8 on the prior cannot be washed out.

TABLE **A.4.** Comparison between limits applying the Bayesian method and those that do not.

k	R_L(uniform [0.8,1] prior)	R_L(no prior)	Exact level
10	0.855	0.79	0.905
9	0.822	0.66	0.904
8	0.812	0.55	0.900
7	0.807	0.45	0.898
6	0.805	0.35	0.905

Case 2: Binomial Confidence Limits — Beta Prior   The prior density of the beta function is

$$g(R) = \frac{(\alpha + \beta + 1)!}{\alpha!\beta!} R^\alpha (1 - R)^\beta.$$

The conditional binomial density function is

$$P(A|R) = \binom{10}{i} R^i (1 - R)^{10-i}.$$

Then we have

$$g(R|A) = \frac{\frac{(\alpha+\beta+1)!}{\alpha!\beta!} R^\alpha (1 - R)^\beta \binom{10}{k} R^k (1 - R)^{10-k}}{\frac{(\alpha+\beta+1)!}{\alpha!\beta!} \int\limits_0^1 R^\alpha (1 - R)^\beta \binom{10}{k} R^k (1 - R)^{10-k} dR}.$$

After simplifications, we obtain

$$g(R|A) = \frac{R^{\alpha+k}(1 - R)^{\beta+10-k}}{\int\limits_0^1 R^{\alpha+k}(1 - R)^{\beta+10-k} dR}.$$

Multiplying and dividing by

$$\frac{(\alpha + \beta + 11)!}{(\alpha + \beta)!(\beta + 10 - k)!}$$

puts the denominator in the form of a beta function with integration over the entire range, and hence, equal to 1. Thus,

$$g(R|A) = \binom{\alpha + \beta + 10}{\alpha + k} R^{\alpha+k}(1 - R)^{\beta+10-k}$$

which again is a beta density function with parameters

$$(\alpha + k) = \alpha' \quad \text{and} \quad (\beta + 10 - k) = \beta'.$$

Integration over $g(R|A)$ from R_L to 1.0 with an integral set to $1 - \alpha$ and a solution of R_L will produce $100(1 - \alpha)\%$ lower confidence bounds on R, that is,

$$\int_{R_L}^{1.0} g(R|A)dR = 1 - \alpha.$$

Case 3: Exponential Confidence Limits — Gamma Prior   For this situation, assume interest is in an upper limit on the exponential parameter λ. The desired statement is of the form

$$p[\lambda < \lambda_U] = 1 - \alpha.$$

If 1,000 hours of test time was accrued with one failure, a 90% upper confidence limit on λ would be

$$p[\lambda < 0.0039] = 0.9.$$

From a study of prior data on the device, assume that λ has a gamma prior density of the form

$$g(\lambda) = \frac{\lambda^{n-1}e^{-\frac{\lambda}{\beta}}}{(n-1)!\beta^n}.$$

With an exponential failure time assumption, the current data in terms of hours of test and failures can be expressed as a Poisson, thus,

$$p(A|\lambda) = \frac{(\lambda T)^n e^{-\lambda T}}{r!},$$

where

r = number of failures,

T = test time, and

A = event which is r failures in T hours of test.

Applying Bayes' results, we have

$$g(\lambda|A) = \frac{\frac{\lambda^{n-1}e^{-\frac{\lambda}{\beta}}}{(n-1)!\beta^n}\frac{(\lambda T)^r e^{-\lambda T}}{r!}}{\int_{\lambda=0}^{\infty} \frac{\lambda^{n-1}e^{-\frac{\lambda}{\beta}}}{(n-1)!\beta^n}\frac{(\lambda T)^r e^{-\lambda T}}{r!}d\lambda}$$

$$= \frac{\lambda^{n+r-1}e^{-\lambda(\frac{1}{\beta}+T)}}{\int_{0}^{\infty} \lambda^{n+r-1}e^{-\lambda(\frac{1}{\beta}+T)d\lambda}}.$$

Note that

$$\int\limits_0^\infty \lambda^{n+r-1} e^{-\lambda(\frac{1}{\beta}+T)} d\lambda = \frac{(n+r-1)!}{\left(\frac{1}{\beta}+T\right)^{n+r}}.$$

Hence,

$$g(\lambda|A) = \frac{\lambda^{n+r-1} e^{-\lambda(\frac{1}{\beta}+T)}(\frac{1}{\beta}+T)^{n+r}}{(n+r-1)!}.$$

Thus, $g(\lambda|A)$ is also a gamma density with parameters $(n+r-1)$ and $1/(1/\beta + T)$. This density can be transformed to the χ^2 density with $2(n+r)$ degree of freedom by the following change of variable. Let

$$\lambda' = 2\lambda(\frac{1}{\beta}+T),$$

then

$$d\lambda = \frac{1}{2}\left(\frac{1}{\frac{1}{\beta}+T}\right) d\lambda'.$$

We have

$$h(\lambda'|A) = \frac{(\lambda')^{\frac{2(n+r)}{2}-1} e^{-\frac{\lambda'}{2}}}{\left[\frac{2(n+r)}{2}-1\right]! 2^{\frac{2(n+r)}{2}}},$$

To obtain a $100(1-\alpha)\%$ upper confidence limit on λ, solve for λ' in the below integral:

$$\int\limits_0^{\lambda'} h(s|A) ds = 1-\alpha$$

and convert λ' to λ via the above transformation.

Example A.22   Given a gamma prior with $n=2$ and $\beta = 0.0001$ and current data as before (i.e., 1,000 hours of test with one failure), the posterior density becomes

$$g(\lambda|A) = \frac{\lambda^2 e^{-\lambda(11,000)}(11,000)^2}{2}$$

converting to χ^2 via the transformation

$$\lambda' = 2\lambda(11{,}000)$$

and

$$h(\lambda'|A) = \frac{(\lambda')^{\frac{6}{2}-1}e^{-\frac{\lambda'}{2}}}{[\frac{6}{2}-1]!2^{\frac{6}{2}}}$$

which is χ^2 with 6 degrees of freedom. Choosing

$$\alpha = 0.1, \qquad \chi_6^2(1-a) = \chi_6^2(0.9) = 10.6$$

and

$$p[\lambda' < 10.6] = 0.9.$$

But $\lambda' = 2\lambda(11{,}000)$, hence,

$$p[2\lambda(11{,}000) < 10.6] = 0.9$$

or

$$p[\lambda < 0.0005] = 0.9.$$

The latter limit conforms to 0.0039 derived without the use of a prior density, i.e., an approximate eight-fold improvement.

The examples shown above involved the development of tighter confidence limits where a prior density of the parameter could be utilized. By the same token, smaller samples for constant and errors could be developed for tests of hypotheses on the parameters. In general, for legitimate applications and where prior data are available, employment of Bayesian methods can reduce cost or give results with less risk for the same dollar value.

Problems

A.1 Let $X_1, X_2, \ldots, X_n$ represent a random sample from the Poisson distribution having pdf

$$f(x; \lambda) = \frac{e^{-\lambda}\lambda^x}{x!} \text{ for } x = 0, 1, 2, \ldots \text{ and } \lambda \geq 0.$$

Find the maximum likelihood estimator $\hat{\lambda}$ of λ.

A.2 Let $X_1, X_2, \ldots, X_n$ be a random sample from the distribution with a discrete pdf

$$P(x) = p^x(1-p)^{1-x} \; x = 0, 1 \quad \text{and } 0 < p < 1.$$

Find the maximum likelihood estimator $\hat{p}$ of p.

A.3 Assume that $X_1, X_2, \ldots, X_n$ represent a random sample from the Pareto distribution, that is,

$$F(x; \lambda, \theta) = 1 - \left(\frac{\lambda}{x}\right)^\theta \text{ for } x \geq \lambda, \lambda > 0, \theta > 0.$$

This distribution is commonly used as a model to study incomes.
Find the maximum likelihood estimators of λ and θ.

A.4 Solve Problem A.1 using the method of moments.

A.5 Let $Y_1 < Y_2 < \cdots < Y_n$ be the order statistics of a random sample $X_1, X_2, \ldots, X_n$ from the distribution with pdf

$$f(x; \theta) = 1$$
$$\text{if } \theta - \frac{1}{2} \leq x \leq \theta + \frac{1}{2}, \quad -\infty < \theta < \infty.$$

Show that any statistic $h(X_1, X_2, \ldots, X_n)$ such that

$$Y_n - \frac{1}{2} \leq h(X_1, X_2, \ldots, X_n) \leq Y_1 + \frac{1}{2}$$

is a maximum likelihood estimator of θ. What can you say about the following functions?

(a) $\dfrac{(4Y_1 + 2Y_n + 1)}{6}$

(b) $\dfrac{(Y_1 + Y_n)}{2}$

(c) $\dfrac{(2Y_1 + 4Y_n - 1)}{6}$

A.6 The lifetime of transistors is assumed to have an exponential distribution with pdf

$$f(x; \theta) = \frac{1}{\theta} e^{-\frac{x}{\theta}} \quad \text{for } x \geq 0 \quad \text{and} \quad \theta > 0.$$

A random sample of size n is observed. Determine the following:

(a) Find an estimator of θ using the method of moments.

(b) Find the maximum likelihood estimator of θ.

(c) Find the MLE of the transistor reliability function, $\hat{R}(t)$, of

$$R(t) = e^{-\frac{t}{\theta}}.$$

References

Bertsbakh, I.B., *Statistical Reliability Theory*, Marcel Dekker, New York, 1989.

Bowker, A.H. and G.J. Lieberman, *Engineering Statistics*, 2nd edition, Prentice-Hall, New Jersey, 1972.

Dai, S.-H. and M.-O. Wang, *Reliability Analysis in Engineering Applications*, Van Nostrand Reinhold, New York, 1992.

Lawless, J.W., *Statistical Models and Methods for Lifetime Data*, John Wiley & Sons, New York, 1982.

Wald, A., *Sequential Analysis*, John Wiley & Sons, New York, 1947.

Appendix B

Stochastic Processes

B.1 Introduction

Stochastic processes are used for the description of a system's operation over time. There are two main types of stochastic processes: continuous and discrete. The complex continuous process is a process describing a system transition from state to state. The simplest process that will be discussed here is a Markov process, i.e., given the current state of the process, its future behavior does not depend on the past.

Among discrete processes, counting processes in reliability engineering are widely used to describe the appearance of events in time, e.g., failures, number of perfect repairs, etc. The simplest counting process is a Poisson process. The Poisson process plays a special role to many applications in reliability engineering. A classic example of such an application is the decay of uranium: Radioactive particles from nuclear material strike a certain target in accordance with a Poisson process of some fixed intensity. A well-known counting process is the so-called renewal process. This process is described as a sequence of events, the intervals between which are independent and identically distributed random variables. In reliability theory, this type of mathematical model is used to describe the number of occurrences of an event in the time interval. In this appendix, we also discuss the quasi-renewal process and the non-homogeneous Poisson process.

A non-negative, integer-valued stochastic process, $N(t)$, is called a counting process if $N(t)$ represents the total number of occurrences of the event in the time interval $[0, t]$ and satisfies these two properties:

(i) if $t_1 < t_2$, then $N(t_1) \le N(t_2)$;
(ii) if $t_1 < t_2$, then $N(t_2) - N(t_1)$ is the number of occurrences of the event in the interval $[t_1, t_2]$.

For example, if $N(t)$ equals the number of persons who have entered a restaurant at or prior to time t, then $N(t)$ is a counting process in which an event occurs whenever a person enters the restaurant.

B.2   Markov Processes

As an introduction to the Markov process, let us examine the following example.

Example B.1   Consider the operation of a system configuration described by the following reliability block diagram as shown in Fig. B.1. From a reliability point of view, the states of the system can be described by

State 1: Full operation (both components operating)
State 2: One component operating — one component failed
State 3: Both components failed.

Define

$$P_i(t) = P[X(t) = i] = P[\text{system is in state } i \text{ at time } t]$$

and

$$P_i(t + dt) = P[X(t + dt) = i$$
$$= P[\text{system is in state } i \text{ at time } t + dt].$$

Define a random variable $X(t)$ which can assume the values 1, 2, or 3 corresponding to the above-mentioned states. Since $X(t)$ is a random variable, one can discuss $P[X(t) = 1]$, $P[X(t) = 2]$ or conditional probability, $P[X(t_1) = 2 | X(t_0) = 1]$. Again, $X(t)$ is defined as a

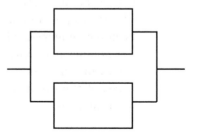

Fɪɢ. **B.1.** Reliability block diagram.

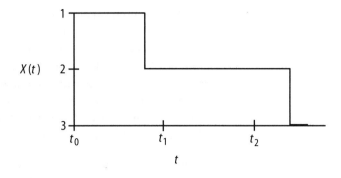

function of time t, the last stated conditional probability, $P[X(t_1) = 2|X(t_0) = 1]$, can be interpreted as the probability of being in state 2 at time t_1, given that the system was in state 1 at time t_0. Figure B.2 is a graphical representation of $X(t)$ as a function of time. In this example, the "stage space" is discrete, i.e., 1, 2, 3, etc., and the parameter space (time) is continuous.

The simple process described above is called a stochastic process, i.e., a process which develops in time (or space) in accordance with some probabilistic (stochastic) laws. There are many types of stochastic processes. In this section, the emphasis will be on Markov processes which are a special type of stochastic process.

Let $t_0 < t_1 < \cdots < t_n$. If

$$P[X(t_n) = A_n|X(t_{n-1}) = A_{n-1}, X(t_{n-2}) = A_{n-2}, \ldots, X(t_0) = A_0]$$
$$= P[X(t_n) = A_n|X(t_{n-1}) = A_{n-1}],$$

then the process is called a Markov process. Given the present state of the process, its future behavior does not depend on past information of the process.

The essential characteristic of a Markov process is that it is a process that has no memory; its future is determined by the present and not the past. If, in addition to having no memory, the process is such that it depends only on the difference $t + dt - t = dt$ and not the value of t, i.e., $P[X(t + dt) = j|X(t) = i]$ is independent of t, then the process is Markov with stationary transition probabilities or homogeneous in time. This is the same property noted in exponential event times, and referring back to the graphical representation of $X(t)$, the times between state changes would in fact be exponential if the process has stationary transition probabilities.

Thus, a Markov process which is time homogeneous can describe processes where events have exponential occurrence times. The random variable of the process is $X(t)$, the state variable rather than the time to failure as in the exponential failure density. To see the types of processes that can be described, a review of the exponential distribution and its properties will be made. Recall that, if $X_1, X_2, \ldots, X_n$ are independent random variables, each with exponential density and a mean equal to $1/\lambda_i$, then $\min\{X_1, X_2, \ldots X_n\}$ has an exponential density with mean $1/\Sigma\lambda_i$.

The significance of the property is as follows:

(1) The failure behavior of the simultaneous operation of components can be characterized by an exponential density with a mean equal to the reciprocal of the sum of the failure rates.
(2) The joint failure/repair behavior of a system where components are operating and/or undergoing repair can be characterized by an exponential density with a mean equal to the reciprocal of the sum of the failure and repair rates.
(3) The failure/repair behavior of a system such as (2) above, but further complicated by active and dormant operating states and sensing and switching, can be characterized by an exponential density.

The above property means that almost all reliability and availability models can be characterized by a time homogeneous Markov process if the various failure times and repair times are exponential. The notation for the Markov process is $\{X(t), t > 0\}$, where $X(t)$ is discrete (state space) and t is continuous (parameter space). By convention, this type of Markov process is called a continuous parameter Markov chain.

From a reliability/availability viewpoint, there are two types of Markov processes. These are defined as follows:

(1) *Absorbing Process*: Contains what is called an "absorbing state" which is a state from which the system can never leave once it has entered, e.g., a failure which aborts a flight or a mission.
(2) *Ergodic Process*: Contains no absorbing states such that $X(t)$ can move around indefinitely, e.g., the operation of a ground power plant where failure only temporarily disrupts the operation.

Figure B.3 gives a summary of the processes to be considered broken down by absorbing and ergodic categories. Both reliability and availability can be described in terms of the probability of the process or system being in defined "up" states, e.g., states 1 and 2 in the initial example. Likewise, meantime between failures (MTBF)

Process	Condition	Measure

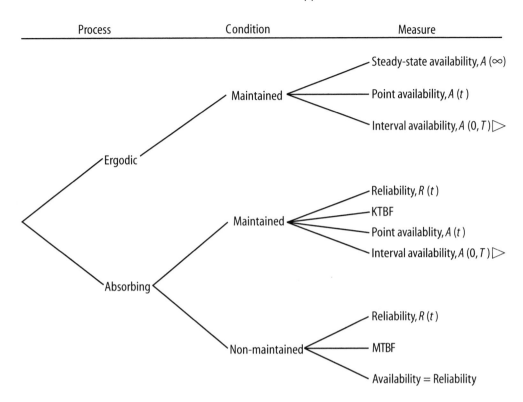

FIG. B.3. Process mode classification.

can be described as the total time in the "up" states before proceeding to the absorbing state or failure state.

Define the incremental transition probability as

$$P_{ij}(dt) = P[X(t + dt) = j | X(t) = i].$$

This is the probability that the process (random variable $X(t)$) will go to state j during the increment t to $t + dt$, given that it was in state i at time t. Since we are dealing with time homogeneous Markov processes, i.e., exponential failure and repair times, the incremental transition probabilities can be derived from an analysis of the exponential hazard function. In Section 2.3, it was shown that the hazard function for the exponential with mean $1/\lambda$ was just λ. This means that the limiting (as $dt \to 0$) conditional probability of an event occurrence between t and $t + dt$, given that an event had not occurred at time t, is just λ, i.e.,

$$h(t) = \lim_{dt \to 0} \frac{P[t < X < t + dt | X > t]}{dt} = \lambda$$

$$h(t)dt = P[t < X < t + dt | X > t] = \lambda dt.$$

The equivalent statement for the random variable $X(t)$ is

$$h(t)dt = P[X(t + dt) = j | X(t) = i] = \lambda dt.$$

Now, $h(t)dt$ is in fact the incremental transition probability, thus the $P_{ij}(dt)$ can be stated in terms of the basic failure and/or repair rates.

Returning to Example B.1, a state transition or flow diagram can be constructed showing the incremental transition probabilities for process or system movement between all possible states (see Fig. B.4).

State 1: Both components operating
State 2: One component up — one component down
State 3: Both components down (absorbing state).

The loops indicate the probability of remaining in the present state during the dt increment.

$$
\begin{array}{lll}
P_{11}(dt) = 1 - 2\lambda dt & P_{12}(dt) = 2\lambda dt & P_{13}(dt) = 0 \\
P_{21}(dt) = 0 & P_{22}(dt) = 1 - \lambda dt & P_{23}(dt) = \lambda dt \\
P_{31}(dt) = 0 & P_{32}(dt) = 0 & P_{33}(dt) = 1.
\end{array}
$$

The zeros on P_{ij}, $i > j$, denote that the process cannot go backwards, i.e., this is not a repair process. The zero on P_{13} denotes that in a process of this type, the probability of more than one event (e.g., failure, repair, etc.) in the incremental time period dt approaches zero as dt approaches zero.

Except for the initial conditions of the process, i.e., the state in which the process starts, the process is completely specified by the incremental transition probabilities. The reason for the latter is

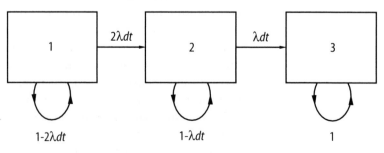

FIG. **B.4.** State-transition diagram.

that the assumption of exponential event (failure or repair) times allows the process to be characterized at any time t since it depends only on what happens between t and $t + dt$. With this in mind and utilizing the same example, determination of the probabilities associated with occupancy in each state will be made.

The incremental transition probabilities can be arranged into a matrix in a way which depicts all possible statewide movements. Thus, for the parallel configurations (see Fig. B.4),

$$[p_{ij}(dt)] = \begin{bmatrix} & 1 & 2 & 3 \\ 1 - 2\lambda dt & 2\lambda dt & 0 \\ 0 & 1 - \lambda dt & \lambda dt \\ 0 & 0 & 1 \end{bmatrix}$$

for $i, j = 1, 2,$ or 3. The matrix $[P_{ij}(dt)]$ is called the incremental, one-step transition matrix. It is a stochastic matrix, i.e., the rows sum to 1.0. As mentioned earlier, this matrix along with the initial conditions completely describes the process.

Now, $[P_{ij}(dt)]$ gives the probabilities for either remaining or moving to all the various states during the interval t to $t + dt$, hence,

$$P_1(t + dt) = (1 - 2\lambda dt)P_1(t)$$
$$P_2(t + dt) = 2\lambda dt P_1(t)(1 - \lambda dt)P_2(t)$$
$$P_3(t + dt) = \lambda dt P_2(t) + P_3(t).$$

By algebraic manipulation, we have

$$\frac{[P_1(t + dt) - P_1(t)]}{dt} = -2\lambda P_1(t)$$
$$\frac{[P_2(t + dt) - P_2(t)]}{dt} = 2\lambda P_1(t) - \lambda P_2(t)$$
$$\frac{[P_3(t + dt) - P_3(t)]}{dt} = \lambda P_2(t).$$

Taking limits of both sides as $dt \to 0$, we obtain

$$P'_1(t) = -2\lambda P_1(t)$$
$$P'_2(t) = 2\lambda P_1(t) - \lambda P_2(t)$$
$$P'_3(t) = \lambda P_2(t).$$

Note that

$$\lim_{dt \to 0} \frac{[P_i(t + dt) - P_i(t)]}{dt} = P'_i(t),$$

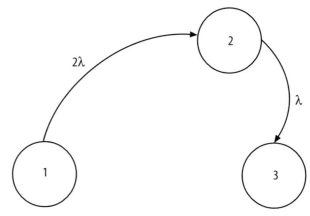

Fig. **B.5.** State-transition diagram.

as a basic definition of derivative. The above system of differential equations can be represented in Fig. B.5. If the above system of linear, first-order differential equations can be solved for $P_1(t)$ and $P_2(t)$, then the reliability of the configuration can be determined.

$$R(t) = P_1(t) + P_2(t).$$

Actually, there is no need to solve all three equations, but only the first two as $P_3(t)$ does not appear and also $P_3(t) = 1 - P_1(t) - P_2(t)$.

Systems of linear, first-order differential equations can be solved by various means including both manual and machine methods. For purposes here, the manual methods employing the LaPlace transform (see Table B.1 at the end of the chapter) will be used.

$$L[P_i(t)] = \int_0^\infty e^{-st} P_i(t)dt = f_i(s)$$

$$L[P'_i(t)] = \int_0^\infty e^{-st} P'_i(t)dt$$

$$= sf_i(s) - P_i(t)|_{t=0} \quad \text{(integration by parts)}.$$

The use of the LaPlace transform will allow transformation of the system of linear, first-order differential equations into a system of linear algebraic equations which can easily be solved, and by means of the inverse transforms, solutions of $P_i(t)$ can be determined.

Returning to the example, the initial condition of the parallel configuration is assumed to be "full-up" such that

$$P_1(t = 0) = 1, \; P_2(t = 0) = 0, \; P_3(t = 0) = 0$$

transforming the equations for $P'_1(t)$ and $P'_2(t)$ gives

$$sf_1(s) - P_1(t)|_{t=0} = -2\lambda f_1(s)$$
$$sf_2(s) - P_2(t)|_{t=0} = 2\lambda f_1(s) - \lambda f_2(s).$$

Evaluating $P_1(t)$ and $P_2(t)$ at $t = 0$ gives

$$sf_1(s) - 1 = -2\lambda f_1(s)$$
$$sf_2(s) - 0 = 2\lambda f_1(s) - \lambda f_2(s)$$

from which we obtain

$$(s + 2\lambda)f_1(s) = 1$$
$$-2\lambda f_1(s) + (s + \lambda)f_2(s) = 0.$$

Solving the above equations for $f_1(s)$ and $f_2(s)$, we have

$$f_1(s) = \frac{1}{(s + 2\lambda)}$$

$$f_2(s) = \frac{2\lambda}{[(s + 2\lambda)(s + \lambda)]}.$$

From Table B.2 of the inverse LaPlace transforms,

$$P_1(t) = e^{-2\lambda t}$$

$$P_2(t) = 2e^{-\lambda t} - 2e^{-2\lambda t}$$

$$R(t) = P_1(t) + P_2(t) = 2e^{-\lambda t} - e^{-2\lambda t}.$$

The example given above is that of a simple absorbing process where we are concerned about reliability. If repair capability in the form of a repair rate μ were added to the model (see Fig. B.6), the methodology would remain the same with only the final result changing.

With a repair rate μ added to the parallel configuration, the incremental transition matrix would be

$$[P_{ij}(dt)] = \begin{bmatrix} & 1 & 2 & 3 \\ 1 - 2\lambda dt & 2\lambda dt & 0 \\ \mu dt & 1 - (\lambda + \mu)dr & \lambda dt \\ 0 & 0 & 1 \end{bmatrix}.$$

The differential equations would become

$$P'_1(t) = -2\lambda P_1(t) + \mu P_2(t)$$
$$P'_2(t) = 2\lambda P_1(t) - (\lambda + \mu)P_2(t),$$

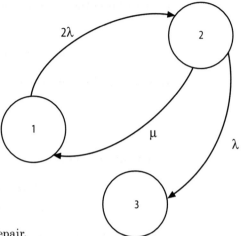

FIG. **B.6.** State-transition diagram with repair.

and the transformed equations would become

$$(s + 2\lambda)f_1(s) - \mu f_2(s) = 1$$
$$-2\lambda f_1(s) + (s + \lambda + \mu)f_2(s) = 0.$$

Hence, we obtain

$$f_1(s) = \frac{(s + \lambda + \mu)}{(s - s_1)(s - s_2)}$$
$$f_2(s) = \frac{2\lambda}{(s - s_1)(s - s_2)},$$

where

$$s_1 = \frac{-(3\lambda + \mu) + \sqrt{(3\lambda + \mu)2 - 8\lambda^2}}{2}$$
$$s_2 = \frac{-(3\lambda + \mu) - \sqrt{(3\lambda + \mu)2 - 8\lambda^2}}{2}.$$

From Table B.1, we obtain

$$P_1(t) = \frac{(s_1 + \lambda + \mu)e^{-s_1 t}}{(s_1 - s_2)} + \frac{(s_2 + \lambda + \mu)e^{-s_2 t}}{(s_2 - s_1)}$$
$$P_2(t) = \frac{2\lambda e^{-s_1 t}}{(s_1 - s_2)} + \frac{2\lambda e^{-s_2 t}}{(s_2 - s_1)}.$$

Thus, the reliability of two-component in a parallel system is given by

$$R(t) = P_1(t) + P_2(t)$$
$$= \frac{(s_1 + 3\lambda + \mu)e^{s_1 t} - (s_2 + 3\lambda + \mu)e^{-s_2 t}}{(s_1 - s_2)}.$$

B.2.1 System Mean Time Between Failures

Another parameter of interest in absorbing Markov processes is the MTBF. Recalling the previous example of a parallel configuration with repair, the differential equations $P'_1(t)$ and $P'_2(t)$ describing the process were

$$P'_1(t) = -2\lambda P_i(t) + \mu P_2(t)$$
$$P'_2(t) = 2\lambda P_1(t) - (\lambda + \mu)P_2(t).$$

Integrating both sides of the above equations yields

$$\int_0^\infty P'_1(t)dt = -2\lambda \int_0^\infty P_1(t)dt + \mu \int_0^\infty P_2(t)dt$$
$$\int_0^\infty P'_2(t)dt = 2\lambda \int_0^\infty P_1(t)dt - (\lambda + \mu) \int_0^\infty P_2(t)dt.$$

Recall from Chapter 2 that

$$\int_0^\infty R(t)dt = \text{MTBF}.$$

Similarly,

$$\int_0^\infty P_1(t)dt = \text{mean time spent in state 1, and}$$
$$\int_0^\infty P_2(t)dt = \text{mean time spent in state 2.}$$

Designating these mean times as T_1 and T_2, respectively, we have

$$P_1(t)|_0^\infty = -2\lambda T_1 + \mu T_2$$
$$P_2(t)|_0^\infty = 2\lambda T_1 - (\lambda + \mu)T_2.$$

But $P_1(t) = 0$ as $t \to \infty$ and $P_1(t) = 1$ for $t = 0$. Likewise, $P_2(t) = 0$ as $t \to \infty$ and $P_2(t) = 0$ for $t = 0$. Thus,

$$-1 = -2\lambda T_1 + \mu T_2$$
$$0 = 2\lambda T_1 - (\lambda + \mu)T_2,$$

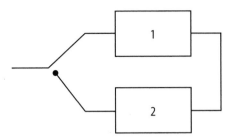

FIG. **B.7.** A cold-standby system.

or, equivalently, that

$$\begin{bmatrix} -1 \\ 0 \end{bmatrix} = \begin{bmatrix} -2\lambda & \mu \\ 2\lambda & -(\lambda + \mu) \end{bmatrix} \begin{bmatrix} T_1 \\ T_2 \end{bmatrix}.$$

Therefore,

$$T_1 = \frac{(\lambda + \mu)}{(2\lambda^2)} \quad T_2 = \frac{1}{\lambda}$$

$$\text{MTBF} = T_1 + T_2 = \frac{(\lambda + \mu)}{(2\lambda^2)} + \frac{1}{\lambda} = \frac{(3\lambda + \mu)}{(2\lambda^2)}.$$

The MTBF for non-maintenance processes is developed exactly the same way as just shown. What remains under absorbing processes is the case for availability for maintained systems. The difference between reliability and availability for absorbing processes is somewhat subtle. A good example is that of a communication system where, if such a system failed temporarily, the mission would continue, but, if it failed permanently, the mission would be aborted. Consider the following cold-standby configuration (see Fig. B.7):

State 1: Main unit operating — spare OK
State 2: Main unit out — restoration underway
State 3: Spare unit installed and operating
State 4: Permanent failure (no spare available).

The incremental transition matrix is given by (see Fig. B.8)

$$[P_{ij}(dt)] = \begin{bmatrix} 1 - \lambda dt & \lambda dt & 0 & 0 \\ 0 & 1 - \mu dt & \mu dt & 0 \\ 0 & 0 & 1 - \lambda dt & \lambda dt \\ 0 & 0 & 0 & 1 \end{bmatrix}.$$

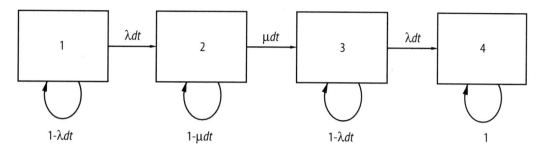

FIG. B.8. State-transition diagram for the cold-standby system.

We obtain

$$P'_1(t) = -\lambda P_1(t)$$
$$P'_2(t) = \lambda P_1(t) - \mu P_2(t)$$
$$P'_3(t) = \mu P_2(t) - \lambda P_3(t).$$

Using the LaPlace transform, we obtain

$$sf_1(s) - 1 = -\lambda f_1(s)$$
$$sf_2(s) = \lambda f_1(s) - \mu f_2(s)$$
$$sf_3(s) = \mu f_2(s) - \lambda f_3(s).$$

After simplifications,

$$f_1(s) = \frac{1}{(s + \lambda)}$$

$$f_2(s) = \frac{\lambda}{[(s + \lambda)(s + \mu)]}$$

$$f_3(s) = \frac{\lambda\mu}{[(s + \lambda)^2(s + \mu)]}.$$

Therefore, the probability of full-up performance, $P_1(t)$, is given by

$$P_1(t) = e^{-\lambda t}.$$

Similarly, the probability of system down and under repair, $P_2(t)$, is

$$P_2(t) = \left[\frac{\lambda}{(\lambda - \mu)}\right][e^{-\mu t} - e^{-\lambda t}]$$

and the probability of system full-up but no spare available, $P_3(t)$, is

$$P_3(t) = \left[\frac{\lambda\mu}{(\lambda - \mu)^2}\right][e^{-\mu t} - e^{-\lambda t} - (\lambda - \mu)te^{-\lambda t}].$$

Hence, the point availability, $A(t)$, is given by

$$A(t) = P_1(t) + P_3(t).$$

If average or interval availability is required, this is achieved by

$$\left(\frac{1}{t}\right) \int_0^T A(t)dt = \left(\frac{1}{t}\right) \int_0^T [P_1(t) + P_3(t)]dt,$$

where T is the interval of concern.

With the above example, cases of the absorbing process (both maintained and non-maintained) have been covered insofar as "manual" methods are concerned. In general, the methodology for treatment of absorbing Markov processes can be "packaged" in a fairly simplified form by utilizing matrix notation. Thus, for example, if the incremental transition matrix is defined as follows:

$$[P_{ij}(dt)] = \begin{bmatrix} 1 - 2\lambda dt & 2\lambda dt & 0 \\ \mu dt & 1 - (\mu + \lambda)dt & \lambda dt \\ 0 & 0 & 1 \end{bmatrix},$$

then if the dt's are dropped and the last row and the last column are deleted, the remainder is designated as the matrix T

$$[T] = \begin{bmatrix} 1 - 2\lambda & 2\lambda \\ \mu & 1 - (\mu + \lambda) \end{bmatrix}.$$

Define $[Q] = [T]' - [I]$, where $[T]'$ is the transposition of $[T]$ and $[I]$ is the unity matrix.

$$[Q] = \begin{bmatrix} 1 - 2\lambda & \mu \\ 2\lambda & 1 - (\mu + \lambda) \end{bmatrix} - \begin{bmatrix} 1 & 0 \\ 0 & 1 \end{bmatrix}$$

$$= \begin{bmatrix} -2\lambda & \mu \\ 2\lambda & -(\mu + \lambda) \end{bmatrix}.$$

Further define $[P(t)]$ and $[P'(t)]$ as column vectors such that

$$[P(t)] = \begin{bmatrix} P_1(t) \\ P_2(t) \end{bmatrix}, \quad [P'(t)] = \begin{bmatrix} P'_1(t) \\ P'_2(t) \end{bmatrix},$$

then

$$[P'(t)] = [Q][P(t)].$$

At the above point, solution of the system of differential equations will produce solutions to $P_1(t)$ and $P_2(t)$. If the MTBF is desired, integration of both sides of the system produces

$$\begin{bmatrix} -1 \\ 0 \end{bmatrix} = [Q] \begin{bmatrix} T_1 \\ T_2 \end{bmatrix}$$

$$\begin{bmatrix} -1 \\ 0 \end{bmatrix} = \begin{bmatrix} -2\lambda & \mu \\ 2\lambda & -(\mu + \lambda) \end{bmatrix} \begin{bmatrix} T_1 \\ T_2 \end{bmatrix} \text{ or}$$

$$[Q]^{-1} \begin{bmatrix} 1 \\ 0 \end{bmatrix} = \begin{bmatrix} T_1 \\ T_2 \end{bmatrix},$$

where $[Q]^{-1}$ is the inverse of $[Q]$ and the mean time to failure is given by

$$\text{MTBF} = T_1 + T_2 = \frac{[3\lambda + \mu]}{(2\lambda)^2}.$$

In the more general MTBF case,

$$[Q]^{-1} \begin{bmatrix} -1 \\ 0 \\ \cdot \\ \cdot \\ \cdot \\ 0 \end{bmatrix} = \begin{bmatrix} T_1 \\ T_2 \\ \cdot \\ \cdot \\ \cdot \\ T_{n-1} \end{bmatrix} \text{ where } \sum_{i=1}^{n-1} T_i = \text{MTBF}$$

and $(n-1)$ is the number of non-absorbing states.

For the reliability/availability case, utilizing the LaPlace transform, the system of linear, first-order differential equations is transformed to

$$s \begin{bmatrix} f_1(s) \\ f_2(s) \end{bmatrix} - \begin{bmatrix} 1 \\ 0 \end{bmatrix} = [Q] \begin{bmatrix} f_1(s) \\ f_2(s) \end{bmatrix}$$

$$[sI - Q] \begin{bmatrix} f_1(s) \\ f_2(s) \end{bmatrix} = \begin{bmatrix} 1 \\ 0 \end{bmatrix}$$

$$\begin{bmatrix} f_1(s) \\ f_2(s) \end{bmatrix} = [sI - Q]^{-1} \begin{bmatrix} 1 \\ 0 \end{bmatrix}$$

$$L^{-1} \begin{bmatrix} f_1(s) \\ f_2(s) \end{bmatrix} = L^{-1} \{ [sI - Q]^{-1} \begin{bmatrix} 1 \\ 0 \end{bmatrix} \}$$

$$\begin{bmatrix} p_1(t) \\ p_2(t) \end{bmatrix} = L^{-1} \{ [sI - Q]^{-1} \begin{bmatrix} 1 \\ 0 \end{bmatrix} \}.$$

Generalization of the latter to the case of $(n-1)$ non-absorbing states is straightforward.

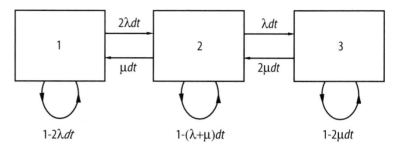

FIG. **B.9.** State-transition diagram with repair.

Ergodic processes, as opposed to absorbing processes, do not have any absorbing states, and hence, movement between states can go on indefinitely. For the latter reason, availability (point, steady-state, or interval) is the only meaningful measure. As an example for ergodic processes, a ground-based power unit configured in parallel will be selected.

The parallel units are identical, each with exponential failure and repair times with means $1/\lambda$ and $1/\mu$, respectively (see Fig. B.9). Assume a two-repairmen capability if required (both units down), then

State 1: Full-up (both units operating)
State 2: One unit down and under repair (other unit up)
State 3: Both units down and under repair.

It should be noted that as in the case of failure events, two or more repairs cannot be made in the dt interval.

$$[P_{ij}(dt)] - \begin{bmatrix} 1 - 2\lambda dt & 2\lambda dt & 0 \\ \mu dt & 1 - (\lambda + \mu)dt & \lambda dt \\ 0 & 2\mu dt & 1 - 2\mu dt \end{bmatrix}.$$

Case I: Point Availability — Ergodic Process   For an ergodic process, as $t \to \infty$, the availability settles down to a constant level. Point availability gives a measure of things before the "settling down" and reflects the initial conditions on the process. Solution of the point availability is similar to the case for absorbing processes except that the last row and column of the transition matrix must be retained and entered into the system of equations. For example, the system of differential equations becomes

$$\begin{bmatrix} P_1'(t) \\ P_2'(t) \\ P_3'(t) \end{bmatrix} = \begin{bmatrix} -2\lambda & \mu & 0 \\ 2\lambda & -(\lambda + \mu) & 2\mu \\ 0 & \lambda & -2\mu \end{bmatrix} \begin{bmatrix} P_1(t) \\ P_2(t) \\ P_3(t) \end{bmatrix}.$$

Similar to the absorbing case, the method of the LaPlace transform can be used to solve for $P_1(t)$, $P_2(t)$, and $P_3(t)$, with the point availability, $A(t)$, given by

$$A(t) = P_1(t) + P_2(t).$$

Case II: Interval Availability — Ergodic Process   This is the same as the absorbing case with integration over time period T of interest. The interval availability, $A(T)$, is

$$A(T) = \frac{1}{T} \int_0^T A(t)dt.$$

Case III: Steady State Availability — Ergodic Process   Here, the process is examined as $t \to \infty$ with complete "washout" of the initial conditions. Letting $t \to \infty$, the system of differential equations can be transformed to linear algebraic equations. Thus,

$$\lim_{t \to \infty} \begin{bmatrix} P_1'(t) \\ P_2'(t) \\ P_3'(t) \end{bmatrix} = \lim_{t \to \infty} \left\{ \begin{bmatrix} -2\lambda & \mu & 0 \\ 2\lambda & -(\lambda + \mu) & 2\mu \\ 0 & \lambda & -2u \end{bmatrix} \begin{bmatrix} P_1(t) \\ P_2(t) \\ P_3(t) \end{bmatrix} \right\}.$$

As $t \to \infty$, $P_i(t) \to$ constant and $P_1'(t) \to 0$. This leads to an unsolvable system, namely,

$$\begin{bmatrix} 0 \\ 0 \\ 0 \end{bmatrix} = \begin{bmatrix} -2\lambda & \mu & 0 \\ 2\lambda & -(\lambda + \mu) & 2\mu \\ 0 & \lambda & -2\mu \end{bmatrix} \begin{bmatrix} P_1(t) \\ P_2(t) \\ P_3(t) \end{bmatrix}.$$

To avoid the above difficulty, an additional equation is introduced:

$$\sum_{i=1}^3 P_i(t) = 1.$$

With the introduction of the new equation, one of the original equations is deleted and a new system is formed:

$$\begin{bmatrix} 1 \\ 0 \\ 0 \end{bmatrix} = \begin{bmatrix} 1 & 1 & 1 \\ -2\lambda & \mu & 0 \\ 2\lambda & -(\lambda + \mu) & 2\mu \end{bmatrix} \begin{bmatrix} P_1(t) \\ P_2(t) \\ P_3(t) \end{bmatrix},$$

or, equivalently, that

$$\begin{bmatrix} P_1(t) \\ P_2(t) \\ P_3(t) \end{bmatrix} = \begin{bmatrix} 1 & 1 & 1 \\ -2\lambda & \mu & 0 \\ 2\lambda & -(\lambda + \mu) & 2\mu \end{bmatrix}^{-1} \begin{bmatrix} 1 \\ 0 \\ 0 \end{bmatrix}.$$

We now obtain the following results:

$$P_1(t) = \frac{\mu^2}{(\mu + \lambda)^2}$$

$$P_2(t) = \frac{2\lambda\mu}{(\mu + \lambda)^2}$$

and

$$P_3(t) = 1 - P_1(t) - P_2(t)$$
$$= \frac{\lambda^2}{(\mu + \lambda)^2}.$$

Therefore, the steady state availability, $A(\infty)$, is given by

$$A(\infty) = P_1(t) + P_2(t)$$
$$= \frac{\mu(\mu + 2\lambda)}{(\mu + \lambda)^2}.$$

Note that Markov methods can also be employed where failure or repair times are not exponential, but can be represented as the sum of exponential times with identical means (Erlang distribution or Gamma distribution with integer valued shape parameters). Basically, the method involves the introduction of "dummy" states which are of no particular interest in themselves, but serve the purpose of changing the hazard function from constant to increasing.

Example B.2   We now present two Markov models (Cases 1 and 2 below) which allow integration of control systems of nuclear power plant reliability and safety analysis. A basic system transition diagram for both models is presented in Fig. B.10. In both models, it is assumed that the control system is composed of a control rod and an associated safety system. The following assumptions are applied in this example.

(a) All failures are statistically independent.
(b) Each unit has a constant rate.
(c) The control system fails when the control rod fails.

The following notations are associated with the system shown in Fig. B.10.

i ith state of the system: $i = 1$ (control and its associated safety system operating normally), $i = 2$ (control operating

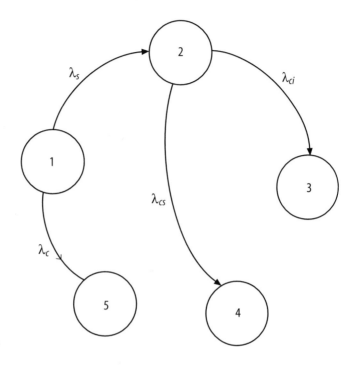

FIG. B.10. System state diagram.

normally, safety system failed), $i = 3$ (control failed with an accident), $i = 4$ (control failed safely), $i = 5$ (control failed but its associated safety system operating normally).

$P_i(t)$ — probability that the control system is in state i at time t, $i = 1, 2, \ldots, 5$.

λ_i — ith constant failure rate: $i = s$ (state 1 to state 2), $i = ci$ (state 2 to state 3), $i = cs$ (state 2 to state 4), $i = c$ (state 1 to state 5)

$P_i(s)$ — LaPlace transform of the probability that the control system is in state i; $i = 1, 2, \ldots, 5$.

s — LaPlace transform variable.

Case 1 The system represented by Model 1 is shown in Fig. B.10. Using the Markov approach, the system of differential equations (associated with Fig. B.10) is given below:

$$P'_1(t) = -(\lambda_s + \lambda_c)P_1(t)$$
$$P'_2(t) = \lambda_s P_1(t) - (\lambda_{ci} + \lambda_{cs})P_2(t)$$
$$P'_3(t) = \lambda_{ci}P_2(t)$$
$$P'_4(t) = \lambda_{cs}P_2(t)$$
$$P'_5(t) = \lambda_c P_1(t).$$

Assume that at time $t = 0$, $P_1(0) = 1$ and $P_2(0) = P_3(0) = P_4(0) = P_5(0) = 0$. Solving the above system of equations, we obtain

$$P_1(t) = e^{-At}$$

$$P_2(t) = \frac{\lambda_s}{B}(e^{-Ct} - e^{-At})$$

$$P_3(t) = \frac{\lambda_s \lambda_{ci}}{AC}\left[1 - \frac{Ae^{-Ct} - Ce^{-At}}{B}\right]$$

$$P_4(t) = \frac{\lambda_s \lambda_{cs}}{AC}\left[1 - \frac{Ae^{-Ct} - Ce^{-At}}{B}\right]$$

$$P_5(t) = \frac{\lambda_c}{A}[1 - e^{-At}],$$

where

$$A = \lambda_s + \lambda_c$$
$$B = \lambda_s + \lambda_c - \lambda_{cs} - \lambda_{ci}$$
$$C = \lambda_{ci} + \lambda_{cs}.$$

The reliability of both the control and its safety system working normally, R_{cs}, is given by

$$R_{cs}(t) = P_1(t) = e^{-At}.$$

The reliability of the control system working normally with or without the safety system functioning successfully is

$$R_{ss}(t) = P_1(t) + P_2(t)$$
$$= e^{-At} + \frac{\lambda_s}{B}(e^{-Ct} - e^{-At}).$$

The mean time to failure (MTTF) of the control with the safety system up is

$$\text{MTTF}_{cs} = \int_0^\infty R_{cs}(t)dt$$
$$= \frac{1}{A}.$$

Similarly, the MTTF of the control with the safety system up or down is

$$\text{MTTF}_{ss} = \int_0^\infty R_{ss}(t)dt$$

$$= \frac{1}{A} + \frac{\lambda_s}{B}\left(\frac{1}{C} - \frac{1}{A}\right).$$

Case 2 This model is the same as Case 1 except that a repair is allowed when the safety system fails with a constant rate μ. The system of differential equations for this model is as follows:

$$P'_1(t) = \mu P_2(t) - AP_1(t)$$
$$P'_2(t) = \lambda_s P_1(t) - (\lambda_{ci} + \lambda_{cs} + \mu)P_2(t)$$
$$P'_3(t) = \lambda_{ci} P_2(t)$$
$$P'_4(t) = \lambda_{cs} P_2(t)$$
$$P'_5(t) = \lambda_c P_1(t).$$

We assume that at time $t = 0$, $P_1(0) = 1$ and $P_2(0) = P_3(0) = P_4(0) = P_5(0) = 0$. Solving the above system of equations, we obtain

$$P_1(t) = e^{-At} + \mu\lambda_s\left[\frac{e^{-At}}{(r_1 + A)(r_2 + A)} + \frac{e^{r_1 t}}{(r_1 + A)(r_1 - r_2)}\right.$$
$$\left. + \frac{e^{r_2 t}}{(r_2 + A)(r_2 - r_1)}\right]$$

$$P_2(t) = \lambda_2 \frac{e^{r_1 t} - e^{r_2 t}}{r_1 - r_2}$$

$$P_3(t) = \left(\frac{r_{ci}\lambda_s}{r_1 r_2}\right)\left[\frac{r_1 e^{r_2 t} - r_2 e^{r_1 t}}{r_2 - r_1} + 1\right]$$

$$P_4(t) = \left(\frac{r_{cs}\lambda_s}{r_1 r_2}\right)\left[\frac{r_1 e^{r_2 t} - r_2 e^{r_1 t}}{r_2 - r_1} + 1\right]$$

$$P_5(t) = \frac{\lambda_c}{A}(1 - e^{-At}) + \mu\lambda_s\lambda_c\left[\frac{1}{r_1 r_2 A} - \frac{e^{-At}}{A(r_1 + A)(r_2 + A)}\right.$$
$$\left. + \frac{e^{r_1 t}}{r_1(r_1 + A)(r_1 - r_2)} + \frac{e^{r_2 t}}{r_2(r_2 + A)(r_2 - r_1)}\right],$$

where

$$r_1, r_2 = \frac{-a \pm \sqrt{a^2 - 4b}}{2},$$
$$a = A + C + u, \text{ and}$$
$$b = \lambda_{ci}\lambda_s + \lambda_{cs}\lambda_s + (\lambda_{ci} + \lambda_{cs} + \mu)\lambda_c.$$

The reliability of both the control and its associated safety system working normally with the safety repairable system is

$$R_{cs}(t) = e^{-At}$$

$$+ \mu\lambda_s \left[\frac{e^{-At}}{(r_1 + A)(r_2 + A)} + \frac{e^{r_1 t}}{(r_1 + A)(r_1 - r_2)} + \frac{e^{r_2 t}}{(r_2 + A)(r_2 - r_1)} \right]$$

The reliability of the control operating normal with or without the safety system operating (but having safety system repair) is

$$R_{ss}(t) = e^{-At} + \frac{\lambda_s(e^{r_1 t} - e^{r_2 t})}{r_1 - r_2}$$

$$+ \mu\lambda_s \left[\frac{e^{-At}}{(r_1 + A)(r_2 + A)} + \frac{e^{r_1 t}}{(r_1 + A)(r_1 - r_2)} + \frac{e^{r_2 t}}{(r_2 + A)(r_2 - r_1)} \right]$$

The MTTF of the control with the safety system operating is

$$\mathrm{MTTF}_{cs} = \int_0^\infty R_{cs}(t)dt$$

$$= \frac{1}{A}(1 + \frac{\mu\lambda_s}{b}).$$

We can see that the repair process has helped to improve the system's MTTF. Similarly, the MTTF of the control with the safety system up or down but with accessible repair is given by

$$\mathrm{MTTF}_{ss} = \int_0^\infty R_{ss}(t)dt$$

$$= \frac{1}{A}(1 + \frac{\mu\lambda_s}{b}) + \frac{\lambda_s}{A}.$$

B.3   Poisson Processes

One of the most important counting processes is the Poisson process.

Definition B.1 A counting process, $N(t)$, is said to be a Poisson process with intensity λ if

(i) the failure process, $N(t)$, has stationary independent increments;

(ii) the number of failures in any time interval of length s has a Poisson distribution with mean λs, that is,

$$P\{N(t+s) - N(t) = n\} = \frac{e^{-\lambda s}(\lambda s)^n}{n!} \quad n = 0, 1, 2, \ldots;$$

(iii) the initial condition is $N(0) = 0$.

This model is also called a homogeneous Poisson process indicating that the failure rate λ does not depend on time t. In other words, the number of failures occurring during the time interval $(t, t + s]$ does not depend on the current time t but only the length of time interval s. A counting process is said to possess independent increments if the number of events in disjoint time intervals are independent.

For a stochastic process with independent increments, the autocovariance function is

$$\text{Cov}[X(t_1), X(t_2)] = \begin{cases} \text{Var}[N(t_1 + s) - N(t_2)] & \text{for } 0 < t_2 - t_1 < s \\ 0 & \text{otherwise,} \end{cases}$$

where

$$X(t) = N(t + s) - N(t).$$

If $X(t)$ is Poisson distributed, then the variance of the Poisson distribution is

$$\text{Cov}[X(t_1), X(t_2)] = \begin{cases} \lambda[s - (t_2 - t_1)] & \text{for } 0 < t_2 - t_1 < s \\ 0 & \text{otherwise.} \end{cases}$$

This result shows that the Poisson increment process is covariance stationary. We now present several properties of the Poisson process.

Property B.1 The sum of independent Poisson processes, $N_1(t)$, $N_2(t)$, $\ldots$, $N_k(t)$, with mean values $\lambda_1 t$, $\lambda_2 t$, $\ldots$, $\lambda_k t$, respectively, is also a Poisson process with mean $(\lambda_1 + \lambda_2 + \cdots + \lambda_k)t$. In other words, the sum of the independent Poisson processes is also a Poisson process with a mean that is equal to the sum of the individual Poisson process' mean.

Property B.2 The difference of two independent Poisson processes, $N_1(t)$ and $N_2(t)$, with mean $\lambda_1 t$ and $\lambda_2 t$, respectively, is not a Poisson process. Instead, it has the probability mass function:

$$P[N_1(t) - N_2(t) = k] = e^{-(\lambda_1+\lambda_2)t} \left(\frac{\lambda_1}{\lambda_2}\right)^{\frac{k}{2}} I_k(2\sqrt{\lambda_1\lambda_2 t}),$$

where $I_k(.)$ is a modified Bessel function of order k [ ].

Property B.3   If the Poisson process, $N(t)$, with mean λt, is filtered such that every occurrence of the event is not completely counted, then the process has a constant probability p of being counted. The result of this process is a Poisson process with mean $\lambda p t$ [ ].

Property B.4   Let $N(t)$ be a Poisson process and Y_n a family of independent and identically distributed random variables which are also independent of $N(t)$. A stochastic process $X(t)$ is said to be a compound Poisson process if it can be represented as

$$X(t) = \sum_{i=1}^{N(t)} Y_i.$$

B.4   Renewal Processes

A renewal process is a more general case of the Poisson process in which the inter-arrival times of the process or the time between failures do not necessarily follow the exponential distribution. For convenience, we will call the occurrence of an event a renewal, the inter-arrival time the renewal period, and the waiting time the renewal time.

Definition B.2   A counting process $N(t)$ that represents the total number of occurrences of an event in the time interval $(0, t]$ is called a renewal process, if the time between failures are independent and identically distributed random variables.

The probability that there are exactly n failures occurring by time t can be written as

$$P\{N(t) = n\} = P\{N(t) \geq n\} - P\{N(t) > n\}. \tag{B.1}$$

Note that the times between the failures are $T_1, T_2, \ldots, T_n$ so the failures occurring at time W_k are

$$W_k = \sum_{i=1}^{k} T_i$$

and

$$T_k = W_k - W_{k-1}.$$

Thus,

$$\begin{aligned}
P\{N(t) = n\} &= P\{N(t) \geq n\} - P\{N(t) > n\} \\
&= P\{W_n \leq t\} - P\{W_{n+1} \leq t\} \\
&= F_n(t) - F_{n+1}(t),
\end{aligned}$$

where $F_n(t)$ is the cumulative distribution function for the time of the nth failure and $n = 0, 1, 2, \ldots$.

Example B.3 Consider a software testing model for which the time to find an error during the testing phase has an exponential distribution with a failure rate of λ. It can be shown that the time of the nth failure follows the gamma distribution with parameters λ and n with probability density function. From Eq. (B.1), we obtain

$$\begin{aligned}
P\{N(t) = n\} &= P\{N(t) \leq n\} - P\{N(t) \geq n - 1\} \\
&= \sum_{k=0}^{n} \frac{(\lambda t)^k}{k!} e^{-\lambda t} - \sum_{k=0}^{n-1} \frac{(\lambda t)^k}{k!} e^{-\lambda t} \\
&= \frac{(\lambda t)^n}{n!} e^{-\lambda t} \quad \text{for } n = 0, 1, 2, \ldots .
\end{aligned}$$

Several important properties of the renewal function are given below.

Property B.5 The mean value function of the renewal process, denoted by $m(t)$, is equal to the sum of the distribution function of all renewal times, that is,

$$\begin{aligned}
m(t) &= E[N(t)] \\
&= \sum_{n=1}^{\infty} F_n(t).
\end{aligned}$$

Proof The renewal function can be obtained as

$$\begin{aligned}
m(t) &= E[N(t)] \\
&= \sum_{n=0}^{\infty} n P\{N(t) = n\} \\
&= \sum_{n=0}^{\infty} n [F_n(t) - F_{n+1}(t)] \\
&= \sum_{n=1}^{\infty} F_n(t).
\end{aligned}$$

The mean value function of the renewal process is also called the renewal function.

Property B.6   The renewal function, $m(t)$, satisfies the following equation:

$$m(t) = F_a(t) + \int_0^t m(t-s)\, dF_a(s),$$

where $F_a(t)$ is the distribution function of the inter-arrival time or the renewal period. The proof is left as an exercise for the reader.

In general, let $y(t)$ be an unknown function to be evaluated and $x(t)$ be any non-negative and integrable function associated with the renewal process. Assume that $F_a(t)$ is the distribution function of the renewal period. We can then obtain the following result.

Property B.7   Let the renewal equation be

$$y(t) = x(t) + \int_0^t y(t-s)\, dF_a(s),$$

then its solution is given by

$$y(t) = x(t) + \int_0^t x(t-s)\, dm(s),$$

where $m(t)$ is the mean value function of the renewal process.

The proof of the above property can be easily derived using the LaPlace transform. It is also noted that the integral equation given in Property B.6 is a special case of Property B.7.

Example B.4   Let $x(t) = a$. Thus, in Property B.7, the solution $y(t)$ is given by

$$y(t) = x(t) + \int_0^t x(t-s)\, dm(s)$$

$$= a + \int_0^t a\, dm(s)$$

$$= a\,(1 + E[N(t)]).$$

B.5   Quasi-Renewal Processes

In this section, a general renewal process, namely, the quasi-renewal process, is discussed. Let $\{N(t), t > 0\}$ be a counting process and let X_n be the time between the $(n - 1)$th and the nth event of this process, $n \geq 1$.

Definition B.3 If the sequence of non-negative random variables $\{X_1, X_2, \ldots\}$ is independent and

$$X_i = aX_{i-1}$$

for $i \geq 2$ where $\alpha > 0$ is a constant, then the counting process $\{N(t), t \geq 0\}$ is said to be a quasi-renewal process with parameter and the first inter-arrival time X_1. When $\alpha = 1$, this process becomes the ordinary renewal process as discussed in Section B.4. This quasi-renewal process can be used to model reliability growth processes in software testing phases and hardware burn-in stages for $\alpha > 1$, and in hardware maintenance processes when $\alpha \leq 1$.

Assume that the probability density function, cumulative distribution function, survival function, and failure rate of random variable X_1 are $f_1(x)$, $F_1(x)$, $s_1(x)$, and $r_1(x)$, respectively. Then the pfd, cdf, survival function, failure rate of X_n for $n = 1, 2, 3, \ldots$ is respectively given below (Wang and Pham, 1996):

$$f_n(x) = \frac{1}{\alpha^{n-1}} f_1\left(\frac{1}{\alpha^{n-1}} x\right)$$

$$F_n(x) = F_1\left(\frac{1}{\alpha^{n-1}} x\right)$$

$$s_n(x) = s_1\left(\frac{1}{\alpha^{n-1}} x\right)$$

$$r_n(x) = \frac{1}{\alpha^{n-1}} r_1\left(\frac{1}{\alpha^{n-1}} x\right).$$

Similarly, the mean and variance of X_n is given as

$$E(X_n) = \alpha^{n-1} E(X_1)$$

$$\text{Var}(X_n) = \alpha^{2n-2} \text{Var}(X_1).$$

Because of the non-negativity of X_1 and the fact that X_1 is not identically 0, we obtain

$$E(X_1) = \mu_1 \neq 0.$$

Proposition B.1 The shape parameters of X_n are the same for $n = 1, 2, 3, \ldots$ for a quasi-renewal process if X_1 follows the gamma, Weibull, or log normal distribution.

This means that after "renewal", the shape parameters of the inter-arrival time will not change. In software reliability, the assumption that the software debugging process does not change the error-free distribution type seems reasonable. Thus, the error-free times of software during the debugging phase modeled by a quasi-renewal process will have the same shape parameters. In this sense, a quasi-renewal process is suitable to model the software reliability growth. It is worthwhile to note that

$$\lim_{n \to \infty} \frac{E(X_1 + X_2 + \cdots + X_n)}{n} = \lim_{n \to \infty} \frac{\mu_1(1 - \alpha^n)}{(1 - \alpha)n}$$
$$= 0 \quad \text{if } \alpha < 1$$
$$= \infty \quad \text{if } \alpha > 1.$$

Therefore, if the inter-arrival time represents the error-free time of a software system, then the average error-free time approaches infinity when its debugging process is occurring for a long debugging time.

Distribution of N(t)

Consider a quasi-renewal process with parameter α and the first inter-arrival time X_1. Clearly, the total number of renewals, $N(t)$, that has occurred up to time t and the arrival time of the nth renewal, SS_n, has the following relationship:

$$N(t) \geq n \text{ if and only if } SS_n \leq t,$$

that is, $N(t)$ is at least n if and only if the nth renewal occurs prior to time t. It is easily seen that

$$SS_n = \sum_{i=1}^{n} X_i = \sum_{i=1}^{n} \alpha^{i-1} X_1 \quad \text{for } n \geq 1.$$

Here, $SS_0 = 0$. Thus, we have

$$P\{N(t) = n\} = P\{N(t) = n\} - P\{N(t) \geq n + 1\}$$
$$= P\{SS_n \leq t\} - P\{SS_{n+1} \leq t\}$$
$$= G_n(t) - G_{n+1}(t),$$

where $G_n(t)$ is the convolution of the inter-arrival times F_1, F_2, $F_3, \ldots, F_n$. In other words,

$$G_n(t) = P\{F_1 + F_2 + \cdots + F_n \leq t\}.$$

If the mean value of $N(t)$ is defined as the renewal function $m(t)$, then

$$m(t) = E[N(t)]$$

$$= \sum_{n=1}^{\infty} P\{N(t) \geq n\}$$

$$= \sum_{n=1}^{\infty} P\{SS_n \leq t\}$$

$$= \sum_{n=1}^{\infty} G_n(t).$$

The derivative of $m(t)$ is known as the renewal density

$$\lambda(t) = m'(t).$$

In renewal theory, random variables representing the inter-arrival distributions only assume non-negative values, and the LaPlace transform of its distribution $F_1(t)$ is defined by

$$\mathcal{L}\{F_1(s)\} = \int_0^{\infty} e^{-sx} dF_1(x).$$

Therefore,

$$\mathcal{L}F_n(s) = \int_0^{\infty} e^{-\alpha^{n-1}st} dF_1(t) = \mathcal{L}F_1(\alpha^{n-1}s)$$

and

$$\mathcal{L}m_n(s) = \sum_{n=1}^{\infty} \mathcal{L}G_n(s)$$

$$= \sum_{n=1}^{\infty} \mathcal{L}F_1(s)\,\mathcal{L}F_1(\alpha s) \cdots \mathcal{L}F_1(\alpha^{n-1}s).$$

Since there is a one-to-one correspondence between distribution functions and its LaPlace transform, it follows that

Proposition B.2 The first inter-arrival distribution of a quasi-renewal process uniquely determines its renewal function.

If the inter-arrival time represents the error-free time (time to first failure), a quasi-renewal process can be used to model reliability growth for both software and hardware. Suppose that all faults of software have the same chance of being detected. If the inter-arrival time of a quasi-renewal process represents the error-free time of a software system, then the expected number of software faults in the time interval $[0, t)$ can be defined by the renewal function, $m(t)$, with parameter $\alpha > 1$. Denoted by $m_r(t)$, the number of remaining software faults at time t, it follows that

$$m_r(t) = m(T_c) - m(t),$$

where $m(T_c)$ is the number of faults that will eventually be detected through a software lifecycle T_c.

B.6   Non-Homogeneous Poisson Processes

The non-homogeneous Poisson process model (NHPP) that represents the number of failures experienced up to time t is a non-homogeneous Poisson process $\{N(t), t \geq 0\}$. The main issue in the NHPP model is to determine an appropriate mean value function to denote the expected number of failures experienced up to a certain time. With different assumptions, the model will end up with different functional forms of the mean value function. Note that in a renewal process, the exponential assumption for the inter-arrival time between failures is relaxed, and in the NHPP, the stationary assumption is relaxed.

The NHPP model is based on the following assumptions:

- The failure process has an independent increment, i.e., the number of failures during the time interval $(t, t + s]$ depends on the current time t and the length of time interval s, and does not depend on the past history of the process.
- The failure rate of the process is given by

$$P\{\text{exactly one failure in } (t, t + \Delta t)\} = P\{N(t + \Delta t) - N(t) = 1\}$$
$$= \lambda(t)\Delta t + o(\Delta t),$$

where $\lambda(t)$ is the intensity function.

- During a small interval Δt, the probability of more than one failure is negligible, that is,

$$P\{\text{two or more failures in } (t, t + \Delta t)\} = o(\Delta t).$$

- The initial condition is $N(0) = 0$.

On the basis of these assumptions, the probability that exactly n failures occurring during the time interval $(0, t)$ for the NHPP is given by

$$Pr\{N(t) = n\} = \frac{[m(t)]^n}{n!} e^{-m(t)} \quad n = 0, 1, 2, \ldots, \tag{B.2}$$

where

$$m(t) = E[N(t)] = \int_0^t \lambda(s)ds \tag{B.3}$$

and $\lambda(t)$ is the intensity function. It can be easily shown that the mean value function $m(t)$ is non-decreasing.

Reliability Function

The reliability $R(t)$, defined as the probability that there are no failures in the time interval $(0, t)$, is given by

$$\begin{aligned} R(t) &= P\{N(t) = 0\} \\ &= e^{-m(t)}. \end{aligned} \tag{B.4}$$

In general, the reliability $R(x/t)$, the probability that there are no failures in the interval $(t, t + x)$, is given by

$$\begin{aligned} R(x|t) &= P\{N(t + x) - N(t) = 0\} \\ &= e^{-[m(t+x)-m(t)]} \end{aligned} \tag{B.5}$$

and its density is given by

$$f(x) = \lambda(t + x) \, e^{-[m(t+x)-m(t)]},$$

where

$$\lambda(x) = \frac{\partial}{\partial x}[m(x)].$$

The variance of the NHPP can be obtained as follows:

$$\text{Var}[N(t)] = \int_0^t \lambda(s)ds$$

and the auto-correlation function is given below:

$$\text{Cor}[s] = E[N(t)]\, E[N(t+s) - N(t)] + E[N^2(t)]$$

$$= \int_0^t \lambda(s)dt \int_0^{t+s} \lambda(s)ds + \int_0^t \lambda(s)ds$$

$$= \int_0^t \lambda(s)ds \left[1 + \int_0^{t+s} \lambda(s)ds \right].$$

Example B.5 Assume that the intensity λ is a random variable with the pdf $f(\lambda)$. Then the probability that exactly n failures occurring during the time interval $(0, t)$ is given as

$$P\{N(t) = n\} = \int_0^\infty e^{-\lambda t} \frac{(\lambda t)^n}{n!} f(\lambda)\, d\lambda$$

It can be shown that if the pdf $f(\lambda)$ is given as the following gamma density function with parameters k and m,

$$f(\lambda) = \frac{1}{\Gamma(m)} k^m\, \lambda^{m-1}\, e^{-k\lambda} \quad \text{for } \lambda \geq 0,$$

then

$$P\{N(t) = n\} = \binom{n + m - 1}{n} p^m q^n \quad n = 0, 1, 2, \ldots$$

is also called a negative binomial density function, where

$$p = \frac{k}{t + k} \quad \text{and} \quad q = \frac{t}{t + k} = 1 - p.$$

Further Reading

The reader interested in a deeper understanding of advanced probability theory and stochastic processes should note the following highly recommended books:

Devore, J.L., *Probability and Statistics for Engineering and the Sciences*, 3rd edition, Brooks/Cole Pub. Co., Pacific Grove, 1991.

Gnedenko, B.V. and I.A. Ushakov, *Probabilistic Reliability Engineering*, John Wiley & Sons, New York, 1995.

Feller, W., *An Introduction to Probability Theory and Its Applications*, 3rd edition, John Wiley & Sons, New York, 1994.

Ross, S.M., *Introduction to Probability Models*, 6th edition, Academic Press, San Diego, 1989.

Problems

B.1   Prove Property B.2.
B.2   Prove Property B.3.
B.3   Prove Property B.6.
B.4   Events occur according to an NHPP in which the mean value function is

$$m(t) = t^3 + 3t^2 + 6t \quad t > 0.$$

What is the probability that n events occur between times $t = 10$ and $t = 15$?

Reference

Wang H., and H. Pham, "A quasi renewal process and its applications in imperfect maintenance," *Int. J. Systems Science*, Vol. 27(10), 1996.

LaPlace Transform

TABLE B.1. The LaPlace transform.

If a function $h(x)$ can be obtained from some prescribed operation on a function $f(x)$, then $h(x)$ is often called a transform of $f(x)$. For example,

$$h(x) = \sqrt{2 + f(x)}$$

$$h(x) = \frac{\partial}{\partial x} f(x).$$

The LaPlace transform of $f(t)$ is the function $f^*(s)$ where

$$f^*(s) = \int_0^\infty e^{-st} f(t) dt.$$

Often the LaPlace transform is denoted as $f^*(s)$ or $\mathscr{L}(f(t))$ or $\mathscr{L}(f)$. The results of the LaPlace transform for a few simple functions are presented below.

Results

1. $\mathscr{L}(1) = \int_0^\infty e^{-st} dt = \dfrac{1}{s}$.

2. $\mathscr{L}(e^{-at}) = \int_0^\infty e^{-st} e^{-at} dt = \int_0^\infty e^{-(s+a)t} dt$.

$$= \frac{1}{s + a}.$$

3. If $f(t) = \dfrac{1}{a}e^{-\frac{t}{a}}$, then

$$\mathcal{L}(f(t)) = \int_0^\infty e^{-st}\frac{1}{a}e^{-\frac{t}{a}}dt = \frac{1}{1+sa}.$$

4. If $f(t) = te^{at}$, then

$$\mathcal{L}(f(t)) = \int_0^\infty e^{-st}te^{at}dt = \frac{1}{(s-a)^2}.$$

5. If $f(t) = \dfrac{1}{a}(e^{at}-1)$, then

$$\mathcal{L}(f(t)) = \int_0^\infty e^{-st}\frac{1}{a}(e^{at}-1)\,dt = \frac{1}{s(s-a)}.$$

6. If $f(t) = (1+at)e^{at}$, then

$$\mathcal{L}(f(t)) = \int_0^\infty e^{-st}(1+at)e^{at}dt = \frac{s}{(s-a)^2}.$$

Similarly, we can obtain the following results.

7. If $f(t) = \dfrac{ae^{at}-be^{bt}}{a-b}$, then

$$\mathcal{L}(f(t)) = \frac{s}{(s-a)(s-b)}\ \text{ for } a \neq b.$$

8. If $f(t) = \dfrac{\alpha^k t^{k-1}e^{-\alpha t}}{\Gamma(k)}$, then

$$\mathcal{L}(f(t)) = \left(\frac{\alpha}{\alpha+s}\right)^k.$$

9. If $f(t) = \dfrac{e^{at}-e^{bt}}{a-b}$ for $a \neq b$, then

$$\mathcal{L}(f(t)) = \frac{1}{(s-a)(s-b)}.$$

10. If $f(t) = \lambda e^{-\lambda t}$, then

$$\mathcal{L}(f(t)) = \frac{\lambda}{\lambda+s}.$$

11. $\mathcal{L}(c_1 f_1(t) + c_2 f_2(t)) = \displaystyle\int_0^\infty e^{-st}[c_1 f_1(t) + c_2 f_2(t)]dt$

$$= c_1\,\mathcal{L}(f_1(t)) + c_2\,\mathcal{L}(f_2(t)).$$

12. If $f_i(t) = \lambda_i e^{-\lambda_i t}$, then

$$\mathscr{L}\left(\sum_{i=1}^{n} f_i(t)\right) = \sum_{i=1}^{n} \frac{\lambda_i}{\lambda_i + s}.$$

13. $\mathscr{L}\left(\sum_{i=1}^{n} f_i(t)\right) = \sum_{i=1}^{n} \mathscr{L}(f_i(t)).$

14. $\mathscr{L}(f'(t)) = \displaystyle\int_0^{\infty} e^{-st} f'(t) dt$

$$= f(t) e^{-st} \Big|_0^{\infty} + s \int_0^{\infty} f(t) e^{-st} dt$$

$$= -f(0^+) + s f^*(s)$$

$$= -f(0^+) + s \, \mathscr{L}(f(t)).$$

Appendix C

Survey of Factors that Affect Software Reliability

The survey form in the following pages is concerned with software reliability and factors involved in the software development process. The factors here include characteristics of the software itself (e.g., size), the development environment (e.g., people and tools), and all other factors during the process. Software reliability models that use testing time as the only influence factor may not be appropriate for the evaluation of software reliability.

Survey Form

Throughout this survey, we hope to obtain your opinion about the impact of factors on software reliability, as well as some background information about your company in order to keep your answers in perspective.

Please read the instructions below and then answer all questions. Section A in the survey deals with the issue of software reliability and factors. The definitions of the factors are provided in Section 8.2. Section B's questions are geared towards illiciting some background information about your company.

Please rank the factors in terms of their importance of inclusion in the software reliability analysis. For example, if you think that program complexity is an extremely important factor, you should rank it 7. On the contrary, if, in your opinion, it will not improve the estimation of software reliability at all, you may rank it 1. Each factor can take an integer value from 0 to 7. Please do not omit a ranking for any factor. Thank you for your cooperation.

Software Engineer

Name (optional) _____ Institution/Company _____

Section A.

Factor name	Not significant					Extremely significant		No opinion
General								
1. Program complexity (e.g., size)	1	2	3	4	5	6	7	0
2. Program categories (e.g., database, operating system, etc.)	1	2	3	4	5	6	7	0
3. Difficulty of programming	1	2	3	4	5	6	7	0
4. Amount of programming effort	1	2	3	4	5	6	7	0
5. Level of programming technologies	1	2	3	4	5	6	7	0
6. Percentage of reused modules	1	2	3	4	5	6	7	0
7. Programming language	1	2	3	4	5	6	7	0
Analysis and Design								
8. Frequency of change of program specifications	1	2	3	4	5	6	7	0
9. Volume of program design documents	1	2	3	4	5	6	7	0
10. Design methodology	1	2	3	4	5	6	7	0
11. Requirements analysis	1	2	3	4	5	6	7	0
12. Relationship of detailed design to requirements	1	2	3	4	5	6	7	0

Factor name	Not significant					Extremely significant		No opinion
13. Work standards	1	2	3	4	5	6	7	0
14. Development management	1	2	3	4	5	6	7	0

Coding

15. Programmer skill	1	2	3	4	5	6	7	0
16. Programmer organization	1	2	3	4	5	6	7	0
17. Development team size	1	2	3	4	5	6	7	0
18. Program workload (stress)	1	2	3	4	5	6	7	0
19. Domain knowledge	1	2	3	4	5	6	7	0
20. Human nature (mistake and work omission)	1	2	3	4	5	6	7	0

Testing

21. Testing environment (duplication of production)	1	2	3	4	5	6	7	0
22. Testing effort	1	2	3	4	5	6	7	0
23. Testing resource allocation	1	2	3	4	5	6	7	0
24. Testing methodologies	1	2	3	4	5	6	7	0
25. Testing coverage	1	2	3	4	5	6	7	0
26. Testing tools	1	2	3	4	5	6	7	0
27. Documentation	1	2	3	4	5	6	7	0

Hardware Systems

28. Processors	1	2	3	4	5	6	7	0
29. Storage devices	1	2	3	4	5	6	7	0
30. Input/output devices	1	2	3	4	5	6	7	0
31. Telecommunication devices	1	2	3	4	5	6	7	0
32. System software	1	2	3	4	5	6	7	0

33. To what extent do you believe that the factors will improve the accuracy of the software reliability? (Circle one)

    10%   20%   40%   60%   80%   100%

Section B. Background Information

1. What kind of applications do you usually develop for?

    ☐ safety-critical   ☐ commercial   ☐ inside users-oriented

2. What type of software development experience do you have?

    ☐ database   ☐ operation system   ☐ communication control   ☐ language processor
    ☐ other _____

3. Number of years have you been working on software development: _____

4. Your title/position
   ☐ manager   ☐ system engineer   ☐ programmer   ☐ tester   ☐ administrator
   ☐ other _____

5. Average percentage of the development time in your group spent on
   analysis phase: _____
   design phase: _____
   coding phase: _____
   testing phase: _____
   total: 100%

6. Average percentage of reusable code in your software applications: _____

Appendix D

Distribution Tables

TABLE D.1. Cumulative areas under the standard normal distribution landscape. (Reproduced from (Lundgren, 1976).)

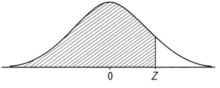

Z	0	1	2	3	4	5	6	7	8	9
−3.0	0.0013	0.0010	0.0007	0.0005	0.0003	0.0002	0.0002	0.0001	0.0001	0.0000
−2.9	0.0019	0.0018	0.0017	0.0017	0.0016	0.0016	0.0015	0.0015	0.0014	0.0014
−2.8	0.0026	0.0025	0.0024	0.0023	0.0023	0.0022	0.0021	0.0021	0.0020	0.0019
−2.7	0.0035	0.0034	0.0033	0.0032	0.0031	0.0030	0.0029	0.0028	0.0027	0.0026
−2.6	0.0047	0.0045	0.0044	0.0043	0.0041	0.0040	0.0039	0.0038	0.0037	0.0036
−2.5	0.0062	0.0060	0.0059	0.0057	0.0055	0.0054	0.0052	0.0051	0.0049	0.0048
−2.4	0.0082	0.0080	0.0078	0.0075	0.0073	0.0071	0.0069	0.0068	0.0066	0.0064
−2.3	0.0107	0.0104	0.0102	0.0099	0.0096	0.0094	0.0091	0.0089	0.0087	0.0084
−2.2	0.0139	0.0136	0.0132	0.0129	0.0126	0.0122	0.0119	0.0116	0.0113	0.0110
−2.1	0.0179	0.0174	0.0170	0.0166	0.0162	0.0158	0.0154	0.0150	0.0146	0.0143
−2.0	0.0228	0.0222	0.0217	0.0212	0.0207	0.0202	0.0197	0.0192	0.0188	0.0183
−1.9	0.0287	0.0281	0.0274	0.0268	0.0262	0.0256	0.0250	0.0244	0.0238	0.0233
−1.8	0.0359	0.0352	0.0344	0.0336	0.0329	0.0322	0.0314	0.0307	0.0300	0.0294
−1.7	0.0446	0.0436	0.0427	0.0418	0.0409	0.0401	0.0392	0.0384	0.0375	0.0367
−1.6	0.0548	0.0537	0.0526	0.0516	0.0505	0.0495	0.0485	0.0475	0.0465	0.0455
−1.5	0.0668	0.0655	0.0643	0.0630	0.0618	0.0606	0.0594	0.0582	0.0570	0.0559
−1.4	0.0808	0.0793	0.0778	0.0764	0.0749	0.0735	0.0722	0.0708	0.0694	0.0681
−1.3	0.0968	0.0951	0.0934	0.0918	0.0901	0.0885	0.0869	0.0853	0.0838	0.0823
−1.2	0.1151	0.1131	0.1112	0.1093	0.1075	0.1056	0.1038	0.1020	0.1003	0.0985
−1.1	0.1357	0.1335	0.1314	0.1292	0.1271	0.1251	0.1230	0.1210	0.1190	0.1170
−1.0	0.1587	0.1562	0.1539	0.1515	0.1492	0.1469	0.1446	0.1423	0.1401	0.1379
−0.9	0.1841	0.1814	0.1788	0.1762	0.1736	0.1711	0.1685	0.1660	0.1635	0.1611
−0.8	0.2119	0.2090	0.2061	0.2033	0.2005	0.1977	0.1949	0.1922	0.1894	0.1867
−0.7	0.2420	0.2389	0.2358	0.2327	0.2297	0.2266	0.2236	0.2206	0.2177	0.2148
−0.6	0.2743	0.2709	0.2676	0.2643	0.2611	0.2578	0.2546	0.2514	0.2483	0.2451

TABLE **D.1.** *continued*

Z	0	1	2	3	4	5	6	7	8	9
−0.5	0.3085	0.3050	0.3015	0.2981	0.2946	0.2912	0.2877	0.2843	0.2810	0.2776
−0.4	0.3446	0.3409	0.3372	0.3336	0.3300	0.3264	0.3228	0.3192	0.3156	0.3121
−0.3	0.3821	0.3783	0.3745	0.3707	0.3669	0.3632	0.3594	0.3557	0.3520	0.3483
−0.2	0.4207	0.4168	0.4129	0.4090	0.4052	0.4013	0.3974	0.3936	0.3897	0.3859
−0.1	0.4602	0.4562	0.4522	0.4483	0.4443	0.4404	0.4364	0.4325	0.4286	0.4247
−0.0	0.5000	0.4960	0.4920	0.4880	0.4840	0.4801	0.4761	0.4721	0.4681	0.4641
0.0	0.5000	0.5040	0.5080	0.5120	0.5160	0.5199	0.5239	0.5279	0.5319	0.5359
0.1	0.5398	0.5438	0.5478	0.5517	0.5557	0.5596	0.5636	0.5675	0.5714	0.5753
0.2	0.5793	0.5832	0.5871	0.5910	0.5948	0.5987	0.6026	0.6064	0.6103	0.6141
0.3	0.6179	0.6217	0.6255	0.6293	0.6331	0.6368	0.6406	0.6443	0.6480	0.6517
0.4	0.6554	0.6591	0.6628	0.6664	0.6700	0.6736	0.6772	0.6808	0.6844	0.6879
0.5	0.6915	0.6950	0.6985	0.7019	0.7054	0.7088	0.7123	0.7157	0.7190	0.7224
0.6	0.7257	0.7291	0.7324	0.7357	0.7389	0.7422	0.7454	0.7486	0.7517	0.7549
0.7	0.7580	0.7611	0.7642	0.7673	0.7703	0.7734	0.7764	0.7794	0.7823	0.7852
0.8	0.7881	0.7910	0.7939	0.7967	0.7995	0.8023	0.8051	0.8078	0.8106	0.8133
0.9	0.8159	0.8186	0.8212	0.8238	0.8264	0.8289	0.8315	0.8340	0.8365	0.8389
1.0	0.8413	0.8438	0.8461	0.8485	0.8508	0.8531	0.8554	0.8577	0.8599	0.8621
1.1	0.8643	0.8665	0.8686	0.8708	0.8729	0.8749	0.8770	0.8790	0.8810	0.8830
1.2	0.8849	0.8869	0.8888	0.8907	0.8925	0.8944	0.8962	0.8980	0.8997	0.9015
1.3	0.9032	0.9049	0.9066	0.9082	0.9099	0.9115	0.9131	0.9147	0.9162	0.9177
1.4	0.9192	0.9207	0.9222	0.9236	0.9251	0.9265	0.9278	0.9292	0.9306	0.9319
1.5	0.9332	0.9345	0.9357	0.9370	0.9382	0.9394	0.9406	0.9418	0.9430	0.9441
1.6	0.9452	0.9463	0.9474	0.9484	0.9495	0.9505	0.9515	0.9525	0.9535	0.9545
1.7	0.9554	0.9564	0.9573	0.9582	0.9591	0.9599	0.9608	0.9616	0.9625	0.9633
1.8	0.9641	0.9648	0.9656	0.9664	0.9671	0.9678	0.9686	0.9693	0.9700	0.9706
1.9	0.9713	0.9719	0.9726	0.9732	0.9738	0.9744	0.9750	0.9756	0.9762	0.9767
2.0	0.9772	0.9778	0.9783	0.9788	0.9793	0.9798	0.9803	0.9808	0.9812	0.9817
2.1	0.9821	0.9826	0.9830	0.9834	0.9838	0.9842	0.9846	0.9850	0.9854	0.9857
2.2	0.9861	0.9864	0.9868	0.9871	0.9874	0.9878	0.9881	0.9884	0.9887	0.9890
2.3	0.9893	0.9896	0.9898	0.9901	0.9904	0.9906	0.9909	0.9911	0.9913	0.9916
2.4	0.9918	0.9920	0.9922	0.9925	0.9927	0.9929	0.9931	0.9932	0.9934	0.9936
2.5	0.9938	0.9940	0.9941	0.9943	0.9945	0.9946	0.9948	0.9949	0.9951	0.9952
2.6	0.9953	0.9955	0.9956	0.9957	0.9959	0.9960	0.9961	0.9962	0.9963	0.9964
2.7	0.9965	0.9966	0.9967	0.9968	0.9969	0.9970	0.9971	0.9972	0.9973	0.9974
2.8	0.9974	0.9975	0.9976	0.9977	0.9977	0.9978	0.9979	0.9979	0.9980	0.9981
2.9	0.9981	0.9982	0.9982	0.9983	0.9984	0.9984	0.9985	0.9985	0.9986	0.9986
3.0	0.9987	0.9990	0.9993	0.9995	0.9997	0.9998	0.9998	0.9999	0.9999	1.0000

TABLE D.2. Percentage points of the t distribution. (Reproduced from (Pearson, 1966).)

ν \ α	0.40	0.25	0.10	0.05	.025	0.01	0.005	0.0025	0.001	0.0005
1	0.325	1.000	3.078	6.314	12.706	31.821	63.657	127.32	318.31	636.620
2	0.289	0.816	1.886	2.920	4.303	6.965	9.925	14.089	23.326	31.598
3	0.277	0.765	1.638	2.353	3.182	4.541	5.841	7.453	10.213	12.924
4	0.271	0.741	1.533	2.132	2.776	3.747	4.604	5.598	7.173	8.610
5	0.267	0.727	1.476	2.015	2.571	3.365	4.032	4.773	5.893	6.869
6	0.265	0.718	1.440	1.943	2.447	3.143	3.707	4.317	5.208	5.959
7	0.263	0.711	1.415	1.895	2.365	2.998	3.499	4.029	4.785	5.408
8	0.262	0.706	1.397	1.860	2.306	2.896	3.355	3.833	4.501	5.041
9	0.261	0.703	1.383	1.833	2.262	2.821	3.250	3.690	4.297	4.781
10	0.260	0.700	1.372	1.812	2.228	2.764	3.169	3.581	4.144	4.587
11	0.260	0.697	1.363	1.796	2.201	2.718	3.106	3.497	4.025	4.437
12	0.259	0.695	1.356	1.782	2.179	2.681	3.055	3.428	3.930	4.318
13	0.259	0.694	1.350	1.771	2.160	2.650	3.012	3.372	3.852	4.221
14	0.258	0.692	1.345	1.761	2.145	2.624	2.977	3.326	3.787	4.140
15	0.258	0.691	1.341	1.753	2.131	2.602	2.947	3.286	3.733	4.073
16	0.258	0.690	1.337	1.746	2.120	2.583	2.921	3.252	3.686	4.015
17	0.257	0.689	1.333	1.740	2.110	2.567	2.898	3.222	3.646	3.965
18	0.257	0.688	1.330	1.734	2.101	2.552	2.878	3.197	3.610	3.922
19	0.257	0.688	1.328	1.729	2.093	2.539	2.861	3.174	3.579	3.883
20	0.257	0.687	1.325	1.725	2.086	2.528	2.845	3.153	3.552	3.850
21	0.257	0.686	1.323	1.721	2.080	2.518	2.831	3.135	3.527	3.819
22	0.256	0.686	1.321	1.717	2.074	2.508	2.819	3.119	3.505	3.792
23	0.256	0.685	1.319	1.714	2.069	2.500	2.807	3.104	3.485	3.767
24	0.256	0.685	1.318	1.711	2.064	2.492	2.797	3.091	3.467	3.745
25	0.256	0.684	1.316	1.708	2.060	2.485	2.787	3.078	3.450	3.725
26	0.256	0.684	1.315	1.706	2.056	2.479	2.779	3.067	3.435	3.707
27	0.256	0.684	1.314	1.703	2.052	2.473	2.771	3.057	3.421	3.690
28	0.256	0.683	1.313	1.701	2.048	2.467	2.763	3.047	3.408	3.674
29	0.256	0.683	1.311	1.699	2.045	2.462	2.756	3.038	3.396	3.659
30	0.256	0.683	1.310	1.697	2.042	2.457	2.750	3.030	3.385	3.646
40	0.255	0.681	1.303	1.684	2.021	2.423	2.704	2.971	3.307	3.551
60	0.254	0.679	1.296	1.671	2.000	2.390	2.660	2.915	3.232	3.460
120	0.254	0.677	1.289	1.658	1.980	2.358	2.617	2.860	3.160	3.373
∞	0.253	0.674	1.282	1.645	1.960	2.326	2.576	2.807	3.090	3.291

TABLE D.3. Percentage points of the chi-squared (χ^2) distribution. (Reproduced from (US Army Material Command, 1968).)

χ^2_α ν	$\chi^2_{0.995}$	$\chi^2_{0.99}$	$\chi^2_{0.975}$	$\chi^2_{0.95}$	$\chi^2_{0.90}$	$\chi^2_{0.80}$	$\chi^2_{0.75}$	$\chi^2_{0.70}$
1	0.0000393	0.000157	0.000982	0.00393	0.0158	0.0642	0.102	0.148
2	0.0100	0.0201	0.0506	0.103	0.211	0.446	0.575	0.713
3	0.0717	0.115	0.216	0.352	0.584	1.005	1.213	1.424
4	0.207	0.297	0.484	0.711	1.064	1.649	1.923	2.195
5	0.412	0.554	0.831	1.145	1.610	2.343	2.675	3.000
6	0.676	0.872	1.237	1.635	2.204	3.070	3.455	3.828
7	0.989	1.239	1.690	2.167	2.833	3.822	4.255	4.671
8	1.344	1.646	2.180	2.733	3.490	4.594	5.071	5.527
9	1.735	2.088	2.700	3.325	4.168	5.380	5.899	6.393
10	2.156	2.558	3.247	3.940	4.865	6.179	6.737	7.267
11	2.603	3.053	3.816	4.575	5.578	6.989	7.584	8.148
12	3.074	3.571	4.404	5.226	6.304	7.807	8.438	9.034
13	3.565	4.107	5.009	5.892	7.042	8.634	9.299	9.926
14	4.075	4.660	5.629	6.571	7.790	9.467	10.165	10.821
15	4.601	5.229	6.262	7.261	8.574	10.307	11.306	11.721
16	5.142	5.812	6.908	7.962	9.312	11.152	11.192	12.624
17	5.697	6.408	7.564	8.672	10.085	12.002	12.792	13.531
18	6.265	7.015	8.231	9.390	10.865	12.857	13.675	14.440
19	6.844	7.633	8.907	10.117	11.651	13.716	14.562	15.352
20	7.434	8.260	9.591	10.851	12.443	14.578	15.452	16.266
21	8.034	8.897	10.283	11.591	13.240	15.445	16.344	17.182
22	8.643	9.542	10.982	12.338	14.041	16.314	17.240	18.101
23	9.260	10.196	11.688	13.091	14.848	17.187	18.137	19.021
24	9.886	10.856	12.401	13.848	15.659	18.062	19.037	19.943
25	10.520	11.524	13.120	14.611	16.473	18.940	19.939	20.867
26	11.160	12.198	13.844	15.379	17.292	19.820	20.843	21.792
27	11.808	12.879	14.573	16.151	18.114	20.703	21.749	22.719
28	12.461	13.565	15.308	16.928	18.939	21.588	22.657	23.647
29	13.121	14.256	16.047	17.708	19.768	22.475	23.567	24.577
30	13.787	14.953	16.791	18.493	20.599	23.364	24.478	25.508
35	17.156	18.484	20.558	22.462	24.812	27.820	29.058	30.181
40	20.674	22.142	24.423	26.507	29.067	32.326	33.664	34.874
45	24.281	25.880	28.356	30.610	33.367	36.863	38.294	39.586
50	27.962	29.687	32.348	34.762	37.706	41.426	42.944	44.314
55	31.708	33.552	36.390	38.956	42.078	46.011	47.612	49.055
60	35.510	37.467	40.474	43.186	46.478	50.614	52.295	53.808
65	39.360	41.427	44.595	47.448	50.902	55.233	56.991	58.572
70	43.253	45.426	48.750	51.737	55.349	59.868	61.698	63.344
75	47.186	49.460	52.935	56.052	59.815	64.515	66.416	68.125
80	51.153	53.526	57.146	60.390	64.299	69.174	71.144	72.913

TABLE D.3. *continued*

ν / χ^2_α	$\chi^2_{0.995}$	$\chi^2_{0.99}$	$\chi^2_{0.975}$	$\chi^2_{.95}$	$\chi^2_{0.90}$	$\chi^2_{0.80}$	$\chi^2_{0.75}$	$\chi^2_{0.70}$
85	55.151	57.621	61.382	64.748	68.799	73.843	75.880	77.707
90	59.179	61.741	65.640	69.124	73.313	78.522	80.623	82.508
95	63.963	65.886	69.919	73.518	77.841	83.210	85.374	87.314
100	67.312	70.053	74.216	77.928	82.381	87.906	90.131	92.125
105	71.414	74.241	78.530	82.352	86.933	92.610	94.894	96.941
110	75.536	78.448	82.861	86.790	91.495	97.321	99.663	101.761
115	79.679	82.672	87.207	91.240	96.067	102.038	104.437	106.585
120	83.839	86.913	91.567	95.703	100.648	106.762	109.216	111.413

	$\chi^2_{0.50}$	$\chi^2_{0.30}$	$\chi^2_{0.25}$	$\chi^2_{0.20}$	$\chi^2_{0.10}$	$\chi^2_{0.05}$	$\chi^2_{0.025}$	$\chi^2_{0.01}$	$\chi^2_{0.005}$
1	0.455	1.074	1.323	1.642	2.706	3.841	5.024	6.635	7.879
2	1.386	2.408	2.773	3.219	4.605	5.991	7.378	9.210	10.597
3	2.366	3.665	4.108	4.642	6.251	7.815	9.348	11.345	12.838
4	3.357	4.878	5.385	5.989	7.779	9.488	11.143	13.277	14.860
5	4.351	6.064	6.626	7.289	9.236	11.070	12.832	15.086	16.750
6	5.348	7.231	7.841	8.558	10.645	12.592	14.449	16.812	18.548
7	6.346	8.383	9.037	9.803	12.017	14.067	16.013	18.475	20.278
8	7.344	9.524	10.219	11.030	13.362	15.507	17.535	20.090	21.955
9	8.343	10.656	11.389	12.242	14.684	16.919	19.023	21.666	23.589
10	9.342	11.781	12.549	13.442	15.987	18.307	20.483	23.209	25.188
11	10.341	12.899	13.701	14.631	17.275	19.675	21.920	24.725	26.757
12	11.340	14.011	14.845	15.812	18.549	21.920	23.337	26.217	28.300
13	12.340	15.119	15.984	16.985	19.812	22.362	24.736	27.688	29.819
14	13.339	16.222	17.117	18.151	21.064	23.685	26.119	29.141	31.319
15	14.339	17.322	18.245	19.311	22.307	24.996	27.488	30.578	32.801
16	15.338	18.418	19.369	20.465	23.542	26.296	28.845	32.000	34.267
17	16.338	19.511	20.489	21.615	24.769	27.587	30.191	33.409	35.718
18	17.338	20.601	21.605	22.760	25.989	28.869	31.526	34.805	37.156
19	18.338	21.689	22.718	23.900	27.204	30.144	32.852	36.191	38.582
20	19.337	22.775	23.828	25.038	28.412	31.410	34.170	37.566	39.997
21	20.337	23.858	24.935	26.171	29.615	32.671	35.479	38.932	41.401
22	21.337	24.939	26.039	27.301	30.813	33.924	36.781	40.289	42.796
23	22.337	26.018	27.141	28.429	32.007	35.172	38.076	41.638	44.181
24	23.337	27.096	28.241	29.553	33.196	36.415	39.364	42.980	45.558
25	24.337	28.172	29.339	30.675	34.382	37.652	40.646	44.314	46.928
26	25.336	29.246	30.434	31.795	35.563	38.885	41.923	45.642	48.290
27	26.336	30.319	31.528	32.912	36.741	40.113	43.194	46.963	49.645
28	27.336	31.391	32.620	34.027	37.916	41.337	44.461	48.278	50.993
29	28.336	32.461	33.711	35.139	39.087	42.557	45.722	49.588	52.336
30	29.336	33.530	34.800	36.250	40.256	43.773	46.979	50.892	53.672

TABLE D.3. *continued*

ν \ χ^2_α	$\chi^2_{0.50}$	$\chi^2_{0.30}$	$\chi^2_{0.25}$	$\chi^2_{0.20}$	$\chi^2_{0.10}$	$\chi^2_{0.05}$	$\chi^2_{0.025}$	$\chi^2_{0.01}$	$\chi^2_{0.005}$
35	34.338	38.860	40.221	41.802	46.034	49.798	53.207	57.359	60.304
40	39.337	44.166	45.615	47.295	51.780	55.755	59.345	63.706	66.792
45	44.337	49.453	50.984	52.757	57.480	61.653	65.141	69.971	73.190
50	49.336	54.725	56.333	58.194	63.141	67.502	71.424	76.167	79.512
55	54.336	59.983	61.665	63.610	68.770	73.309	77.384	82.305	85.769
60	59.336	65.229	66.982	69.006	74.370	79.080	83.301	88.391	91.970
65	64.336	70.466	72.286	74.387	79.946	84.819	89.181	94.433	93.122
70	69.335	75.693	77.578	79.752	85.500	90.530	95.027	100.436	104.230
75	74.335	80.912	82.860	85.105	91.034	96.216	100.843	106.403	110.300
80	79.335	86.124	88.132	90.446	96.550	101.879	106.632	112.338	116.334
85	84.335	91.329	93.396	95.777	102.050	107.521	112.397	118.244	122.337
90	89.335	96.529	98.653	101.097	107.536	113.145	118.139	124.125	128.310
95	94.335	101.723	103.902	106.409	113.008	118.751	123.861	129.980	134.257
100	99.335	106.911	109.145	111.713	118.468	124.342	129.565	135.814	140.179
105	104.335	112.095	114.381	117.009	123.917	129.918	135.250	141.627	146.078
110	109.335	117.275	119.612	112.299	129.355	135.480	140.920	147.421	151.956
115	114.335	122.451	124.838	127.581	134.782	141.030	146.574	153.197	157.814
120	119.335	127.623	130.059	132.858	140.201	146.568	152.215	158.956	163.654

TABLE D.4. Percentage points of the F distribution. (Reproduced from (Pearson, 1966)).

$F_{0.25, \nu_1, \nu_2}$

ν_2 \ ν_1	1	2	3	4	5	6	7	8	9	10	12	15	20	24	30	40	60	120	∞
1	5.83	7.50	8.20	8.58	8.82	8.98	9.10	9.19	9.26	9.32	9.41	9.49	9.58	9.63	9.67	9.71	9.76	9.80	9.85
2	2.57	3.00	3.15	3.23	3.28	3.31	3.34	3.35	3.37	3.38	3.39	3.41	3.43	3.43	3.44	3.45	3.46	3.47	3.48
3	2.02	2.28	2.36	2.39	2.41	2.42	2.43	2.44	2.44	2.44	2.45	2.46	2.46	2.46	2.47	2.47	2.47	2.47	2.47
4	1.81	2.00	2.05	2.06	2.07	2.08	2.08	2.08	2.08	2.08	2.08	2.08	2.08	2.08	2.08	2.08	2.08	2.08	2.08
5	1.69	1.85	1.88	1.89	1.89	1.89	1.89	1.89	1.89	1.89	1.89	1.89	1.88	1.88	1.88	1.88	1.87	1.87	1.87
6	1.62	1.76	1.78	1.79	1.79	1.78	1.78	1.78	1.77	1.77	1.77	1.76	1.76	1.75	1.75	1.75	1.74	1.74	1.74
7	1.57	1.70	1.72	1.72	1.71	1.71	1.70	1.70	1.70	1.69	1.68	1.68	1.67	1.67	1.66	1.66	1.65	1.65	1.65
8	1.54	1.66	1.67	1.66	1.66	1.65	1.64	1.64	1.63	1.63	1.62	1.62	1.61	1.60	1.60	1.59	1.59	1.58	1.58
9	1.51	1.62	1.63	1.63	1.62	1.61	1.60	1.60	1.59	1.59	1.58	1.57	1.56	1.56	1.55	1.54	1.54	1.53	1.53
10	1.49	1.60	1.60	1.59	1.59	1.58	1.57	1.56	1.56	1.55	1.54	1.53	1.52	1.52	1.51	1.51	1.50	1.49	1.48
11	1.47	1.58	1.58	1.57	1.56	1.55	1.54	1.53	1.53	1.52	1.51	1.50	1.49	1.49	1.48	1.47	1.47	1.46	1.45
12	1.46	1.56	1.56	1.55	1.54	1.53	1.52	1.51	1.51	1.50	1.49	1.48	1.47	1.46	1.45	1.45	1.44	1.43	1.42
13	1.45	1.55	1.55	1.53	1.52	1.51	1.50	1.49	1.49	1.48	1.47	1.46	1.45	1.44	1.43	1.42	1.42	1.41	1.40
14	1.44	1.53	1.53	1.52	1.51	1.50	1.49	1.48	1.47	1.46	1.45	1.44	1.43	1.42	1.41	1.41	1.40	1.39	1.38
15	1.43	1.52	1.52	1.51	1.49	1.48	1.47	1.46	1.46	1.45	1.44	1.43	1.41	1.41	1.40	1.39	1.38	1.37	1.36
16	1.42	1.51	1.51	1.50	1.48	1.47	1.46	1.45	1.44	1.44	1.43	1.41	1.40	1.39	1.38	1.37	1.36	1.35	1.34
17	1.42	1.51	1.50	1.49	1.47	1.46	1.45	1.44	1.43	1.43	1.41	1.40	1.39	1.38	1.37	1.36	1.35	1.34	1.33
18	1.41	1.50	1.49	1.48	1.46	1.45	1.44	1.43	1.42	1.42	1.40	1.39	1.38	1.37	1.36	1.35	1.34	1.33	1.32
19	1.41	1.49	1.49	1.47	1.46	1.44	1.43	1.42	1.41	1.41	1.40	1.38	1.37	1.36	1.35	1.34	1.33	1.32	1.30
20	1.40	1.49	1.48	1.47	1.45	1.44	1.43	1.42	1.41	1.40	1.39	1.37	1.36	1.35	1.34	1.33	1.32	1.31	1.29
21	1.40	1.48	1.48	1.46	1.44	1.43	1.42	1.41	1.40	1.39	1.38	1.37	1.35	1.34	1.33	1.32	1.31	1.30	1.28
22	1.40	1.48	1.47	1.45	1.44	1.42	1.41	1.40	1.39	1.39	1.37	1.36	1.34	1.33	1.32	1.31	1.30	1.29	1.28
23	1.39	1.47	1.47	1.45	1.43	1.42	1.41	1.40	1.39	1.38	1.37	1.35	1.34	1.33	1.32	1.31	1.30	1.28	1.27
24	1.39	1.47	1.46	1.44	1.43	1.41	1.40	1.39	1.38	1.38	1.36	1.35	1.33	1.32	1.31	1.30	1.29	1.28	1.26
25	1.39	1.47	1.46	1.44	1.42	1.41	1.40	1.39	1.38	1.37	1.36	1.34	1.33	1.32	1.31	1.29	1.28	1.27	1.25
26	1.38	1.46	1.45	1.44	1.42	1.41	1.39	1.38	1.37	1.37	1.35	1.34	1.32	1.31	1.30	1.29	1.28	1.26	1.25
27	1.38	1.46	1.45	1.43	1.42	1.40	1.39	1.38	1.37	1.36	1.35	1.33	1.32	1.31	1.30	1.28	1.27	1.26	1.24
28	1.38	1.46	1.45	1.43	1.41	1.40	1.39	1.38	1.37	1.36	1.34	1.33	1.31	1.30	1.29	1.28	1.27	1.25	1.24
29	1.38	1.45	1.45	1.43	1.41	1.40	1.38	1.37	1.36	1.35	1.34	1.32	1.31	1.30	1.29	1.27	1.26	1.25	1.23
30	1.38	1.45	1.44	1.42	1.41	1.39	1.38	1.37	1.36	1.35	1.34	1.32	1.30	1.29	1.28	1.27	1.26	1.24	1.23
40	1.36	1.44	1.42	1.40	1.39	1.37	1.36	1.35	1.34	1.33	1.31	1.30	1.28	1.26	1.25	1.24	1.22	1.21	1.19
60	1.35	1.42	1.41	1.38	1.37	1.35	1.33	1.32	1.31	1.30	1.29	1.27	1.25	1.24	1.22	1.21	1.19	1.17	1.15
120	1.34	1.40	1.39	1.37	1.35	1.33	1.31	1.30	1.29	1.28	1.26	1.24	1.22	1.21	1.19	1.18	1.16	1.13	1.10
∞	1.32	1.39	1.37	1.35	1.33	1.31	1.29	1.28	1.27	1.25	1.24	1.22	1.19	1.18	1.16	1.14	1.12	1.08	1.00

Degrees of freedom for the numerator (ν_1)

Degrees of freedom for the denominator (ν_2)

TABLE D.4. *continued*

$$F_{0.10,\,\nu_1,\,\nu_2}$$

		Degrees of freedom for the numerator (ν_1)																	
ν_1	1	2	3	4	5	6	7	8	9	10	12	15	20	24	30	40	60	120	∞
1	39.86	49.50	53.59	55.83	57.24	58.20	58.91	59.44	59.86	60.19	60.71	61.22	61.74	62.00	62.26	62.53	62.79	63.06	63.33
2	8.53	9.00	9.16	9.24	9.29	9.33	9.35	9.37	9.38	9.39	9.41	9.42	9.44	9.45	9.46	9.47	9.47	9.48	9.49
3	5.54	5.46	5.39	5.34	5.31	5.28	5.27	5.25	5.24	5.23	5.22	5.20	5.18	5.18	5.17	5.16	5.15	5.14	5.13
4	4.54	4.32	4.19	4.11	4.05	4.01	3.98	3.95	3.94	3.92	3.90	3.87	3.84	3.83	3.82	3.80	3.79	3.78	3.76
5	4.06	3.78	3.62	3.52	3.45	3.40	3.37	3.34	3.32	3.30	3.27	3.24	3.21	3.19	3.17	3.16	3.14	3.12	3.10
6	3.78	3.46	3.29	3.18	3.11	3.05	3.01	2.98	2.96	2.94	2.90	2.87	2.84	2.82	2.80	2.78	2.76	2.74	2.72
7	3.59	3.26	3.07	2.96	2.88	2.83	2.78	2.75	2.72	2.70	2.67	2.63	2.59	2.58	2.56	2.54	2.51	2.49	2.47
8	3.46	3.11	2.92	2.81	2.73	2.67	2.62	2.59	2.56	2.54	2.50	2.46	2.42	2.40	2.38	2.36	2.34	2.32	2.29
9	3.36	3.01	2.81	2.69	2.61	2.55	2.51	2.47	2.44	2.42	2.38	2.34	2.30	2.28	2.25	2.23	2.21	2.18	2.16
10	3.29	2.92	2.73	2.61	2.52	2.46	2.41	2.38	2.35	2.32	2.28	2.24	2.20	2.18	2.16	2.13	2.11	2.08	2.06
11	3.23	2.86	2.66	2.54	2.45	2.39	2.34	2.30	2.27	2.25	2.21	2.17	2.12	2.10	2.08	2.05	2.03	2.00	1.97
12	3.18	2.81	2.61	2.48	2.39	2.33	2.28	2.24	2.21	2.19	2.15	2.10	2.06	2.04	2.01	1.99	1.96	1.93	1.90
13	3.14	2.76	2.56	2.43	2.35	2.28	2.23	2.20	2.16	2.14	2.10	2.05	2.01	1.98	1.96	1.93	1.90	1.88	1.85
14	3.10	2.73	2.52	2.39	2.31	2.24	2.19	2.15	2.12	2.10	2.05	2.01	1.96	1.94	1.91	1.89	1.86	1.83	1.80
15	3.07	2.70	2.49	2.36	2.27	2.21	2.16	2.12	2.09	2.06	2.02	1.97	1.92	1.90	1.87	1.85	1.82	1.79	1.76
16	3.05	2.67	2.46	2.33	2.24	2.18	2.13	2.09	2.06	2.03	1.99	1.94	1.89	1.87	1.84	1.81	1.78	1.75	1.72
17	3.03	2.64	2.44	2.31	2.22	2.15	2.10	2.06	2.03	2.00	1.96	1.91	1.86	1.84	1.81	1.78	1.75	1.72	1.69
18	3.01	2.62	2.42	2.29	2.20	2.13	2.08	2.04	2.00	1.98	1.93	1.89	1.84	1.81	1.78	1.75	1.72	1.69	1.66
19	2.99	2.61	2.40	2.27	2.18	2.11	2.06	2.02	1.98	1.96	1.91	1.86	1.81	1.79	1.76	1.73	1.70	1.67	1.63
20	2.97	2.59	2.38	2.25	2.16	2.09	2.04	2.00	1.96	1.94	1.89	1.84	1.79	1.77	1.74	1.71	1.68	1.64	1.61
21	2.96	2.57	2.36	2.23	2.14	2.08	2.02	1.98	1.95	1.92	1.87	1.83	1.78	1.75	1.72	1.69	1.66	1.62	1.59
22	2.95	2.56	2.35	2.22	2.13	2.06	2.01	1.97	1.93	1.90	1.86	1.81	1.76	1.73	1.70	1.67	1.64	1.60	1.57
23	2.94	2.55	2.34	2.21	2.11	2.05	1.99	1.95	1.92	1.89	1.84	1.80	1.74	1.72	1.69	1.66	1.62	1.59	1.55
24	2.93	2.54	2.33	2.19	2.10	2.04	1.98	1.94	1.91	1.88	1.83	1.78	1.73	1.70	1.67	1.64	1.61	1.57	1.53
25	2.92	2.53	2.32	2.18	2.09	2.02	1.97	1.93	1.89	1.87	1.82	1.77	1.72	1.69	1.66	1.63	1.59	1.56	1.52
26	2.91	2.52	2.31	2.17	2.08	2.01	1.96	1.92	1.88	1.86	1.81	1.76	1.71	1.68	1.65	1.61	1.58	1.54	1.50
27	2.90	2.51	2.30	2.17	2.07	2.00	1.95	1.91	1.87	1.85	1.80	1.75	1.70	1.67	1.64	1.60	1.57	1.53	1.49
28	2.89	2.50	2.29	2.16	2.06	2.00	1.94	1.90	1.87	1.84	1.79	1.74	1.69	1.66	1.63	1.59	1.56	1.52	1.48
29	2.89	2.50	2.28	2.15	2.06	1.99	1.93	1.89	1.86	1.83	1.78	1.73	1.68	1.65	1.62	1.58	1.55	1.51	1.47
30	2.88	2.49	2.28	2.14	2.03	1.98	1.93	1.88	1.85	1.82	1.77	1.72	1.67	1.64	1.61	1.57	1.54	1.50	1.46
40	2.84	2.44	2.23	2.09	2.00	1.93	1.87	1.83	1.79	1.76	1.71	1.66	1.61	1.57	1.54	1.51	1.47	1.42	1.38
60	2.79	2.39	2.18	2.04	1.95	1.87	1.82	1.77	1.74	1.71	1.66	1.60	1.54	1.51	1.48	1.44	1.40	1.35	1.29
120	2.75	2.35	2.13	1.99	1.90	1.82	1.77	1.72	1.68	1.65	1.60	1.55	1.48	1.45	1.41	1.37	1.32	1.26	1.19
∞	2.71	2.30	2.08	1.94	1.85	1.77	1.72	1.67	1.63	1.60	1.55	1.49	1.42	1.38	1.34	1.30	1.24	1.17	1.00

Degrees of freedom of the denominator (ν_2)

TABLE D.4. continued

$$F_{0.05,\ v_1,\ v_2}$$

Degrees of freedom for the numerator (v_1)

v_2	1	2	3	4	5	6	7	8	9	10	12	15	20	24	30	40	60	120	∞
1	161.40	199.50	215.70	224.60	230.20	234.00	236.80	238.90	240.50	241.90	243.90	245.90	248.00	249.10	250.10	251.10	252.20	253.30	254.30
2	18.51	19.00	19.16	19.25	19.30	19.33	19.35	19.37	19.38	19.40	19.41	19.43	19.45	19.45	19.45	19.47	19.48	19.49	19.50
3	10.13	9.55	9.28	9.12	9.01	8.94	8.89	8.85	8.81	8.79	8.74	8.70	8.66	8.64	8.62	8.59	8.57	8.55	8.53
4	7.71	6.94	6.59	6.39	6.26	6.16	6.09	6.04	6.00	5.96	5.91	5.86	5.80	5.77	5.75	5.72	5.69	5.66	5.63
5	6.61	5.79	5.41	5.19	5.05	4.95	4.88	4.82	4.77	4.74	4.68	4.62	4.56	4.53	4.50	4.46	4.43	4.40	4.36
6	5.99	5.14	4.76	4.53	4.39	4.28	4.21	4.15	4.10	4.06	4.00	3.94	3.87	3.84	3.81	3.77	3.74	3.70	3.67
7	5.59	4.74	4.35	4.12	3.97	3.87	3.79	3.73	3.68	3.64	3.57	3.51	3.44	3.41	3.38	3.34	3.30	3.27	3.23
8	5.32	4.46	4.07	3.84	3.69	3.58	3.50	3.44	3.39	3.35	3.28	3.22	3.15	3.12	3.08	3.04	3.01	2.97	2.93
9	5.12	4.26	3.86	3.63	3.48	3.37	3.29	3.23	3.18	3.14	3.07	3.01	2.94	2.90	2.86	2.83	2.79	2.75	2.71
10	4.95	4.10	3.71	3.48	3.33	3.22	3.14	3.07	3.02	2.98	2.91	2.85	2.77	2.74	2.70	2.66	2.62	2.58	2.54
11	4.84	3.98	3.59	3.36	3.20	3.09	3.01	2.95	2.90	2.85	2.79	2.72	2.65	2.61	2.57	2.53	2.49	2.45	2.40
12	4.75	3.89	3.49	3.26	3.11	3.00	2.91	2.85	2.80	2.75	2.69	2.62	2.54	2.51	2.47	2.43	2.38	2.34	2.30
13	4.67	3.81	3.41	3.18	3.03	2.92	2.83	2.77	2.71	2.67	2.60	2.53	2.46	2.42	2.38	2.34	2.30	2.25	2.21
14	4.60	3.74	3.34	3.11	2.96	2.85	2.76	2.70	2.65	2.60	2.53	2.46	2.39	2.35	2.31	2.27	2.22	2.18	2.13
15	4.54	3.68	3.29	3.06	2.90	2.79	2.71	2.64	2.59	2.54	2.48	2.40	2.33	2.29	2.25	2.20	2.16	2.11	2.07
16	4.49	3.63	3.24	3.01	2.85	2.74	2.66	2.59	2.54	2.49	2.42	2.35	2.28	2.24	2.19	2.15	2.11	2.06	2.01
17	4.45	3.59	3.20	2.96	2.81	2.70	2.61	2.55	2.49	2.45	2.38	2.31	2.23	2.19	2.15	2.10	2.06	2.01	1.96
18	4.41	3.55	3.16	2.93	2.77	2.66	2.58	2.51	2.46	2.41	2.34	2.27	2.19	2.15	2.11	2.06	2.02	1.97	1.92
19	4.38	3.52	3.13	2.90	2.74	2.63	2.54	2.48	2.42	2.38	2.31	2.23	2.16	2.11	2.07	2.03	1.98	1.93	1.88
20	4.35	3.49	3.10	2.87	2.71	2.60	2.51	2.45	2.39	2.35	2.28	2.20	2.12	2.08	2.04	1.99	1.95	1.90	1.84
21	4.32	3.47	3.07	2.84	2.68	2.57	2.49	2.42	2.37	2.32	2.25	2.18	2.10	2.05	2.01	1.96	1.92	1.87	1.81
22	4.30	3.44	3.05	2.82	2.66	2.55	2.46	2.40	2.34	2.30	2.23	2.15	2.07	2.03	1.98	1.94	1.89	1.84	1.78
23	4.28	3.42	3.03	2.80	2.64	2.53	2.44	2.37	2.32	2.27	2.20	2.13	2.05	2.01	1.96	1.91	1.86	1.81	1.76
24	4.26	3.40	3.01	2.78	2.62	2.51	2.42	2.36	2.30	2.25	2.18	2.11	2.03	1.98	1.94	1.89	1.84	1.79	1.73
25	4.24	3.39	2.99	2.76	2.60	2.49	2.40	2.34	2.28	2.24	2.16	2.09	2.01	1.96	1.92	1.87	1.82	1.77	1.71
26	4.23	3.37	2.98	2.74	2.59	2.47	2.39	2.32	2.27	2.22	2.15	2.07	1.99	1.95	1.90	1.85	1.80	1.75	1.69
27	4.21	3.35	2.96	2.73	2.57	2.46	2.37	2.31	2.25	2.20	2.13	2.06	1.97	1.93	1.88	1.84	1.79	1.73	1.67
28	4.20	3.34	2.95	2.71	2.56	2.45	2.36	2.29	2.24	2.19	2.12	2.04	1.96	1.91	1.87	1.82	1.77	1.71	1.65
29	4.18	3.33	2.93	2.70	2.55	2.43	2.35	2.28	2.22	2.18	2.10	2.03	1.94	1.90	1.85	1.81	1.75	1.70	1.64
30	4.17	3.32	2.92	2.69	2.53	2.42	2.33	2.27	2.21	2.16	2.09	2.01	1.93	1.89	1.84	1.79	1.74	1.68	1.62
40	4.08	3.23	2.84	2.61	2.45	2.34	2.25	2.18	2.12	2.08	2.00	1.92	1.84	1.79	1.74	1.69	1.64	1.58	1.51
60	4.00	3.15	2.76	2.53	2.37	2.25	2.17	2.10	2.04	1.99	1.92	1.84	1.75	1.70	1.65	1.59	1.53	1.47	1.39
120	3.92	3.07	2.68	2.45	2.29	2.17	2.09	2.02	1.96	1.91	1.83	1.75	1.66	1.61	1.55	1.50	1.43	1.35	1.25
∞	3.84	3.00	2.60	2.37	2.21	2.10	2.01	1.94	1.88	1.83	1.75	1.67	1.57	1.52	1.46	1.39	1.32	1.22	1.00

Degrees of freedom for the denominator (v_2)

TABLE D.4. *continued*

$$F_{0.25, \nu_1, \nu_2}$$

Degrees of freedom for the numerator (ν_1)

ν_2 \ ν_1	1	2	3	4	5	6	7	8	9	10	12	15	20	24	30	40	60	120	∞
1	647.80	799.50	864.20	899.60	921.80	937.10	948.20	956.70	963.30	968.60	976.70	984.90	993.10	997.20	1001.00	1006.00	1010.00	1014.00	1018.00
2	38.51	39.00	39.17	39.25	39.30	39.33	39.36	39.37	39.39	39.40	39.41	39.43	39.45	39.46	39.46	39.47	39.48	39.49	39.50
3	17.44	16.04	15.44	15.10	14.88	14.73	14.62	14.54	14.47	14.42	14.34	14.25	14.17	14.12	14.08	14.04	13.99	13.95	13.90
4	12.22	10.65	9.98	9.60	9.36	9.20	9.07	8.98	8.90	8.84	8.75	8.66	8.56	8.51	8.46	8.41	8.36	8.31	8.26
5	10.01	8.43	7.76	7.39	7.15	6.98	6.85	6.76	6.68	6.62	6.52	6.43	6.33	6.28	6.23	6.18	6.12	6.07	6.02
6	8.81	7.26	6.60	6.23	5.99	5.82	5.70	5.60	5.52	5.46	5.37	5.27	5.17	5.12	5.07	5.01	4.96	4.90	4.85
7	8.07	6.54	5.89	5.52	5.29	5.12	4.99	4.90	4.82	4.76	4.67	4.57	4.47	4.42	4.36	4.31	4.25	4.20	4.14
8	7.57	6.06	5.42	5.05	4.82	4.65	4.53	4.43	4.36	4.30	4.20	4.10	4.00	3.95	3.89	3.84	3.78	3.73	3.67
9	7.21	5.71	5.08	4.72	4.48	4.32	4.20	4.10	4.03	3.96	3.87	3.77	3.67	3.61	3.56	3.51	3.45	3.39	3.33
10	6.94	5.46	4.83	4.47	4.24	4.07	3.95	3.85	3.78	3.72	3.62	3.52	3.42	3.37	3.31	3.26	3.20	3.14	3.08
11	6.72	5.26	4.63	4.28	4.04	3.88	3.76	3.66	3.59	3.53	3.43	3.33	3.23	3.17	3.12	3.06	3.00	2.94	2.88
12	6.55	5.10	4.47	4.12	3.89	3.73	3.61	3.51	3.44	3.37	3.28	3.18	3.07	3.02	2.96	2.91	2.85	2.79	2.72
13	6.41	4.97	4.35	4.00	3.77	3.60	3.48	3.39	3.31	3.25	3.15	3.05	2.95	2.89	2.84	2.78	2.72	2.66	2.60
14	6.30	4.86	4.24	3.89	3.66	3.50	3.38	3.29	3.21	3.15	3.05	2.95	2.84	2.79	2.73	2.67	2.61	2.55	2.49
15	6.20	4.77	4.15	3.80	3.58	3.41	3.29	3.20	3.12	3.06	2.96	2.86	2.76	2.70	2.64	2.59	2.52	2.46	2.40
16	6.12	4.69	4.08	3.73	3.50	3.34	3.22	3.12	3.05	2.99	2.89	2.79	2.68	2.63	2.57	2.51	2.45	2.38	2.32
17	6.04	4.62	4.01	3.66	3.44	3.28	3.16	3.06	2.98	2.92	2.82	2.72	2.62	2.56	2.50	2.44	2.38	2.32	2.25
18	5.98	4.56	3.95	3.61	3.38	3.22	3.10	3.01	2.93	2.87	2.77	2.67	2.56	2.50	2.44	2.38	2.32	2.26	2.19
19	5.92	4.51	3.90	3.56	3.33	3.17	3.05	2.96	2.88	2.82	2.72	2.62	2.51	2.45	2.39	2.33	2.27	2.20	2.13
20	5.87	4.46	3.86	3.51	3.29	3.13	3.01	2.91	2.84	2.77	2.68	2.57	2.46	2.41	2.35	2.29	2.22	2.16	2.09
21	5.83	4.42	3.82	3.48	3.25	3.09	2.97	2.87	2.80	2.73	2.64	2.53	2.42	2.37	2.31	2.25	2.18	2.11	2.04
22	5.79	4.38	3.78	3.44	3.22	3.05	2.93	2.84	2.76	2.70	2.60	2.50	2.39	2.33	2.27	2.21	2.14	2.08	2.00
23	5.75	4.35	3.75	3.41	3.18	3.02	2.90	2.81	2.73	2.67	2.57	2.47	2.36	2.30	2.24	2.18	2.11	2.04	1.97
24	5.72	4.32	3.72	3.38	3.15	2.99	2.87	2.78	2.70	2.64	2.54	2.44	2.33	2.27	2.21	2.15	2.08	2.01	1.94
25	5.69	4.29	3.69	3.35	3.13	2.97	2.85	2.75	2.68	2.61	2.51	2.41	2.30	2.24	2.18	2.12	2.05	1.98	1.91
26	5.66	4.27	3.67	3.33	3.10	2.94	2.82	2.73	2.65	2.59	2.49	2.39	2.28	2.22	2.16	2.09	2.03	1.95	1.88
27	5.63	4.24	3.65	3.31	3.08	2.92	2.80	2.71	2.63	2.57	2.47	2.36	2.25	2.19	2.13	2.07	2.00	1.93	1.85
28	5.61	4.22	3.63	3.29	3.06	2.90	2.78	2.69	2.61	2.55	2.45	2.34	2.23	2.17	2.11	2.05	1.98	1.91	1.83
29	5.59	4.20	3.61	3.27	3.04	2.88	2.76	2.67	2.59	2.53	2.43	2.32	2.21	2.15	2.09	2.03	1.96	1.89	1.81
30	5.57	4.18	3.59	3.25	3.03	2.87	2.75	2.65	2.57	2.51	2.41	2.31	2.20	2.14	2.07	2.01	1.94	1.87	1.79
40	5.42	4.05	3.46	3.13	2.90	2.74	2.62	2.53	2.45	2.39	2.29	2.18	2.07	2.01	1.94	1.88	1.80	1.72	1.64
60	5.29	3.93	3.34	3.01	2.79	2.63	2.51	2.41	2.33	2.27	2.17	2.06	1.94	1.88	1.82	1.74	1.67	1.58	1.48
120	5.15	3.80	3.23	2.89	2.67	2.52	2.39	2.30	2.22	2.16	2.05	1.94	1.82	1.76	1.69	1.61	1.53	1.43	1.31
∞	5.02	3.69	3.12	2.79	2.57	2.41	2.29	2.19	2.11	2.05	1.94	1.83	1.71	1.64	1.57	1.48	1.39	1.27	1.00

Degrees of freedom for the denominator (ν_2)

TABLE D.4. continued

$$F_{0.01, v_1, v_2}$$

$v_2 \backslash v_1$	Degrees of freedom for the numerator (v_1)																		
	1	2	3	4	5	6	7	8	9	10	12	15	20	24	30	40	60	120	∞
1	4052	4999.5	5403	5625	5764	5859	5928	5982	6022	6056	6106	6157	6209	6235	6261	6287	6313	6339	6366
2	98.50	99.00	99.17	99.25	99.30	99.33	99.36	99.37	99.39	99.40	99.42	99.43	99.45	99.46	99.47	99.47	99.48	99.49	99.50
3	34.12	30.82	29.46	28.71	28.24	27.91	27.67	27.49	27.35	27.23	27.05	26.87	26.69	26.60	26.50	26.41	26.32	26.22	26.13
4	21.20	18.00	16.69	15.98	15.52	15.21	14.98	14.80	14.66	14.55	14.37	14.20	14.02	13.93	13.84	13.75	13.65	13.56	13.46
5	16.26	13.27	12.06	11.39	10.97	10.67	10.46	10.29	10.16	10.05	9.89	9.72	9.55	9.47	9.38	9.29	9.20	9.11	9.02
6	13.75	10.92	9.78	9.15	8.75	8.47	8.26	8.10	7.98	7.87	7.72	7.56	7.40	7.31	7.23	7.14	7.06	6.97	6.88
7	12.25	9.55	8.45	7.85	7.46	7.19	6.99	6.84	6.72	6.62	6.47	6.31	6.16	6.07	5.99	5.91	5.82	5.74	5.65
8	11.26	8.65	7.59	7.01	6.63	6.37	6.18	6.03	5.91	5.81	5.67	5.52	5.36	5.28	5.20	5.12	5.03	4.95	4.86
9	10.56	8.02	6.99	6.42	6.06	5.80	5.61	5.47	5.35	5.26	5.11	4.96	4.81	4.73	4.65	4.57	4.48	4.40	4.31
10	10.04	7.56	6.55	5.99	5.64	5.39	5.20	5.06	4.94	4.85	4.71	4.56	4.41	4.33	4.25	4.17	4.08	4.00	3.91
11	9.65	7.21	6.22	5.67	5.32	5.07	4.89	4.74	4.63	4.54	4.40	4.25	4.10	4.02	3.94	3.86	3.78	3.69	3.60
12	9.33	6.93	5.95	5.41	5.06	4.82	4.64	4.50	4.39	4.30	4.16	4.01	3.86	3.78	3.70	3.62	3.54	3.45	3.36
13	9.07	6.70	5.74	5.21	4.86	4.62	4.44	4.30	4.19	4.10	3.96	3.82	3.66	3.59	3.51	3.43	3.34	3.25	3.17
14	8.86	6.51	5.56	5.04	4.69	4.46	4.28	4.14	4.03	3.94	3.80	3.66	3.51	3.43	3.35	3.27	3.18	3.09	3.00
15	8.68	6.36	5.42	4.89	4.56	4.32	4.14	4.00	3.89	3.80	3.67	3.52	3.37	3.29	3.21	3.13	3.05	2.96	2.87
16	8.53	6.23	5.29	4.77	4.44	4.20	4.03	3.89	3.78	3.69	3.55	3.41	3.26	3.18	3.10	3.02	2.93	2.84	2.75
17	8.40	6.11	5.18	4.67	4.34	4.10	3.93	3.79	3.68	3.59	3.46	3.31	3.16	3.08	3.00	2.92	2.83	2.75	2.65
18	8.29	6.01	5.09	4.58	4.25	4.01	3.84	3.71	3.60	3.51	3.37	3.23	3.08	3.00	2.92	2.84	2.75	2.66	2.57
19	8.18	5.93	5.01	4.50	4.17	3.94	3.77	3.63	3.52	3.43	3.30	3.15	3.00	2.92	2.84	2.76	2.67	2.58	2.49
20	8.10	5.85	4.94	4.43	4.10	3.87	3.70	3.56	3.46	3.37	3.23	3.09	2.94	2.86	2.78	2.69	2.61	2.52	2.42
21	8.02	5.78	4.87	4.37	4.04	3.81	3.64	3.51	3.40	3.31	3.17	3.03	2.88	2.80	2.72	2.64	2.55	2.46	2.36
22	7.95	5.72	4.82	4.31	3.99	3.76	3.59	3.45	3.35	3.26	3.12	2.98	2.83	2.75	2.67	2.58	2.50	2.40	2.31
23	7.88	5.66	4.76	4.26	3.94	3.71	3.54	3.41	3.30	3.21	3.07	2.93	2.78	2.70	2.62	2.54	2.45	2.35	2.26
24	7.82	5.61	4.72	4.22	3.90	3.67	3.50	3.36	3.26	3.17	3.03	2.89	2.74	2.66	2.58	2.49	2.40	2.31	2.21
25	7.77	5.57	4.68	4.18	3.85	3.63	3.46	3.32	3.22	3.13	2.99	2.85	2.70	2.62	2.54	2.45	2.36	2.27	2.17
26	7.72	5.53	4.64	4.14	3.82	3.59	3.42	3.29	3.18	3.09	2.96	2.81	2.66	2.58	2.50	2.42	2.33	2.23	2.13
27	7.68	5.49	4.60	4.11	3.78	3.56	3.39	3.26	3.15	3.06	2.93	2.78	2.63	2.55	2.47	2.38	2.29	2.20	2.10
28	7.64	5.45	4.57	4.07	3.75	3.53	3.36	3.23	3.12	3.03	2.90	2.75	2.60	2.52	2.44	2.35	2.26	2.17	2.06
29	7.60	5.42	4.54	4.04	3.73	3.50	3.33	3.20	3.09	3.00	2.87	2.73	2.57	2.49	2.41	2.33	2.23	2.14	2.03
30	7.56	5.39	4.51	4.02	3.70	3.47	3.30	3.17	3.07	2.98	2.84	2.70	2.55	2.47	2.39	2.30	2.21	2.11	2.01
40	7.31	5.18	4.31	3.83	3.51	3.29	3.12	2.99	2.89	2.80	2.66	2.52	2.37	2.29	2.20	2.11	2.02	1.92	1.80
60	7.08	4.98	4.13	3.65	3.34	3.12	2.95	2.82	2.72	2.63	2.50	2.35	2.20	2.12	2.03	1.94	1.84	1.73	1.60
120	6.85	4.79	3.95	3.48	3.17	2.96	2.79	2.66	2.56	2.47	2.34	2.19	2.03	1.95	1.86	1.76	1.66	1.53	1.38
∞	6.63	4.61	3.78	3.32	3.02	2.80	2.64	2.51	2.41	2.32	2.18	2.04	1.88	1.79	1.70	1.59	1.47	1.32	1.00

Degrees of freedom for the denominator (v_2)

TABLE D.5. Cumulative poisson distribution.

χ	0.01	0.05	0.10	$\alpha = \lambda t$ 0.20	0.30	0.40	0.50	0.60
0	0.990	0.951	0.904	0.818	0.740	0.670	0.606	0.548
1	0.999	0.998	0.995	0.982	0.963	0.938	0.909	0.878
2		0.999	0.999	0.998	0.996	0.992	0.985	0.976
3				0.999	0.999	0.999	0.998	0.996
4					0.999	0.999	0.999	0.999
5							0.999	0.999

χ	.70	.80	.90	$\alpha = \lambda t$ 1.00	1.10	1.20	1.30	1.40
0	0.496	0.449	0.406	0.367	0.332	0.301	0.272	0.246
1	0.844	0.808	0.772	0.735	0.699	0.662	0.626	0.591
2	0.965	0.952	0.937	0.919	0.900	0.879	0.857	0.833
3	0.994	0.990	0.986	0.981	0.974	0.966	0.956	0.946
4	0.999	0.998	0.997	0.996	0.994	0.992	0.989	0.985
5	0.999	0.999	0.999	0.999	0.999	0.998	0.997	0.996
6		0.999	0.999	0.999	0.999	0.999	0.999	0.999
7				0.999	0.999	0.999	0.999	0.999
8							0.999	0.999

χ	1.50	1.60	1.70	$\alpha = \lambda t$ 1.80	1.90	2.00	2.10	2.20
0	0.223	0.201	0.182	0.165	0.149	0.135	0.122	0.110
1	0.557	0.524	0.493	0.462	0.433	0.406	0.379	0.354
2	0.808	0.783	0.757	0.730	0.703	0.676	0.649	0.622
3	0.934	0.921	0.906	0.891	0.874	0.857	0.838	0.819
4	0.981	0.976	0.970	0.963	0.955	0.947	0.937	0.927
5	0.995	0.993	0.992	0.989	0.986	0.983	0.979	0.975
6	0.999	0.998	0.998	0.997	0.996	0.995	0.994	0.992
7	0.999	0.999	0.999	0.999	0.999	0.998	0.998	0.998
8	0.999	0.999	0.999	0.999	0.999	0.999	0.999	0.999
9			0.999	0.999	0.999	0.999	0.999	0.999
10							0.999	0.999

TABLE D.5. *continued*

				$\alpha = \lambda t$				
χ	2.30	2.40	2.50	2.60	2.70	2.80	2.90	3.00
0	0.100	0.090	0.082	0.074	0.067	0.060	0.055	0.049
1	0.330	0.308	0.287	0.267	0.248	0.231	0.214	0.199
2	0.596	0.569	0.543	0.518	0.493	0.469	0.445	0.423
3	0.799	0.778	0.757	0.736	0.714	0.691	0.669	0.647
4	0.916	0.904	0.891	0.877	0.862	0.847	0.831	0.815
5	0.970	0.964	0.957	0.950	0.943	0.934	0.925	0.916
6	0.990	0.988	0.985	0.982	0.979	0.975	0.971	0.966
7	0.997	0.996	0.995	0.994	0.993	0.991	0.990	0.988
8	0.999	0.999	0.998	0.998	0.998	0.997	0.996	0.996
9	0.999	0.999	0.999	0.999	0.999	0.999	0.999	0.998
10	0.999	0.999	0.999	0.999	0.999	0.999	0.999	0.999
11			0.999	0.999	0.999	0.999	0.999	0.999
12							0.999	0.999

				$\alpha = \lambda t$				
χ	3.50	4.00	4.50	5.00	5.50	6.00	6.50	7.00
0	0.030	0.018	0.011	0.006	0.004	0.002	0.001	0.000
1	0.135	0.091	0.061	0.040	0.026	0.017	0.011	0.007
2	0.320	0.238	0.173	0.124	0.088	0.061	0.043	0.029
3	0.536	0.433	0.342	0.265	0.201	0.151	0.111	0.081
4	0.725	0.628	0.532	0.440	0.357	0.285	0.223	0.172
5	0.857	0.785	0.702	0.615	0.528	0.445	0.369	0.300
6	0.934	0.889	0.831	0.762	0.686	0.606	0.526	0.449
7	0.973	0.948	0.913	0.866	0.809	0.743	0.672	0.598
8	0.990	0.978	0.959	0.931	0.894	0.847	0.791	0.729
9	0.996	0.991	0.982	0.968	0.946	0.916	0.877	0.830
10	0.998	0.997	0.993	0.986	0.974	0.957	0.933	0.901
11	0.999	0.999	0.997	0.994	0.989	0.979	0.966	0.946
12	0.999	0.999	0.999	0.997	0.995	0.991	0.983	0.973
13	0.999	0.999	0.999	0.999	0.998	0.996	0.992	0.987
14		0.999	0.999	0.999	0.999	0.998	0.997	0.994
15			0.999	0.999	0.999	0.999	0.998	0.997
16				0.999	0.999	0.999	0.999	0.999
17					0.999	0.999	0.999	0.999
18						0.999	0.999	0.999
19							0.999	0.999

TABLE **D.5.** *continued*

χ	7.50	8.00	8.50	α = λt 9.00	9.50	10.00	15.00	20.00
0	0.000	0.000	0.000	0.000	0.000	0.000	0.000	0.000
1	0.004	0.003	0.001	0.001	0.000	0.000	0.000	0.000
2	0.020	0.013	0.009	0.006	0.004	0.002	0.000	0.000
3	0.059	0.042	0.030	0.021	0.014	0.010	0.000	0.000
4	0.132	0.099	0.074	0.054	0.040	0.029	0.000	0.000
5	0.241	0.191	0.149	0.115	0.088	0.067	0.002	0.000
6	0.378	0.313	0.256	0.206	0.164	0.130	0.007	0.000
7	0.524	0.452	0.385	0.323	0.268	0.220	0.018	0.000
8	0.661	0.592	0.523	0.455	0.391	0.332	0.037	0.002
9	0.776	0.716	0.652	0.587	0.521	0.457	0.069	0.005
10	0.862	0.815	0.763	0.705	0.645	0.583	0.118	0.010
11	0.920	0.888	0.848	0.803	0.751	0.696	0.184	0.021
12	0.957	0.936	0.909	0.875	0.836	0.791	0.267	0.039
13	0.978	0.965	0.948	0.926	0.898	0.864	0.363	0.066
14	0.989	0.982	0.972	0.958	0.940	0.916	0.465	0.104
15	0.995	0.991	0.986	0.977	0.966	0.951	0.568	0.156
16	0.998	0.996	0.993	0.988	0.982	0.972	0.664	0.221
17	0.999	0.998	0.997	0.994	0.991	0.985	0.748	0.297
18	0.999	0.999	0.998	0.997	0.995	0.992	0.819	0.381
19	0.999	0.999	0.999	0.998	0.998	0.996	0.875	0.470
20	0.999	0.999	0.999	0.999	0.999	0.998	0.917	0.559
21	0.999	0.999	0.999	0.999	0.999	0.999	0.946	0.643
22		0.999	0.999	0.999	0.999	0.999	0.967	0.720
23			0.999	0.999	0.999	0.999	0.980	0.787
24					0.999	0.999	0.988	0.843
25						0.999	0.993	0.887
26							0.996	0.922
27							0.998	0.947
28							0.999	0.965
29							0.999	0.978
30							0.999	0.986
31							0.999	0.991
32							0.999	0.995
33							0.999	0.997
34								0.998

TABLE D.6. Percentage points of the binomial distribution.

$$P(X \le k) = \sum_{x=0}^{k} \binom{n}{k} p^x (1-p)^{n-x}$$

n	k	0.05	0.10	0.15	0.20	p 0.25	0.30	0.35	0.40	0.45	0.50
2	0	0.9025	0.8100	0.7225	0.6400	0.5265	0.4900	0.4225	0.3600	0.3025	0.2500
	1	0.9975	0.9900	0.9775	0.9600	0.9375	0.9100	0.8775	0.8400	0.7975	0.7500
3	0	0.8574	0.7290	0.6141	0.5120	0.4219	0.3430	0.2746	0.2160	0.1664	0.1250
	1	0.9927	0.9720	0.9393	0.8960	0.8438	0.7840	0.7183	0.6480	0.5748	0.5000
	2	0.9999	0.9990	0.9966	0.9920	0.9844	0.9730	0.9571	0.9360	0.9089	0.8750
4	0	0.8145	0.6561	0.5220	0.4096	0.3164	0.2401	0.1785	0.1296	0.0915	0.0625
	1	0.9860	0.9477	0.8905	0.8192	0.7383	0.6517	0.5630	0.4752	0.3910	0.3125
	2	0.9995	0.9963	0.9880	0.9728	0.9492	0.9163	0.8735	0.8208	0.7585	0.6875
	3	1.0000	0.9999	0.9995	0.9984	0.9961	0.9919	0.9850	0.9744	0.9590	0.9375
5	0	0.7738	0.5905	0.4437	0.3277	0.2373	0.1681	0.1160	0.0778	0.0503	0.0313
	1	0.9774	0.9185	0.8352	0.7373	0.6328	0.5282	0.4284	0.3370	0.2562	0.1875
	2	0.9988	0.9914	0.9734	0.9421	0.8965	0.8369	0.7648	0.6826	0.5931	0.5000
	3	1.0000	0.9995	0.9978	0.9933	0.9844	0.9692	0.9460	0.9130	0.8688	0.8125
	4	1.0000	1.0000	0.9999	0.9997	0.9990	0.9976	0.9947	0.9898	0.9815	0.9688
6	0	0.7351	0.5314	0.3771	0.2621	0.1780	0.1176	0.0754	0.0467	0.0277	0.0156
	1	0.9672	0.8857	0.7765	0.6554	0.5339	0.4202	0.3191	0.2333	0.1636	0.1094
	2	0.9978	0.9841	0.9527	0.9011	0.8306	0.7443	0.6471	0.5443	0.4415	0.3438
	3	0.9999	0.9987	0.9941	0.9830	0.9624	0.9295	0.8826	0.8208	0.7447	0.6563
	4	1.0000	0.9999	0.9996	0.0084	0.9954	0.9891	0.9777	0.9590	0.9308	0.8906
	5	1.0000	1.0000	1.0000	0.9999	0.9998	0.9993	0.9982	0.9959	0.9917	0.9844
7	0	0.6983	0.4783	0.3206	0.2097	0.1335	0.0824	0.0490	0.0280	0.0152	0.0078
	1	0.9556	0.8503	0.7166	0.5767	0.4449	0.3294	0.2338	0.1586	0.1024	0.0625
	2	0.9962	0.9743	0.9262	0.8520	0.7564	0.6471	0.5323	0.4199	0.3164	0.2266
	3	0.9998	0.9973	0.9879	0.9667	0.9294	0.8740	0.8002	0.7102	0.6983	0.5000
	4	1.0000	0.9998	0.9988	0.9953	0.9871	0.9712	0.9444	0.9037	0.8471	0.7734
	5	1.0000	1.0000	0.9999	0.9996	0.9987	0.9962	0.9910	0.9812	0.9643	0.9375
	6	1.0000	1.0000	1.0000	1.0000	0.9999	0.9998	0.9994	0.9984	0.9963	0.9922
8	0	0.6634	0.4305	0.2725	0.1678	0.1001	0.0567	0.0319	0.0168	0.0084	0.0039
	1	0.9428	0.8131	0.6572	0.5033	0.3671	0.2553	0.1691	0.1064	0.0632	0.0352
	2	0.9942	0.9619	0.8948	0.7969	0.6785	0.5518	0.4278	0.3154	0.2201	0.1445
	3	0.9996	0.9950	0.9786	0.9437	0.8862	0.8059	0.7064	0.5941	0.4770	0.3633
	4	1.0000	0.9996	0.9971	0.9896	0.9727	0.9420	0.8939	0.8263	0.7396	0.6367
	5	1.0000	1.0000	0.9998	0.9988	0.9958	0.9887	0.9747	0.9502	0.9115	0.8555
	6	1.0000	1.0000	1.0000	0.9999	0.9996	0.9987	0.9964	0.9915	0.9819	0.9648
	7	1.0000	1.0000	1.0000	1.0000	1.0000	0.9999	0.9998	0.9993	0.9983	0.9961

TABLE **D.6.** *continued*

n	k	0.05	.0.10	0.15	0.20	*p* 0.25	0.30	0.35	0.40	0.45	0.50
9	0	0.6302	0.3874	0.2316	0.1342	0.0751	0.0404	0.0207	0.0101	0.0046	0.0020
	1	0.9299	0.7748	0.5995	0.4362	0.3003	0.1960	0.1211	0.0705	0.0385	0.0195
	2	0.9916	0.9470	0.8591	0.7382	0.6007	0.4628	0.3373	0.2318	0.1495	0.0898
	3	0.9994	0.9917	0.9661	0.9144	0.8343	0.7297	0.6089	0.4826	0.3614	0.2539
	4	1.0000	0.9991	0.9944	0.9804	0.9511	0.9012	0.8283	0.7334	0.6214	0.5000
	5	1.0000	0.9999	0.9994	0.9969	0.9900	0.9747	0.9464	0.9006	0.8342	0.7461
	6	1.0000	1.0000	1.0000	0.9997	0.9987	0.9957	0.9888	0.9750	0.9502	0.9102
	7	1.0000	1.0000	1.0000	1.0000	0.9999	0.9996	0.9986	0.9962	0.9909	0.9805
	8	1.0000	1.0000	1.0000	1.0000	1.0000	1.0000	0.9999	0.9997	0.9992	0.9980
10	0	0.5987	0.3487	0.1969	0.1074	0.0563	0.0282	0.0135	0.0060	0.0025	0.0010
	1	0.9139	0.7361	0.5443	0.3758	0.2440	0.1493	0.0860	0.0464	0.0233	0.0107
	2	0.9885	0.9298	0.8202	0.6778	0.5256	0.3828	0.2616	0.1673	0.0996	0.0547
	3	0.9990	0.9872	0.9500	0.8791	0.7759	0.6496	0.5138	0.3823	0.2660	0.1719
	4	0.9999	0.9984	0.9901	0.9672	0.9219	0.8497	0.7515	0.6331	0.5044	0.3770
	5	1.0000	0.9999	0.9986	0.9936	0.9803	0.9527	0.9051	0.8338	0.7384	0.6230
	6	1.0000	1.0000	0.9999	0.9991	0.9965	0.9894	0.9740	0.9452	0.8980	0.8281
	7	1.0000	1.0000	1.0000	0.9999	0.9996	0.9984	0.9952	0.9877	0.9726	0.9453
	8	1.0000	1.0000	1.0000	1.0000	1.0000	0.9999	0.9995	0.9983	0.9955	0.9893
	9	1.0000	1.0000	1.0000	1.0000	1.0000	1.0000	1.0000	0.9999	0.9997	0.9990
11	0	0.5688	0.3138	0.1673	0.0859	0.0422	0.0198	0.0088	0.0036	0.0014	0.0005
	1	0.8981	0.6974	0.4922	0.3221	0.1971	0.1130	0.0606	0.0302	0.0139	0.0059
	2	0.9848	0.9104	0.7788	0.6174	0.4552	0.3127	0.2001	0.1189	0.0652	0.0327
	3	0.9984	0.9815	0.9306	0.8389	0.7133	0.5696	0.4256	0.2963	0.1911	0.1133
	4	0.9999	0.9972	0.9841	0.9496	0.8854	0.7897	0.6683	0.5328	0.3971	0.2744
	5	1.0000	0.9997	0.9973	0.9883	0.9657	0.9218	0.8513	0.7535	0.6331	0.5000
	6	1.0000	1.0000	0.9997	0.9980	0.9924	0.9784	0.9499	0.9006	0.8262	0.7256
	7	1.0000	1.0000	1.0000	0.9998	0.9988	0.9957	0.9878	0.9707	0.9390	0.8867
	8	1.0000	1.0000	1.0000	1.0000	0.9999	0.9994	0.9980	0.9941	0.9852	0.9673
	9	1.0000	1.0000	1.0000	1.0000	1.0000	1.0000	0.9998	0.9993	0.9978	0.9941
	10	1.0000	1.0000	1.0000	1.0000	1.0000	1.0000	1.0000	1.0000	0.9998	0.9995
12	0	0.5404	0.2824	0.1422	0.0687	0.0317	0.0138	0.0057	0.0022	0.0008	0.0002
	1	0.8816	0.6590	0.4435	0.2749	0.1584	0.0850	0.0424	0.0196	0.0083	0.0032
	2	0.9804	0.8891	0.7358	0.5583	0.3907	0.2528	0.1513	0.0834	0.0421	0.0193
	3	0.9978	0.9744	0.9078	0.7946	0.6488	0.4925	0.3467	0.2253	0.1345	0.0730
	4	0.9998	0.9957	0.9761	0.9274	0.8424	0.7237	0.5833	0.4382	0.3044	0.1938
	5	1.0000	0.9995	0.9954	0.9806	0.9456	0.8822	0.7873	0.6652	0.5269	0.3872
	6	1.0000	0.9999	0.9993	0.9961	0.9857	0.9614	0.9154	0.8418	0.7393	0.6128
	7	1.0000	1.0000	0.9999	0.9994	0.9972	0.9905	0.9745	0.9427	0.8883	0.8062
	8	1.0000	1.0000	1.0000	0.9999	0.9996	0.9983	0.9944	0.9847	0.9644	0.9270
	9	1.0000	1.0000	1.0000	1.0000	1.0000	0.9998	0.9992	0.9972	0.9921	0.9807
	10	1.0000	1.0000	1.0000	1.0000	1.0000	1.0000	0.9999	0.9997	0.9989	0.9968
	11	1.0000	1.0000	1.0000	1.0000	1.0000	1.0000	1.0000	1.0000	0.9999	0.9998

TABLE D.6. *continued*

n	k	0.05	.0.10	0.15	0.20	p 0.25	0.30	0.35	0.40	0.45	0.50
13	0	0.5133	0.2542	0.1209	0.0550	0.0238	0.0097	0.0037	0.0013	0.0004	0.0001
	1	0.8646	0.6213	0.3983	0.2336	0.1267	0.0637	0.0296	0.0126	0.0049	0.0017
	2	0.9755	0.8661	0.6920	0.5017	0.3326	0.2025	0.1132	0.0579	0.0268	0.0112
	3	0.9969	0.9658	0.8820	0.7473	0.5843	0.4206	0.2783	0.1686	0.0929	0.0461
	4	0.9997	0.9935	0.9658	0.9009	0.7940	0.6543	0.5005	0.3530	0.2279	0.1334
	5	1.0000	0.9991	0.9925	0.9700	0.9198	0.8346	0.7159	0.5744	0.4268	0.2905
	6	1.0000	0.9999	0.9987	0.9930	0.9757	0.9376	0.8705	0.7712	0.6437	0.5000
	7	1.0000	1.0000	0.9998	0.9988	0.9944	0.9818	0.9538	0.9023	0.8212	0.7095
	8	1.0000	1.0000	1.0000	0.9998	0.9990	0.9960	0.9874	0.9679	0.9302	0.8666
	9	1.0000	1.0000	1.0000	1.0000	0.9999	0.9993	0.9975	0.9922	0.9797	0.9539
	10	1.0000	1.0000	1.0000	1.0000	1.0000	0.9999	0.9997	0.9987	0.9959	0.9888
	11	1.0000	1.0000	1.0000	1.0000	1.0000	1.0000	1.0000	1.0000	1.0000	1.0000
14	0	0.4877	0.2288	0.1028	0.0440	0.0178	0.0068	0.0024	0.0008	0.0002	0.0001
	1	0.8470	0.5846	0.3567	0.1979	0.1010	0.0475	0.0205	0.0081	0.0029	0.0009
	2	0.9699	0.8416	0.6479	0.4481	0.2811	0.1608	0.0838	0.0398	0.0170	0.0065
	3	0.9958	0.9559	0.8535	0.6982	0.5213	0.3552	0.2205	0.1243	0.0632	0.0287
	4	0.9996	0.9908	0.9533	0.8702	0.7415	0.5842	0.4227	0.2793	0.1672	0.0898
	5	1.0000	0.9985	0.9885	0.9561	0.8883	0.7805	0.6405	0.4859	0.3373	0.2120
	6	1.0000	0.9998	0.9978	0.9884	0.9617	0.9067	0.8164	0.6925	0.5461	0.3953
	7	1.0000	1.0000	0.9997	0.9976	0.9897	0.9685	0.9247	0.8499	0.7414	0.6047
	8	1.0000	1.0000	1.0000	0.9996	0.9978	0.9917	0.9757	0.9417	0.8811	0.7880
	9	1.0000	1.0000	1.0000	1.0000	0.9997	0.9983	0.9940	0.9825	0.9574	0.9102
	10	1.0000	1.0000	1.0000	1.0000	1.0000	0.9998	0.9989	0.9961	0.9886	0.9713
	11	1.0000	1.0000	1.0000	1.0000	1.0000	1.0000	0.9999	0.9994	0.9978	0.9935
	12	1.0000	1.0000	1.0000	1.0000	1.0000	1.0000	1.0000	1.0000	1.0000	1.0000
15	0	0.4633	0.2059	0.0874	0.0352	0.0134	0.0047	0.0016	0.0005	0.0001	0.0000
	1	0.8290	0.5490	0.3186	0.1671	0.0802	0.0353	0.0142	0.0052	0.0017	0.0005
	2	0.9638	0.8159	0.6042	0.3980	0.2361	0.1268	0.0617	0.0271	0.0107	0.0037
	3	0.9945	0.9444	0.8227	0.6482	0.4613	0.2969	0.1729	0.0905	0.0424	0.0176
	4	0.9994	0.9873	0.9383	0.8358	0.6865	0.5155	0.3519	0.2173	0.1204	0.0592
	5	0.9999	0.9977	0.9832	0.9389	0.8516	0.7216	0.5643	0.4032	0.2608	0.1509

TABLE **D.7.** Critical values d_n, α for Kolmogorov–Smirnov test. (Reproduced from (Dai, 1992).)

n \ α	0.20	0.10	0.05	0.02	0.01
1	0.90000	0.95000	0.97500	0.99000	0.99500
2	0.68377	0.77639	0.84189	0.90000	0.92929
3	0.56481	0.63604	0.70760	0.78456	0.82900
4	0.49265	0.56522	0.62394	0.68887	0.73424
5	0.44698	0.50945	0.56328	0.62718	0.66853
6	0.41037	0.46799	0.51926	0.57741	0.61661
7	0.38148	0.43607	0.48342	0.53844	0.57581
8	0.35831	0.40962	0.45427	0.50654	0.54179
9	0.33910	0.38746	0.43001	0.47960	0.51332
10	0.32260	0.36866	0.40925	0.45662	0.48893
11	0.30829	0.35242	0.39122	0.43670	0.46770
12	0.29577	0.33815	0.37543	0.41918	0.44905
13	0.28470	0.32549	0.36143	0.40362	0.43247
14	0.27481	0.31417	0.34890	0.38970	0.41762
15	0.26588	0.30397	0.33760	0.37713	0.40420
16	0.25778	0.29472	0.32733	0.36571	0.39201
17	0.25039	0.28627	0.31796	0.35528	0.38086
18	0.24360	0.27851	0.30936	0.34569	0.37062
19	0.23735	0.27136	0.30143	0.33685	0.36117
20	0.23156	0.26473	0.29408	0.32866	0.35241
21	0.22617	0.25858	0.28724	0.32104	0.34427
22	0.22115	0.25283	0.28087	0.31394	0.33666
23	0.21645	0.24746	0.27490	0.30728	0.32954
24	0.21205	0.24242	0.26391	0.30104	0.32286
25	0.20790	0.23768	0.26404	0.29518	0.31657
26	0.20399	0.23320	0.25907	0.28962	0.31064
27	0.20030	0.22898	0.25438	0.28438	0.30502
28	0.19680	0.22497	0.24993	0.27942	0.29971
29	0.19348	0.22117	0.24571	0.27471	0.29466
30	0.19032	0.21756	0.24170	0.27023	0.28087

TABLE D.8. Tolerance factors for normal distribution. (Reproduced from (Eisenhart, 1947).)

Factors K such that the probability is y that at least a proportion $1 - \alpha$ of the distribution will be included between $\overline{X} \pm KS$, where $\overline{X}$ and S are estimators of the mean and the standard deviation computed from a random sample of size n.

			$\gamma = 0.75$					$\gamma = 0.90$		
α n	0.25	0.10	0.05	0.01	0.001	0.25	0.10	0.05	0.01	0.001
2	4.498	6.301	7.414	9.531	11.920	11.407	15.978	18.800	24.167	30.227
3	2.501	3.538	4.187	5.431	6.844	4.132	5.847	6.919	8.974	11.309
4	2.035	2.892	3.431	4.471	5.657	2.932	4.166	4.943	6.440	8.149
5	1.825	2.599	3.088	4.033	5.117	2.454	3.494	4.152	5.423	6.879
6	1.704	2.429	2.889	3.779	4.802	2.196	3.131	3.723	4.870	6.188
7	1.624	2.318	2.757	3.611	4.593	2.034	2.902	3.452	4.521	5.750
8	1.568	2.238	2.663	3.491	4.444	1.921	2.743	3.264	4.278	5.446
9	1.525	2.178	2.593	3.400	4.330	1.839	2.626	3.125	4.098	5.220
10	1.492	2.131	2.537	3.328	4.241	1.775	2.535	3.018	3.959	5.046
11	1.465	2.093	2.493	3.271	4.169	1.724	2.463	2.933	3.849	4.906
12	1.443	2.062	2.456	3.223	4.110	1.683	2.404	2.863	3.758	4.792
13	1.425	2.036	2.424	3.183	4.059	1.648	2.355	2.805	3.682	4.697
14	1.409	2.013	2.398	3.148	4.016	1.619	2.314	2.756	3.618	4.615
15	1.395	1.994	2.375	3.118	3.979	1.594	2.278	2.713	3.562	4.545
16	1.383	1.977	2.355	3.092	3.946	1.572	2.246	2.676	3.514	4.484
17	1.372	1.962	2.337	3.069	3.917	1.552	2.219	2.643	3.471	4.430
18	1.363	1.948	2.321	3.048	3.891	1.535	2.194	2.614	3.433	4.382
19	1.355	1.936	2.307	3.030	3.867	1.520	2.172	2.588	3.399	4.339
20	1.347	1.925	2.294	3.013	3.846	1.506	2.152	2.564	3.368	4.300
21	1.340	1.915	2.282	2.998	3.827	1.493	2.135	2.543	3.340	4.264
22	1.334	1.906	2.271	2.984	3.809	1.482	2.118	2.524	3.315	4.232
23	1.328	1.898	2.261	2.971	3.793	1.471	2.103	2.506	3.292	4.203
24	1.322	1.891	2.252	2.959	3.778	1.462	2.089	2.489	3.270	4.176
25	1.317	1.883	2.244	2.948	3.764	1.453	2.077	2.474	3.251	4.151
26	1.313	1.877	2.236	2.938	3.751	1.444	2.065	2.460	3.232	4.127
27	1.309	1.871	2.229	2.929	3.740	1.437	2.054	2.447	3.215	4.106
28	1.305	1.865	2.222	2.920	3.728	1.430	2.044	2.435	3.199	4.085
29	1.301	1.860	2.216	2.911	3.718	1.423	2.034	2.424	3.184	4.066
30	1.297	1.855	2.210	2.904	3.708	1.417	2.025	2.413	3.170	4.049
31	1.294	1.850	2.204	2.896	3.699	1.411	2.017	2.403	3.157	4.032
32	1.291	1.846	2.199	2.890	3.690	1.405	2.009	2.393	3.145	4.016
33	1.288	1.842	2.194	2.883	3.682	1.400	2.001	2.385	3.133	4.001
34	1.285	1.838	2.189	2.877	3.674	1.395	1.994	2.376	3.122	3.987
35	1.283	1.834	2.185	2.871	3.667	1.390	1.988	2.368	3.112	3.974
36	1.280	1.830	2.181	2.866	3.660	1.386	1.981	2.361	3.102	3.961
37	1.278	1.827	2.177	2.860	3.653	1.381	1.975	2.353	3.092	3.949
38	1.275	1.824	2.173	2.855	3.647	1.377	1.969	2.346	3.083	3.938

TABLE **D.8.** *continued*

α / n	$\gamma = 0.75$					$\gamma = 0.90$				
n	0.25	0.10	0.05	0.01	0.001	0.25	0.10	0.05	0.01	0.001
39	1.273	1.821	2.169	2.850	3.641	1.374	1.964	2.340	3.075	3.927
40	1.271	1.818	2.166	2.846	3.635	1.370	1.959	2.334	3.066	3.917
41	1.269	1.815	2.162	2.841	3.629	1.366	1.954	2.328	3.059	3.907
42	1.267	1.812	2.159	2.837	3.624	1.363	1.949	2.322	3.051	3.897
43	1.266	1.810	2.156	2.833	3.619	1.360	1.944	2.316	3.044	3.888
44	1.264	1.807	2.153	2.829	3.614	1.357	1.940	2.311	3.037	3.879
45	1.262	1.805	2.150	2.826	3.609	1.354	1.935	2.306	3.030	3.871
46	1.261	1.802	2.148	2.822	3.605	1.351	1.931	2.301	3.024	3.863
47	1.259	1.800	2.145	2.819	3.600	1.348	1.927	2.297	3.018	3.855
48	1.258	1.798	2.143	2.815	3.596	1.345	1.924	2.292	3.012	3.847
49	1.256	1.796	2.140	2.812	3.592	1.343	1.920	2.288	3.006	3.840
50	1.255	1.794	2.138	2.809	3.588	1.340	1.916	2.284	3.001	3.833

α / n	$\gamma = 0.95$					$\gamma = 0.99$				
n	0.25	0.10	0.05	0.01	0.001	0.25	0.10	0.05	0.01	0.001
2	22.858	32.019	37.674	48.430	60.573	114.363	160.193	188.491	242.300	303.054
3	5.922	8.380	9.916	12.861	16.208	13.378	18.930	22.401	29.055	36.616
4	3.779	5.369	6.370	8.299	10.502	6.614	9.398	11.150	14.527	18.383
5	3.002	4.275	5.079	6.634	8.415	4.643	6.612	7.855	10.260	13.015
6	2.604	3.712	4.414	5.775	7.337	3.743	5.337	6.345	8.301	10.548
7	2.361	3.369	4.007	5.248	6.676	3.233	4.613	5.488	7.187	9.142
8	2.197	3.136	3.732	4.891	6.226	2.905	4.147	4.936	6.468	8.234
9	2.078	2.967	3.532	4.631	5.899	2.677	3.822	4.550	5.966	7.600
10	1.987	2.839	3.379	4.433	5.649	2.508	3.582	4.265	5.594	7.129
11	1.916	2.737	3.259	4.277	5.452	2.378	3.397	4.045	5.308	6.766
12	1.858	2.655	3.162	4.150	5.291	2.274	3.250	3.870	5.079	6.477
13	1.810	2.587	3.081	4.044	5.158	2.190	3.130	3.727	4.893	6.240
14	1.770	2.529	3.012	3.955	5.045	2.120	3.029	3.608	4.737	6.043
15	1.735	2.480	2.954	3.878	4.949	2.060	2.945	3.507	4.605	5.876
16	1.705	2.437	2.903	3.812	4.865	2.009	2.872	3.421	4.492	5.732
17	1.679	2.400	2.858	3.754	4.791	1.965	2.808	3.345	4.393	5.607
18	1.655	2.366	2.819	3.702	4.725	1.926	2.753	3.279	4.307	5.497
19	1.635	2.337	2.784	3.656	4.667	1.891	2.703	3.221	4.230	5.399
20	1.616	2.310	2.752	3.615	4.614	1.860	2.659	3.168	4.161	5.312
21	1.599	2.286	2.723	3.577	4.567	1.833	2.620	3.121	4.100	5.234
22	1.584	2.264	2.697	3.543	4.523	1.808	2.584	3.078	4.044	5.163
23	1.570	2.244	2.673	3.512	4.484	1.785	2.551	3.040	3.993	5.098
24	1.557	2.225	2.651	3.483	4.447	1.764	2.522	3.004	3.947	5.039
25	1.545	2.208	2.631	3.457	4.413	1.745	2.494	2.972	3.904	4.985
26	1.534	2.193	2.612	3.432	4.382	1.727	2.469	2.941	3.865	4.935

TABLE **D.8.** *continued*

α n	$\gamma = 0.95$					$\gamma = 0.99$				
	0.25	0.10	0.05	0.01	0.001	0.25	0.10	0.05	0.01	0.001
27	1.523	2.178	2.595	3.409	4.353	1.711	2.446	2.914	3.828	4.888
28	1.514	2.164	2.579	3.388	4.326	1.695	2.424	2.888	3.794	4.845
29	1.505	2.152	2.554	3.368	4.301	1.681	2.404	2.864	3.763	4.805
30	1.497	2.140	2.549	3.350	4.278	1.668	2.385	2.841	3.733	4.768
31	1.489	2.129	2.536	3.332	4.256	1.656	2.367	2.820	3.706	4.732
32	1.481	2.118	2.524	3.316	4.235	1.644	2.351	2.801	3.680	4.699
33	1.475	2.108	2.512	3.300	4.215	1.633	2.335	2.782	3.655	4.668
34	1.468	2.099	2.501	3.286	4.197	1.623	2.320	2.764	3.632	4.639
35	1.462	2.090	2.490	3.272	4.179	1.613	2.306	2.748	3.611	4.611
36	1.455	2.081	2.479	3.258	4.161	1.604	2.293	2.732	3.590	4.585
37	1.450	2.073	2.470	3.246	4.146	1.595	2.281	2.717	3.571	4.560
38	1.446	2.068	2.464	3.237	4.134	1.587	2.269	2.703	3.552	4.537
39	1.441	2.060	2.455	3.226	4.120	1.579	2.257	2.690	3.534	4.514
40	1.435	2.052	2.445	3.213	4.104	1.571	2.247	2.677	3.518	4.493
41	1.430	2.045	2.437	3.202	4.090	1.564	2.236	2.665	3.502	4.472
42	1.426	2.039	2.429	3.192	4.077	1.557	2.227	2.653	3.486	4.453
43	1.422	2.033	2.422	3.183	4.065	1.551	2.217	2.642	3.472	4.434
44	1.418	2.027	2.415	3.173	4.053	1.545	2.208	2.631	3.458	4.416
45	1.414	2.021	2.408	3.165	4.042	1.539	2.200	2.621	3.444	4.399
46	1.410	2.016	2.402	3.156	4.031	1.533	2.192	2.611	3.431	4.383
47	1.406	2.011	2.396	3.148	4.021	1.527	2.184	2.602	3.419	4.367
48	1.403	2.006	2.390	3.140	4.011	1.522	2.176	2.593	3.407	4.352
49	1.399	2.001	2.384	3.133	4.002	1.517	2.169	2.584	3.396	4.337
50	1.396	1.969	2.379	3.126	3.993	1.512	2.162	2.576	3.385	4.323

TABLE **D.9.** One-sided tolerance factors for normal distribution. (Reproduced from (Eisenhart, 1947).)

Factors K such that the Probability is γ that at least a Proportion 1 − α of the distribution will be less than
$\overline{X} + KS$ *(or greater than* $\overline{X} − KS$*), where* $\overline{X}$ *and S are estimators of the mean and the standard deviation computed from a random sample of size n.*

n \ α	γ = 0.75 0.25	0.10	0.05	0.01	0.001	γ = 0.90 0.25	0.10	0.05	0.01	0.001
3	1.464	2.501	3.152	4.396	5.805	2.602	4.258	5.310	7.340	9.651
4	1.256	2.134	2.680	3.726	4.910	1.972	3.187	3.957	5.437	7.128
5	1.152	1.961	2.463	3.421	4.507	1.698	2.742	3.400	4.666	6.112
6	1.087	1.860	2.336	3.243	4.273	1.540	2.494	3.091	4.242	5.556
7	1.043	1.791	2.250	3.126	4.118	1.435	2.333	2.894	3.972	5.201
8	1.010	1.740	2.190	3.042	4.008	1.360	2.219	2.755	3.783	4.955
9	0.984	1.702	2.141	2.977	3.924	1.302	2.133	2.649	3.641	4.772
10	0.964	1.671	2.103	2.927	3.858	1.257	2.065	2.568	3.532	4.629
11	0.947	1.646	2.073	2.885	3.804	1.219	2.012	2.503	3.444	4.515
12	0.933	1.624	2.048	2.851	3.760	1.188	1.966	2.448	3.371	4.420
13	0.919	1.606	2.026	2.822	3.722	1.162	1.928	2.403	3.310	4.341
14	0.909	1.591	2.007	2.796	3.690	1.139	1.895	2.363	3.257	4.274
15	0.899	1.577	1.991	2.776	3.661	1.119	1.866	2.329	3.212	4.215
16	0.891	1.566	1.977	2.756	3.637	1.101	1.842	2.299	3.172	4.164
17	0.883	1.554	1.964	2.739	3.615	1.085	1.820	2.272	3.136	4.118
18	0.876	1.544	1.951	2.723	3.595	1.071	1.800	2.249	3.106	4.078
19	0.870	1.536	1.942	2.710	3.577	1.058	1.781	2.228	3.078	4.041
20	0.865	1.528	1.933	2.697	3.561	1.046	1.765	2.208	3.052	4.009
21	0.859	1.520	1.923	2.686	3.545	1.035	1.750	2.190	3.028	3.979
22	0.854	1.514	1.916	2.675	3.532	1.025	1.736	2.174	3.007	3.952
23	0.849	1.508	1.907	2.665	3.520	1.016	1.724	2.159	2.987	3.927
24	0.845	1.502	1.901	2.656	3.509	1.007	1.712	2.145	2.969	3.904
25	0.842	1.496	1.895	2.647	3.497	0.999	1.702	2.132	2.952	3.882
30	0.825	1.475	1.869	2.613	3.454	0.966	1.657	2.080	2.884	3.794
35	0.812	1.458	1.849	2.588	3.421	0.942	1.623	2.041	2.833	3.730
40	0.803	1.445	1.834	2.568	3.395	0.923	1.598	2.010	2.793	3.679
45	0.795	1.435	1.821	2.552	3.375	0.908	1.577	1.986	2.762	3.638
50	0.788	1.426	1.811	2.538	3.358	0.894	1.560	1.965	2.735	3.604

TABLE **D.9.** *continued*

	$\gamma = 0.95$					$\gamma = 0.99$				
α n	0.25	0.10	0.05	0.01	0.001	0.25	0.10	0.05	0.01	0.001
3	3.804	6.158	7.655	10.552	13.857					
4	2.619	4.163	5.145	7.042	9.215					
5	2.149	3.407	4.202	5.741	7.501					
6	1.895	3.006	3.707	5.062	6.612	2.849	4.408	5.409	7.334	9.540
7	1.732	2.755	3.399	4.641	6.061	2.490	3.856	4.730	6.411	8.348
8	1.617	2.582	3.188	4.353	5.686	2.252	3.496	4.287	5.811	7.566
9	1.532	2.454	3.031	4.143	5.414	2.085	3.242	3.971	5.389	7.014
10	1.465	2.355	2.911	3.981	5.203	1.954	3.048	3.739	5.075	6.603
11	1.411	2.275	2.815	3.852	5.036	1.854	2.897	3.557	4.828	6.284
12	1.366	2.210	2.736	3.747	4.900	1.771	2.773	3.410	4.633	6.032
13	1.329	2.155	2.670	3.659	4.787	1.702	2.677	3.290	4.472	5.826
14	1.296	2.108	2.614	3.585	4.690	1.645	2.592	3.189	4.336	5.651
15	1.268	2.068	2.566	3.520	4.607	1.596	2.521	3.102	4.224	5.507
16	1.242	2.032	2.523	3.463	4.534	1.553	2.458	3.028	4.124	5.374
17	1.220	2.001	2.486	3.415	4.471	1.514	2.405	2.962	4.038	5.268
18	1.200	1.974	2.453	3.370	4.415	1.481	2.357	2.906	3.961	5.167
19	1.183	1.949	2.423	3.331	4.364	1.450	2.315	2.855	3.893	5.078
20	1.167	1.926	2.396	3.295	4.319	1.424	2.275	2.807	3.832	5.003
21	1.152	1.905	2.371	3.262	4.276	1.397	2.241	2.768	3.776	4.932
22	1.138	1.887	2.350	3.233	4.238	1.376	2.208	2.729	3.727	4.866
23	1.126	1.869	2.329	3.206	4.204	1.355	2.179	2.693	3.680	4.806
24	1.114	1.853	2.309	3.181	4.171	1.336	2.154	2.663	3.638	4.755
25	1.103	1.838	2.292	3.158	4.143	1.319	2.129	2.632	3.601	4.706
30	1.059	1.778	2.220	3.064	4.022	1.249	2.029	2.516	3.446	4.508
35	1.025	1.732	2.166	2.994	3.934	1.195	1.957	2.431	3.334	4.364
40	0.999	1.697	2.126	2.941	3.866	1.154	1.902	2.365	3.250	4.255
45	0.978	1.669	2.092	2.897	3.811	1.122	1.857	2.313	3.181	4.168
50	0.961	1.646	2.065	2.863	3.766	1.096	1.821	2.296	3.124	4.096

References

Dai, S.-H. and M.-O. Wang, *Reliability Analysis in Engineering Applications*, Van Nostrand Reinhold, New York, 1992.

Eisenhart, C., M.W. Hastay, and W.A. Wallis, *Techniques of Statistical Analysis*, McGraw–Hill Book Company, New York, 1947, Chapter 2.

Lindgren, B.W., *Statistical Theory*, Macmillan Publishing Company, New York, 1976, pp. 391–393; 4th edition. CRC Press, Boca Raton, 1993.

Pearson, E.S. and H.O. Hartley (eds.), *Biometrica Tables for Statisticians*, 3rd edition, Vol. 1, Lubrecht & Cramer Ltd., New York, 1976.

Reliability Handbook, AMCP 702–3, Headquarters, US Army Material Command, Washington, DC, October 1968.

Solutions to Selected Problems

Chapter 4

4.2 The MLE of N and ϕ can be obtained by solving the following two equations simultaneously:

$$\sum_{i=1}^{n}\frac{1}{N-i+1} = \phi\sum_{i=1}^{n}t_i$$

$$\sum_{i=1}^{n}(N-i+1)t_i = \frac{n}{\phi}.$$

The above two equations can be solved and obtained as follows:

$$\left(\sum_{i=1}^{n}(N-i+1)t_i\right)\left(\sum_{i=1}^{n}\frac{1}{N-i+1}\right) = n\sum_{i=1}^{n}t_i$$

and

$$\phi = \frac{n}{\displaystyle\sum_{i=1}^{n}(N-i+1)t_i}.$$

4.5 The probability of removing k induced errors and $r-k$ indigenous errors in m tests is a combination of binomial and hypergeometric distributions and is given by

$$P(k; N+n_1, n_1, r, m) = \binom{m}{r}(1-q)^r q^{m-r}\frac{\dbinom{n_1}{k}\dbinom{N}{r-k}}{\dbinom{N+n_1}{r}}$$

$$N \geq r-k \geq 0, n_1 \geq k \geq 0, \quad \text{and} \quad m \geq r.$$

Chapter 5

5.3 (a) Note that the mean value function of the Goel–Okumoto (G–O) model (Eq. (5.6)) is given as

$$m(t) = a(1 - e^{-bt}).$$

By Eq. (5.7), the estimate of parameters a and b, using the MLE, are given below:

$$a = 99$$
$$b = 0.28.$$

(b) The mean value function is

$$m(t) = 99(1 - e^{-0.28t})$$

and the reliability function is

$$R(x|t) = e^{-[m(t+x)-m(t)]}$$
$$= e^{-99[e^{-0.28(t+x)} - e^{-0.28t}]}.$$

(c) Assume $x = 2$ and $t = 10$, then

$$R(2|10) = e^{-99e^{-0.28(12)} + 99e^{-0.28(10)}}$$
$$= 0.077.$$

Hence, the probability that a software failure does not occur in the time interval [10, 12] is 0.077.

(d) Let us consider the logarithmic NHPP model, also called the Musa–Okumoto (M–O) model (Musa, 1984), in which the failure rate is the reciprocal of another function that is linear with respect to time:

$$\lambda(t) = \frac{ab}{(1 + bt)}.$$

Hence, the expected cumulative number of failures, or the mean value function, is a logarithmic function of time:

$$m(t) = a\ln(1 + bt).$$

The log likelihood function is as follows:

$$\ln L$$
$$= \sum_{i=1}^{n} \left[(y_i - y_{i-1})\ln[m(t_i) - m(t_{i-1})] - [m(t_i)m(t_{i-1})] - \ln[(y_i - y_{i-1})!] \right].$$

Taking the partial derivatives of the above function with respect to the unknown parameters a and b and set:

$$\frac{\partial}{\partial a} \ln L = 0$$
$$\frac{\partial}{\partial b} \ln L = 0,$$

we obtain

$$a = \frac{y_n}{\ln(1 + bt_n)}$$

$$\frac{at_n}{(1 + bt_n)} = \sum_{i=1}^{n}(y_i - y_{i-1})\frac{\left(\frac{t_i}{1+bt_i} - \frac{t_{i-1}}{1+bt_{i-1}}\right)}{\ln(1 + bt_i) - \ln(1 + bt_{i-1})}.$$

The estimate of a and b can be obtained by solving the above equations simultaneously. Given the data, we obtain

$$a = 94$$
$$b = 0.17.$$

The mean value function and the reliability function are

$$m(t) = a \ln(1 + bt)$$
$$= 94 \ln(1 + 0.17t)$$

and

$$R(t) = e^{-94 \ln(1+0.17t)},$$

respectively.

The probability that a software failure does not occur during the interval $[10, 12]$ is

$$R(x = 2|t = 10) = e^{-[m(12)-m(10)]}$$
$$= 1.44 \times 10^{-5}.$$

From Table S.1 and Fig. S.1, we observe that the G–O model fits the actual data better than the logarithmic NHPP M–O model, and therefore, the G–O model is a better model for this application data.

TABLE S.1. Comparison of data.

Hour	Cumulative $m(t) = 99(1 - e^{-0.28t})$	Cumulative $m(t) = 94\ln(1 + 0.17t)$	Cumulative actual data
1	24	15	27
2	42	28	43
3	56	39	54
4	67	49	64
5	75	59	75
6	81	66	82
7	85	74	84
8	88	81	89
9	91	87	92
10	93	93	93

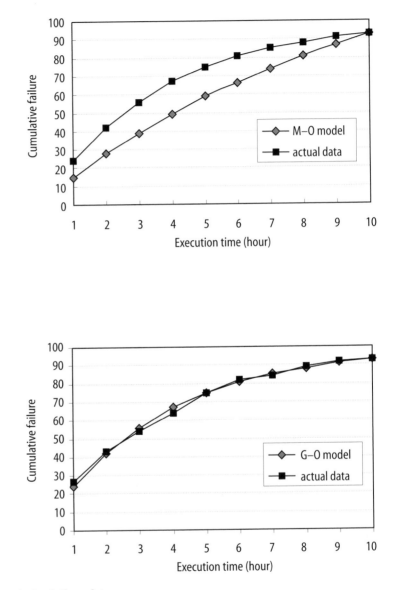

FIG. S.1. Comparison of cumulative failure data
between the Musa–Okumoto (M–O) and
Goel–Okumoto (G–O) models.

Chapter 6

6.4   Taking the first differentiate of the cost function $E(T)$, as given in Eq. (6.10), with respect to T, we obtain

$$\frac{\partial}{\partial T}[E(T)] = [c_3 - \sum_{i=1}^{3}(C_{i2} - C_{i1})ap_ib_ie^{-(1-\beta_i)b_iT}]\int_{T}^{\infty}g(t)dt$$

$$= [c_3 - \sum_{i=1}^{3}(C_{i2} - C_{i1})\lambda_i(T)]\int_{T}^{\infty}g(t)dt$$

$$= [c_3 - h(T)]\int_{T}^{\infty}g(t)dt,$$

where $h(T)$ is defined as in Eq. (6.11). Note that the function $\lambda_i(T)$ is, defined as in Eq. (5.23), decreasing in T for any T, and $C_{i2} \geq C_{i1}$ for all i. Therefore, the function $h(T)$ is continuous and decreasing in T for all T.

Case 1   If $h(0) \leq C_3$, then $h(T) \leq C_3$ for all T. This implies that

$$\frac{d}{dt}[E(T)] = C_3 - h(T) \geq 0 \quad \text{for all } T,$$

that is, the function $E(T)$ is increasing in T. Therefore, the optimal testing time T^* is 0 that minimizes $E(T)$.

Case 2   If $h(0) > C_3$, then there exists a unique value T^* such that $h(T^*) = C_3$. Since the function $h(T)$ is continuous and decreasing in T, therefore, $T^* = h^{-1}(C_3)$.

Chapter 7

7.2

TABLE S.2. System reliability and cost of configurations (a)–(e).

Scheme	System reliability	Total cost ($)
(a) 7 RB	0.9978	117,750
(b) 7 NVP	0.9980	106,500
(c) 3 RB 2 NVP	0.9995	109,500
(d) 2 NVP 3 RB	0.9980	116,500
(e) 2 RB 3 NVP	0.9991	111,000

Appendix A

A.1 The likelihood function is given by

$$L(\lambda; \underline{x}) = \prod_{i=1}^{n} f(x_i, \lambda)$$

$$= \prod_{i=1}^{n} \frac{e^{-\lambda} \lambda^{x_i}}{x_i!} = \frac{e^{-n\lambda} \sum_{i=1}^{n} x_i}{\prod_{i=1}^{n} x_i!}. \tag{S.1}$$

We wish to find the value of λ that maximizes (5.1). First, taking logarithms,

$$\ln L(\underline{X}; \lambda) = -n\lambda + \sum_{i=1}^{n} x_i[\ln \lambda] - \sum_{i=1}^{n} \ln x_i!$$

Taking the partial derivative of $\ln L(\lambda; \underline{X})$ with respect to λ and setting the result equal to zero,

$$\frac{\partial}{\partial \lambda}[\ln L(\lambda; \underline{X})] = -n + \frac{1}{\lambda} \sum_{i=1}^{n} x_i \equiv 0$$

$$\hat{\lambda} = \frac{1}{n} \sum_{i=1}^{n} x_i.$$

A.2 Let $X_1, X_2, \ldots, X_n$ be a random sample from the distribution with a discrete pdf

$$P(x) = p^x(1-p)^{1-x} \quad x = 0, 1 \quad \text{and} \quad 0 < p < 1.$$

The likelihood function is

$$L(X, p) = p^{\sum_{i=1}^{n} x_i} (1-p)^{n - \sum_{i=1}^{n} x_i}.$$

Then

$$\ln L = \sum_{i=1}^{n} x_i \log p + (n - \sum_{i=1}^{n} x_i) \log(1-p)$$

and

$$\frac{\partial \ln L}{\partial p} = \frac{\sum_{i=1}^{n} x_i}{p} - \frac{(n - \sum_{i=1}^{n} x_i)}{1-p} = 0.$$

The solution for $\hat{p}$ is

$$p = \frac{\sum_{i=1}^{n} x_i}{n}.$$

A.3 The pdf of Pareto distribution is given by

$$f(x; \lambda, \theta) = \frac{\partial}{\partial x} F(x; \lambda, \theta)$$
$$= \theta \, \lambda^{\theta} x^{-(\theta+1)} \quad \text{for } x \geq \lambda, \quad \text{and} \quad \lambda, \theta > 0.$$

The likelihood function is

$$L(\lambda, \theta; \underline{X}) = \prod_{i=1}^{n} f(x_i; \lambda, \theta) = \theta^n \lambda^{n\theta} \prod_{i=1}^{n} x_i^{-(\theta+1)}.$$

Taking log on both sides,

$$\mathcal{L} \equiv \ln L(\lambda, \theta; \underline{X}) = n \ln\theta + n\,\theta \ln\lambda - (\theta+1) \ln\left(\prod_{i=1}^{n} x_i\right).$$

Then we obtain

$$\frac{\partial \mathcal{L}}{\partial \lambda} = \frac{n\theta}{\lambda} \equiv 0 \qquad\qquad\qquad \text{(S.2)}$$

$$\frac{\partial \mathcal{L}}{\partial \theta} = \frac{n}{\theta} + n \ln\lambda - \ln\left(\prod_{i=1}^{n} x_i\right) \equiv 0. \qquad \text{(S.3)}$$

From Eq. (S.2), we obtain

$$\hat{\lambda} = \min\{X_i\}$$

and from Eq. (S.2)

$$\hat{\theta} = \frac{n}{\ln\left(\dfrac{\prod\limits_{i=1}^{n} x_i}{\hat{\lambda}^n}\right)}$$

$$= \frac{n}{\ln\left[\dfrac{\prod x_i}{(\min\{x_i\}^n)}\right]}.$$

A.4 Using the method of moments, the expected value of the random variable X having a Poisson distribution is λ. Hence, the estimator μ of is the first sample moment. Therefore,

$$\hat{\lambda} = \frac{1}{n} \sum_{i=1}^{n} x_i = \underline{x}.$$

A.5 The pdf is

$$f(x; \theta) = 1 \quad \text{if} \quad \theta - \frac{1}{2} \leq x \leq \theta + \frac{1}{2}$$

and let $Y_1 < Y_2 < \cdots < Y_n$ be the order statistic. Then we can rewrite:

$$\theta - \frac{1}{2} \le x \le \theta + \frac{1}{2}$$

as

$$x - \frac{1}{2} \le \theta \le x + \frac{1}{2}.$$

The likelihood function is

$$L(\theta; X) = 1 \quad \text{for} \quad x_i - \frac{1}{2} \le \theta \le x_i + \frac{1}{2} \quad \text{for all } i.$$

This implies that

(a) Assume

$$h(\underline{X}) = \frac{4Y_1 + 2Y_n + 1}{6}$$

$$\max\{X_i\} - \frac{1}{2} \le \theta \le \min\{X_i\} + \frac{1}{2}$$

or

$$Y_n - \frac{1}{2} \le \theta \le Y_1 + \frac{1}{2}$$

is an MLE of θ, then

$$Y_n - \frac{1}{2} \le \frac{4Y_i + 2Y_n + 1}{6} \le Y_1 + \frac{1}{2}.$$

After simplifications, we obtain

$$6Y_n - 3 \le 4Y_1 + 2Y_n + 1 \le 6Y_1 + 3$$

or

$$Y_n \le Y_1 + 1.$$

(b) Similarly,

$$Y_n - \frac{1}{2} \le \frac{Y_1 + Y_n}{2} \le Y_1 + \frac{1}{2}$$

or

$$Y_n \le Y_1 + 1.$$

(c)

$$Y_n - \frac{1}{2} \le \frac{2Y_1 + 4Y_n - 1}{6} \le Y_1 + \frac{1}{2}$$

or

$$Y_n \le Y_1 + 1.$$

This is also fine!

A.6

$$f(x; \theta) = \frac{1}{\theta} e^{-\frac{x}{\theta}} \qquad x \geq 0.$$

(a) Using the method of moments, the estimator of mean is the first sample moment. Thus,

$$\hat{\theta} = \frac{\sum\limits_{i=1}^{n} X_i}{n} = \bar{x}.$$

(b) The likelihood function is

$$L(\theta; X) = \frac{1}{\theta^n} e^{-\frac{1}{\theta} \sum\limits_{i=1}^{n} x_i}.$$

Taking log, we obtain

$$\mathcal{L} \equiv \ln L(\theta; \underline{X}) = -n \ln\theta - \frac{1}{\theta} \sum\limits_{i=1}^{n} x_i.$$

Then

$$\frac{\partial \mathcal{L}}{\partial \theta} = -\frac{n}{\theta} + \frac{\sum\limits_{i=1}^{n} x_i}{\theta^2} = 0.$$

This implies that

$$\hat{\theta} = \frac{1}{n} \sum\limits_{i=1}^{n} x_i = \bar{x}. \tag{S.4}$$

(c)

$$\hat{R}(t) = e^{-\frac{t}{\hat{\theta}}}$$

where $\hat{\theta}$ is given in Eq. (S.4).

Appendix B

B.1 (Property B.2) Define $N(t) = N_1(t) - N_2(t)$. We have

$$P[N(t) = k] = \sum\limits_{i=0}^{\infty} P[N_1(t) = k + i]P[N_2(t) = i].$$

Since $N_i(t)$ for $i = 1, 2$, is a Poisson process with mean $\lambda_i t$, therefore,

$$P[N(t) = k] = \sum\limits_{i=0}^{\infty} \frac{e^{-\lambda_1 t}(\lambda_1 t)^{k+i}}{(k+i)!} \cdot \frac{e^{-\lambda_2 t}(\lambda_2 t)^i}{i!}$$

$$= e^{-(\lambda_1 + \lambda_2)t} \left(\frac{\lambda_1}{\lambda_2}\right)^{\frac{k}{2}} \sum\limits_{i=0}^{\infty} \frac{(\sqrt{\lambda_1 \lambda_2} t)^{2n+k}}{i!(k+i)!}$$

$$= e^{-(\lambda_1 + \lambda_2)t} \left(\frac{\lambda_1}{\lambda_2}\right)^{\frac{k}{2}} I_k(2\sqrt{\lambda_1 \lambda_2} t).$$

B.2 (Property B.3) Let $N_a(t)$ be the number of events actually counted in time $[0, t]$. Assuming that each filtering is independent,

$$P[N_a(t) = n|N(t) = n + k] = \binom{n+k}{n}p^n(1-p)^k.$$

Note that

$$P[N(t) = n + k] = \frac{e^{-\lambda t}(\lambda t)^{n+k}}{(n+k)!}.$$

Therefore,

$$P[N_a(t) = n + k] = \sum_{k=0}^{\infty} P[N_a(t) = n|N(t) = n + k]\, P[N(t) = n + k]$$

$$= \sum_{k=0}^{\infty} \binom{n+k}{n}p^n(1-p)^k \frac{e^{-\lambda t}(\lambda t)^{n+k}}{(n+k)!}$$

$$= e^{-\lambda t}\frac{(p\lambda t)^n}{n!} \sum_{k=0}^{\infty} \frac{((1-p)\lambda t)^k}{k!}$$

$$= \frac{e^{-p\lambda t}(p\lambda t)^n}{n!}.$$

This result shows that the process $N_a(t)$ is a Poisson process with mean $p\lambda t$.

B.3 (Property B.6) The renewal function can be obtained in the form of a conditional expectation, given the first inter-arrival time $T_1 = s$, as follows:

$$m(t) = \int_0^\infty E[N(t)|T_1 = s]dF_a(s)$$

Note that

$$E[N(t)|T_1 = s] = 1 + E[N(t - s)]$$
$$= 1 + m(t - s).$$

Therefore, we obtain

$$m(t) = \int_0^t [1 + m(t - s)]\, dF_a(s)$$

$$= \int_0^t dF_a(s) + \int_0^t m(t - s)\, dF_a(s)$$

$$= F_a(t) + \int_0^t m(t - s)\, dF_a(s).$$

Reference

Musa, J.D. and K. Okumoto, "A logarithmic Poisson execution time model for software reliability measurement," in *Proc. 7th International Conference on Software Engineering*, Orlando, 1984, IEEE Computer Society Press, Los Angeles, 1984, pp. 230–238.

Index

Absorbing process, 264, 272
AHP (see analytic hierarchy process)
Analytic hierarchy process, 62
Analysis phase, 50
Auto-correlation function, 292
Availability, 32, 272, 273

Bayesian, 253
Beta distribution, 29
Binomial distribution, 19
Bug, 7

Chilled water system, 38
Coding phase, 56
Common-cause failures, 192
Complexity
    program size, 204
    measures, 95
    metric, 74
Compound Poisson process, 284
Computer aided traffic control, 179
Conditional probability, 253, 262
Conditional reliability, 21, 139
Confidence interval, 235, 238, 240, 241
Constant failure rate, 279
Continuous distribution, 21
Control flow, 79
    complexity, 80
Correlated failure, 192
Correlation analysis, 205, 208
Correlation factors, 208
Cost model, 162, 169
Counting process, 262, 282
Coutinho model, 96
Coverage definition, 203
Criticality category, 8, 200
Cumulative distribution function, 25

Curve fitting model, 94
Customer demand, 52
Cyclomatic complexity, 78, 200

Data
    analysis, 69
    collection, 48, 69
Debugging, 120
Decision-making process, 245
Defect
    definition, 7
    type, 8
Delayed S-shaped model, 133
Density function, 14, 20
Dependence of faults/failures, 192
Design fault, 53
Design phase, 53
Discrete distribution, 19
Distribution, 19
    beta, 29
    binomial, 19, 316
    chi-square, 237, 304
    continuous, 21, 264
    exponential, 21, 285
    gamma, 28, 285
    log normal, 24
    negative binomial, 292
    normal, 22, 320, 322
    Poisson, 20, 92, 283, 313
    Rayleigh, 133
    Weibull, 26

Environment condition, 44
Environmental factors, 199, 216
    data collection and analysis, 205
Ergodic process, 264, 276
Error
    definition, 1
    detected, 85

Error estimation, 82
Error model, 94
Error seeding, 81
Error types, 120
Estimation
    goodness of fit, 230
    least square, 233
    maximum likelihood, 106, 211
    maximum partial likelihood, 211
    method of moments, 230
Execution time, 48
Expected life of system, 153
Expected cost, 163
Explantory variables, 218
Exponential
    distribution, 21, 99, 263, 285
    failure time model, 264

Factor analysis, 205
Factors affecting software
    reliability, 206
Failure
    incident, 180, 192
    correlated, 192
    intensity, 88, 213
Failure data
    types, 69
    interfailure times (see
    time domain)
Failure independence, 180
Failure rate model, 87, 95
Failure rate, 18, 85, 264
Fault
    common-cause, 192
    counting, 262
    definition, 8
    removal, 47
    seeding, 81

Fault-tolerant software, 179
  advanced techniques, 183
  definition, 180
  N-version programming, 182
  N self-checking programming, 184
  recovery block, 182
  self-checking duplex, 184
Flow graph, 46

Gamma distribution, 28
Geometric Poisson model, 92
Generalized NHPP model, 148
G–O model, 108
goodness of fit test, 139, 230
Growth model, 95

Halstead metrics, 74
Hardware-related software
  failures, 7, 39
Hardware reliability, 39
Hazard rate, 19, 29
Homogeneous Poisson process, 283
Homogeneous Markov process, 264
Hossain–Dahiya model, 145
Hyperexponential model, 113
Hypergeometric distribution, 85

Imperfect debugging, 93, 98, 120,
  126, 145
Independence
  assumption, 87, 108, 180, 181, 188, 282
  failure, 87
  fault, 108
Induced errors, 81
Inflection S-shaped model, 133
Inherent errors, 81
Input space, 45
Input state, 46
Inter-arrival time, 284
Interval availability, 274, 277
Interval-domain data, 69
Interval estimation, 235

Jelinski–Moranda model, 87

Kolmogorov–Smirnov test, 232, 320
K–S test (see Kolmogorov–Smirnov
  test)

LaPlace transform, 273, 292
Least squared estimation, 233
Lifecycle cost, 170
Life distribution, 19
Likelihood function, 106

Lines of code, 6
Littlewood Markov model, 98

Maintainability, 30
Markov chain, 264
Markov model, 98, 278
Markov process, 97, 262
Maximum likelihood estimation,
  106, 211, 223
McCabe's cyclomatic, 78
Mean time between failures, 32, 152,
  271
Mean time to failure, 16
Mean time to repair, 31
Mean value function, 97, 149, 285
Model classification, 74
Model comparison, 137, 138
Model estimation
  EPJM, 213
  Jelinski–Moranda, 87, 213, 216
  Goel–Okumoto, 108
  Moranda geometric, 92
  Musa, 112
  NHPP, 115
  S-shaped, 115, 118, 128
  Pham–Nordmann, 129
  Pham–Zhang, 132
  Pham exponential imperfect
    debugging, 132
Moranda geometric model, 92
MTBF (see mean time between
  failures)
MTTF (see mean time to failure)
MTTR (see mean time to repair)
Musa model, 112

NASA space shuttle, 180
Negative binomial function, 292
NVP (see N-version programming)
N-version programming, 182, 187
NHPP model, 96, 108, 115, 129, 132
Non-homogeneous Poisson process,
  96, 133, 290
Normal distribution, 22, 320, 322
NTDS data, 134
Null hypothesis, 245
Number of errors, 81, 94, 96

Operating phase, 60, 169
Optimal release policy, 162, 163
Optimization, 171, 173

Parallel system, 270
Parameter estimation, 106, 123, 130,
  221

Pareto model, 145
Partial likelihood approach, 212
Pham exponential imperfect
  debugging, 129
Pham–Nordmann model, 129, 133
Pham–Zhang model, 132, 133
Point availability, 274, 276
Poisson
  distribution, 20, 92, 283, 313
  process, 99, 282
Posterior distribution, 254
Prior distribution, 254
Probability density function, 14
Proportional hazard model, 210, 213
Proportional hazard J–M model, 213

Quasi-renewal process, 287

Real-time control software, 155
Recovery block, 187
Redundancy, 180
Regression coefficient, 94
Regression model, 94
Release policy, 162
Release time, 166
Reliability
  definitions, 13, 14
  function, 291
  of hardware, 43
  of software, 8
Reliability block diagram, 42
Reliability estimation, 110
Reliability evaluation, 270
Reliability growth, 95
Reliability requirement, 105
Reliability model
  Delayed S-shaped, 133
  Goel–Okumoto model, 108
  Hybrid fault-tolerant software,
    189
  Inflection S-shaped, 133
  Jelinski–Moranda model, 87
  Moranda geometric model, 92
  Musa model, 96
  NHPP model, 96, 108, 132
  N-version programming, 187
  Pham exponential imperfect
    model, 129
  Pham–Nordmann model, 129
  Pham–Zhang model, 132
  Recovery block, 187
  S-shaped model, 115, 118, 128
Reliability prediction (see
  reliability model)

Reliability subject to cost, 173
Renewal process, 284
Repairable system, 282
Resource allocation, 160, 203
Risk, 2
    definition, 7
    cost model, 162

S-shaped NHPP model, 115, 118, 128
Safety integrity, 7
Safety system, 278
Schick–Wolverton model, 90
Seeding of faults, 81
Self-checking, 184
Sequential sampling, 245
Severity classification of failure, 8
Software
    complexity, 8
    defects, 8
    metrics, 74
    quality, 37
    validation, 68
    verification, 67
Software cost model, 162, 165, 169
Software development process,
    50, 62
Software diversity, 180
Software environmental factors, 211

Software evaluation, 63
Software failure, 1, 160, 162
    hardware-related, 7, 39
Software fault, 1
Software fault tolerant (see
    Fault-tolerant software)
Software lifecycle, 48, 170
Software quality, 159
Software redundancy, 180
Software reliability, 39
    definition, 8
    function, 45
    factors affecting, 171, 210
Software reliability and cost model,
    162
Software reliability growth, 99
Software reliability engineering, 4
Software safety, 98, 99
Software testing, 45, 165, 285
Specification, 52
Standby system, 272
State space, 99
Steady state availability, 277
Stochastic processes, 261, 284
Sum of squared errors, 134
Superposition of NHPP, 115
System failure, 40
System reliability, 43

System reliability model
    hardware and software, 43
    hybrid fault-tolerant, 189
    N-version programming, 187
    N self-checking duplex, 184
    recovery blocks, 187

t-distribution, 237, 303
Test, 230
Testing cost, 165, 169
Testing effort, 162
Testing phase, 56
Time domain, 69, 107
Type of faults, 120

Validation, 56, 68
Verification and validation, 56, 67
Versions of software, 181
Voting system, 181

Weibull distribution, 26
Wall and Ferguson model, 96

Yamada software reliability model,
    127, 128
Yamada imperfect debugging
    model, 133

Software Reliability
NHPP Software™
Hoang Pham
© Springer-Verlag Singapore Pte. Ltd. 2000